ELECTROSTATICS AND
ITS APPLICATIONS

ELECTROSTATICS AND ITS APPLICATIONS

A. D. MOORE, Editor

Electrical and Computer Engineering Department
University of Michigan, Ann Arbor

A WILEY-INTERSCIENCE PUBLICATION

JOHN WILEY & SONS, New York · London · Sydney · Toronto

Library of Congress Cataloging in Publication Data:

Moore, Arthur Dearth, 1895–
  Electrostatics and its applications.

  "A Wiley-Interscience publication."
  Includes bibliographical references.
  1. Electrostatics. I. Title.

QC571.M663   537.1   72-13945

ISBN 0-471-61450-5

Printed in the United States of America

10 9 8 7 6 5 4 3 2

Contributors

W. R. Bell, *Ion Physics Corporation, Burlington, Massachusetts*

James D. Cobine, *Atmospheric Sciences Research Center, State University of New York at Albany, Albany, New York*

Joe S. Crane, *Department of Physics, Cameron State Agricultural College, Lawton, Oklahoma*

W. P. Dyrenforth, *Senior Vice President, Carpco Research and Engineering, Inc., Jacksonville, Florida*

H. Frank Eden, *Program Director for Meteorology, Atmospheric Sciences Section, National Science Foundation, Washington, D. C.*

Noel J. Felici, *Laboratoire D'electrostatique, Grenoble, France*

Charles D. Hendricks, *Director, Charged Particle Research Laboratory, Electrical Engineering Department, University of Illinois, Urbana, Illinois*

Ion I. Inculet, *Chairman, Electrical Engineering Group, The University of Western Ontario, London, Ontario, Canada*

Oleg D. Jefimenko, *Department of Physics, West Virginia University, Morgantown, West Virginia*

James E. Lawver, *Director, Mineral Resources Research Center, University of Minnesota, Minneapolis, Minnesota*

Emery P. Miller, *Vice President, Technical Operations, Ransburg Corporation, Indianapolis, Indiana*

A. D. Moore, *Electrical and Computer Engineering Department, University of Michigan, Ann Arbor, Michigan*

John H. Moran, *Chief Electrical Engineer, Lapp Insulator Division— Interpace, LeRoy, New York*

M. J. Mulcahy, *Director, High Voltage Consulting and Testing, Ion Physics Corporation, Burlington, Massachusetts*

Herbert Ackland Pohl, *Department of Physics, Oklahoma State University, Stillwater, Oklahoma (NATO Senior Science Fellow on leave to Cavendish Laboratory, University of Cambridge, Cambridge, England)*

Myron Robinson, *Health and Safety Laboratory, U.S. Atomic Energy Commission, New York, New York*

Donald S. Swatik, *Manager, Advanced Development Engineering, Computer Peripherals, Inc., (A joint venture of Control Data Corporation and The National Cash Register Company), Rochester, Michigan*

Bernard Vonnegut, *Atmospheric Sciences Research Center, State University of New York at Albany, Albany, New York*

This book is dedicated to a man who stood
before a hostile Parliament in 1766.
When asked for his name and place of abode,
he replied . . .

FRANKLIN, OF PHILADELPHIA

Foreword

The field of electrostatics is a strange one. Although it was known to the ancients, later studied by Gilbert, von Guericke, van Musschenbroek, the Abbé Nollet, Charles Dufay, and Faraday, and popularized by Franklin, many of the phenomena considered by these investigators are still not adequately understood!

Electrostatics is a field of physical science replete with opportunities for new discovery. On all sides are effects that need better explanation.

A few pioneers have utilized electrostatics to make themselves as well as some of their stockholders very rich. Too often the electrostatic mechanisms utilized are closely guarded proprietary secrets, although in a number of instances the physical understandings are so unclear that their utilization must be classified as an art.

At an Adirondack Conference held under the auspices of our Atmospheric Sciences Research Center in 1969, top experts in this field spent three days of informal discussion at our Field Station on the slopes of Whiteface Mountain. For the first time some of the scientists and engineers dealing with electrostatic phenomena came together and laid the groundwork for subsequent scientific cooperation, as well as for the holding of the Albany Conference on Electrostatics, the formation of the Electrostatics Society of America, and the preparation of this book.

In scanning its table of contents I am very much impressed with the remarkable group of electrostatics experts which A. D. Moore has selected and persuaded to cooperate as contributors. I believe these writings provide a focus establishing electrostatics as an important field in the physical sciences which deserves the intensive research of faculty and students in our universities, as well as scientists and engineers of industrial organizations throughout the world.

New industries will develop from the discoveries resulting from this effort, as this woefully neglected area of science is further explored, understood, and utilized.

I predict that many exciting, satisfying, and important new developments will follow the publication of this book.

VINCENT J. SCHAEFER, *Director*
Atmospheric Sciences Research Center
Albany, New York

Preface

What is electrostatics? To the mathematician it is an absorbing line of theory. To the physicist it is a science of many and varied manifestations. To the engineer it is a nuisance or hazard he must eliminate, or again, it represents the perfecting of an application the world needs. To the ecologist it is a major means of preventing pollution of the atmosphere. To the secretary, it is liberation from coping with batches of carbon copies. To the housewife it is a nuisance when it makes dust stick to her walls. To me it is an endlessly fascinating mixture of science and art, highly diversified in character, waiting to be trapped between the covers of a book that is needed by many who are already "in the game" and the many newcomers who are constantly joining it.

When the newcomer, with little or no background in electrostatics, has a need to eliminate a hazard or a desire to develop a possible application, he starts by "looking up the literature." Time and again he is frustrated and discouraged, for although the literature *in toto* is enormous, it is too often of little or no help. The newcomer's problem has not been neglected in planning this book. He should receive an introduction to what he has to know, and the selected references may take him the rest of the way.

No one person could write this book and do so authoritatively. An individual could write with authority only in his area; then he would have to fill in as a reporter. From the start, I planned for multiple authorship, so that each major area would be covered by those who are among the top experts in the various areas.

The book is not a collection of papers already given at conferences or symposia. It is all original writing. And, it is the first book of its kind.

From the book-planning standpoint, electrostatics presents problems. It refuses to be neatly compartmentalized into chapters, with little overlap from one to another. What is possible, and what was done, was to devote whole chapters to the major applications; but again, and inherently, some of the same effects and phenomena are bound to be found in common among several chapters. There is some repetition. This can be helpful, as when a point covered by one writer is not comprehended, but understanding comes when another deals with the same point with different language and a different approach.

Each of these experts was responsible for the content of his contribution. I was not qualified to tell them what to include and what to pass over. I did, however, decide on what chapters to have, and their headings, and I rounded up the writers qualified for each job. Nothing was done to influence their style. Everyone has his favorite way of organizing and presenting a subject. To interfere in this respect is to risk disturbing a writer's approach.

This being a real world, with mounting publication costs, inevitably I had to use my judgment in making approximate space allocations to each contributor. If it turns out that some areas should have had more pages, the fault lies with me and not the writer.

A part of my plan was to have some chapters begin, not with the typical brief introductions, but with *overviews*. Let me give my reasons. Electrostatics is a very odd activity, in that the door to discovery or invention may in one case consent to open only to the practiced touch of an expert. But in another, that door may be leaned against and opened by a surprised and pleased amateur experimenter whose knowledge of electrostatics is meager and perhaps even partly wrong. There are gradations in between. Hence the overview plan. The overviews can be read by the newcomers and "electrostatics amateurs"—including interested laymen, technicians, patent attorneys, and engineers and scientists not familiar with electrostatics—for general understanding. Depending on what the reader learns there, he may (or may not) decide to go on through the chapter.

Electrostatics is an "uneven" activity to a very unusual degree. Parts of it are ancient; other parts are as modern as yesterday. The phenomena needed to explain it range from very simple, to so complex that full understanding is yet to be achieved. Likewise, the chapters are uneven with respect to length, theory needed, mathematics required, and so on. This is the nature of electrostatics, and the book reflects it.

Experts will promptly find that for the most part, vector analysis has been avoided. Scalar treatment largely prevails. Since this was my decision, I should remind the experts that the advantages of neatness and space-saving qualities of vector treatment may be outweighed by another consideration: we must not exclude the readers who do not know vector analysis or who have forgotten it. Some latitude did have to be allowed, especially in terms of the space-saving needs in the material on dielectrophoresis.

Modern electrostatics is almost completely neglected in our colleges and universities. When a teacher becomes interested in initiating an elective course in it, he too begins by looking up the literature. He is faced with a difficult task, but with this book available, he will have more than enough to keep his students both busy and interested.

A. D. MOORE
Ann Arbor, Michigan

September 1972

Acknowledgments

This book has many origins. Beginning with retirement, my Electrostatics Lecture–Demonstration was given during the 1964 Western Tour of 16 colleges and universities; and another 16 were covered in the 1965 South-and-East Tour. By now it has covered many more schools, together with conferences, firms, and research entities, to the extent of 62,000 miles of station wagon travel with the equipment, from border to border and coast to coast, and into Canada. The book owes much to the hundreds of contacts thus made with faculty members, by which it became clear to me that nearly all our teachers are quite unaware of the scope of electrostatics and the great services it is rendering. Also at this point let me express my gratitude to my own university for the rare privilege of maintaining my laboratory to use as a travel base and otherwise to enable me to continue with experimentation.

There is a debt to Doubleday for bringing out my nontechnical book on electrostatics in 1968, for the correspondence it brought pointed to the need for a professional book of the present kind. The book is now available in Italian, German, and Japanese editions.

There has been a close and invaluable association with Dr. Vincent J. Schaefer, Director of the Atmospheric Sciences Research Center at the State University of New York at Albany, and Dr. Bernard Vonnegut of that Center. We three organized the Adirondack Conference on Electrostatics, which was attended by about half of the eventual contributors to this book. More stimulation for furthering the work on the book came from the charter members who joined me in founding the Electrostatics Society of America in 1970. And nearly all the writing team attended when the ESA held the Albany Conference on Electrostatics in 1971. My appreciation goes to Dennis Flanagan, Editor of *Scientific American,* for the invitation to write on electrostatics for the March 1972 issue. The ensuing correspondence gave further proof that this book is needed. I wish to thank my younger colleague, Professor Edward A. Martin: with his fine grasp of theory, he has often helped my editorial work by discussing various points as the chapters came in.

I am deeply grateful for the enthusiasm with which the members of the writing team accepted the invitation to join, and for their contributions. It meant a great deal of work for men who are always fully occupied with their regular assignments and with extras that such men inevitably attract. Finally, my gratitude goes to Beatrice Shube, Editor at Wiley-Interscience who, for these three years, has been a constant source of sound advice and encouragement. The book owes her much.

                                                                             A.D.M.

Contents

Electrostatics and
Its Applications

CHAPTER 1

Introduction

A. D. Moore

Electrical and Computer Engineering Department
University of Michigan
Ann Arbor, Michigan

1.1 HISTORY

At least as early as 600 B.C. the Greeks had the natural magnet, lodestone, and knew some of its properties. They had amber, and when they rubbed it, it did strange things. Thus man's knowledge of electricity *began* with electrostatics.

Little was added to the world knowledge of electricity until 1600, when William Gilbert, a remarkable man who was physician to Queen Elizabeth, conducted many ingenious experiments and published his famous *De Magnete*. The next man to get busy was Otto von Guericke of Magdeburg, perhaps best known for his crude vacuum pump and his demonstration that partially evacuated hemispheres could not be pulled apart by teams of horses. He made the world's first electrostatic generator, a friction-type machine. It was a sulfur ball on a shaft, with a turning crank. When von Guericke rubbed his dry hand on it, there were electrical manifestations.

Slowly, various types of friction machines appeared; many ardent experimenters joined the game and made exciting discoveries. There was much interchange by letters, and the first literature began to appear. The nature of electric charge was much debated. Although it is hardly recognized today, all these early scientists were amateurs, and even those who managed to obtain a higher education learned what was really amateur science.

A fairly complete history of those exciting decades would fill volumes and do honor to many famous names. Here we must skip them all but two: those great experimenters, Michael Faraday and Benjamin Franklin. Both were amateurs in the best sense. Faraday, with little schooling, was self-trained. Franklin did not even begin to study electrostatics until middle age,

1

when he bought the equipment of another, but he soon was world famous. And now, for a historical note. Franklin got into electrostatics in the 1740s. By 1752 he had gained a tremendous reputation throughout the Western world for his electrical discoveries, and especially for showing that lightning was electrical and that rods would protect buildings from it. In those days most men of education—doctors, ministers, lawyers, politicians—were avid followers of science. Thus, when Franklin later began crossing the Atlantic in our behalf, he was universally known and admired, not as a statesman but as a great man of science. And much of that science was electrostatics. Without that high standing, it seems doubtful that he would have enjoyed the political acceptance that enabled him to play so potent a part in securing aid for the Colonies in the Revolutionary War.

Some able historians believe that England might have lost anyway, or again, that France would have entered on our side of her own accord. We must respect those opinions. On the other hand, historians should not ignore the degree in which Franklin's prior reputation in electrostatics was responsible for his acceptance as a statesman. Thus I like to claim that had it not been for electrostatics, we might today be a part of the British Empire.

The fact is, then, that up to 1800 the origins of electrical science were in the hands of electrostatics experimenters. Then, just two centuries after Gilbert's book appeared, there came a landmark development: Volta's wet cell, or battery.

For the first time, men had *current*: steady electrical current, and plenty of it. Any broad history of the advance of science would rate Volta's contribution as one of the great turning points. Electrochemistry could get a real start. Laws, such as Ohm's law, could be formulated. Faraday could insulate wire, use it knowingly, and discover electromagnetic induction in 1831. The following decades came alive with developments at an ever-increasing rate. There were induction coils. Morse gave us the telegraph, but not in an easy fashion: he began with a dream, and an almost complete ignorance of electricity. He wound his first electromagnet with bare wire! Edison, along with his lamp, produced a complete system of generation and distribution. Bell brought in his telephone. The arc light brightened streets and parks, and many other wondrous things have since come.

The frenetic activity in applied electromagnetics did not completely shut off interest in electrostatics in the last century. There was a renewed spurt of interest around 1860, when the unreliable friction generators were replaced by "influence machines." Charges produced by influence (now called induction) were carried along to collectors. We now call them induction machines.

These culminated in 1878 in the famous Wimshurst generator, with two or more glass discs rotating in opposite directions. Some of these generators were giants. Here at last was a far more vigorous source of high voltage electrostatically produced, and hopes rose for successful applications. Some patents were taken out prior to 1900, but there was not enough output, and disappointment prevailed. Apparently, the only major use of the Wimshurst was to operate the early x-ray tubes.

1.2 ELECTROSTATICS AFTER 1900

By 1900 electromagnetics had virtually eclipsed electrostatics, and hopes for the older art began to wane. Worse yet, it was acquiring a bad reputation. Natural electrostatics in the thunderstorm, admired perhaps by some, had always been feared. The increasing use of central heating kept the indoor atmosphere dryer in winter, and people just did not like the shocks they were getting from walking on rugs and then reaching for door knobs. With the proliferation of technology, new nuisances and hazards began to appear; for example, as the printing of paper called for higher speeds, friction effects began to build up charges, sometimes turning sparks loose that caused volatiles to burn or explode. It began to be recognized that charge accumulations could set off dust explosions.

Along with all this, renewed efforts to tame electrostatics and put it to work tended to give it the reputation of being tricky and unreliable. Humidity, or the lack of it, might turn a process into a failure. Or again, a process involving a complex mixture of phenomena might refuse to come under control until more understanding in later days could make it a success. Moreover, until the transformer–rectifier combination came along, there simply was not enough energy available to do what needed to be done.

It is then no wonder that electrostatics acquired a bad reputation, which has persisted even in the face of a gradual growth of useful and even indispensable processes, for most of these are hidden away somewhere and are little known except to the groups directly concerned with them.

The first major break came around 1906 at the hands of an extremely able physical chemist, Frederick G. Cottrell. Smelters and cement mills were ruining the ecological balance of their areas. The success of Cottrell precipitators in the western United States was a breakthrough of great impact. By 1923 the Detroit Edison Company was able to install the first precipitators to trap fly ash made in coal-burning powerhouses. The pollution picture today would be far worse than it is, were it not for the 20 million tons of fly ash trapped annually, as well as for the use of a number of other precipitator applications.

More developments, such as electrostatic separation of minerals and other mixtures, and paint spraying, came later. The first process, now working all over the world, separates immense quantities of materials for the mining industry or otherwise cleans food products. Spray application speeds up painting and saves great amounts of paint, and in so doing relieves the pollution problem.

Along with these developments came the taming of electrostatics to apply dry coatings. Better sandpaper and grit cloth products are made to the extent of $200 million per year. Dry coating has extended to flocking, and powder application. Powders applied electrostatically and later fused to refrigerators and many other products are rapidly being adopted.

In 1935 Chester Carlson set out to make a better copier. After considering various other methods, he settled on electrostatics. His first success was crude enough to discourage anyone else, but he never gave up. In 1944 he had the Battelle Memorial Institute take on the development of what later became the Xerox process. When the machines at last came to market, they brought on the most drastic and rapid revolution ever to affect the business machines area. The Xerox Corporation now does about $2 billion worth of business a year. Other important electrostatic applications are covered in the appropriate chapters.

The present century has also seen an ever-increasing interest in natural electrostatics, or atmospheric electricity. Many very able workers in meteorology have been busy at research for years. A large literature has accumulated, and it constantly grows, as new findings are made and new theories are evolved. Good scientists cannot resist a challenge, such as that offered by trying to explain the enormous charge accumulations built up in thunderstorms. Along with trying to understand such phenomena, there is and must be a paramount interest displayed in protecting planes and helicopters and their communications and control systems from lightning hits, and from interference due to discharges from planes themselves. The chapter on atmospheric electrostatics will be of interest to many readers.

Electrostatics is at last on the march. The rich variety of its effects is fully exemplified in the following chapters, and there is simply no telling where it will go and what it will be doing for us in new ways in the coming decades.

1.3   AND NOW, THE BOOK—

First come lead-off chapters to develop the basic theory of electric fields and particle charging. Another fundamental chapter follows on dielectrics. After treating electrets, motors, and generators, there are eight chapters on various areas of application. The chapter on atmospheric electrostatics is followed by one on nuisances and hazards and by a chapter giving brief mention to many promising effects and phenomena, some of which will eventually grow to full-chapter status. And finally, we have the charming chapter by Felici, on the status abroad.

In closing this introduction, let us again remind ourselves that in olden times, electrostatics was wholly in the hands of amateurs, and that even though great advances have been made in theory and understanding, the day of the experimenter and amateur is by no means over. Two reasons can be cited. The first lies in the history of creativity. It tells us that time and time again, an amateur (newcomer) has entered an area that had become professional, and invented or discovered what everyone else has missed. The second lies in the nature of most areas of electrostatics, which depend on making corona in order to make ions that will have useful effects. But to make ions is to make space charge. And space charge has a puckish way of making hash out of our precise mathematical formulations. The result is that we are glad to have theory and analysis take us as far as can be; but eventually the experimenter has to take over and finish the job.

That is, electrostatics is both a science and an art. As far as anyone can see, it will remain a fascinating mixture of the two.

SYMBOLS USED IN CHAPTERS 2 TO 4

SYMBOL	QUANTITY	UNIT
m	mass	kilograms (kg)
L	length	meters (m)
t	time	seconds (sec)
Q, q	charge	coulombs (C)
F	force	newtons (N)
E	electric field intensity	volts/meter (V/m)
W	energy	joules (J)
g	gravitational acceleration	meters/sec^2 (m/sec^2)
v	velocity	meters/sec (m/sec)
I	electric current	amperes (A)
J	electric current density	amperes/meter^2 (A/m^2)
R	resistance	ohms (Ω)
V	electric potential difference	volts (V)
n	number density	number/meter^3
T	temperature	degrees Centigrade or Kelvin ($^\circ$C or $^\circ$K)
C	capacitance	farads (F) or coulombs/volt (C/V)

R, r	distance	meters (m)
D	Maxwell displacement	coulombs/meter^2 (C/m^2)
A	area	meters^2 (m^2)
x	distance	meters (m)
ln	natural logarithm	none
a, b	radius	meters (m)
f	frequency	hertz (Hz)
h	Planck's constant	
	6.6×10^{-34}	joule-second (J-sec)

e electron charge 1.6×10^{-19} coulombs (C)

A_0 universal constant sometimes referred to as the Richardson–Dushman constant, 1.2×10^6 A/(m^2) ($^\circ$K^2)

x, y, z	Cartesian coordinates	meters (m)
f_e	force of electric origin	newtons (N)
k	Boltzmann constant,	
	1.38×10^{-23}	joules/degree (J/$^\circ$K)

$\int$ integral sign

$\sum_{i=1}^{N} x_i \equiv$ $x_1 + x_2 + x_3 + \cdots + x_{N-1} + x_N$ summation of N terms

$\equiv$ identically equal or "defined as"

dv, dw, dq, etc.   differential elements or very small changes in v, w, q, etc.

$\dfrac{\partial v}{\partial x}$ the rate of change of v with respect to only the variable x; the partial derivative of v with respect to x.

$\perp$ ground connection symbol

———o/ o———   single pole, single throw switch

GREEK LETTERS

SYMBOL	QUANTITY	UNITS
σ (l.c. sigma)	electrical conductivity	mho/meter (mho/m) or ampere/volt-meter (A/V-m)
ρ (l.c. rho)	electrical resistivity	ohm-meter (Ω-m) or volt-meter/ampere (V-m/A)

μ (l.c. mu)   mobility   meter^2/volt-second (m^2/V-sec)

σ_s (l.c. sigma)   surface charge density   coulombs/meter^2 (C/m^2)

ε_0 (l.c. epsilon)   permittivity of free space   farad/meter (F/m)

ε_r (l.c. epsilon)   dielectric constant relative permittivity   no units, just a number

ε (l.c. epsilon)   permittivity   farads/meter (F/m)

τ (l.c. tau)   time constant   seconds (sec)

ρ_m (l.c. rho)   mass density   kilograms/meter^3 (kg/m^3)

π (l.c. pi)   3.1415...   none

ϕ (l.c. phi)   work function   volts (V)

λ (l.c. lambda)   linear charge density   coulombs/meter (C/m)

η (l.c. eta)   viscosity   kilogram/meter-second (kg/m-sec)

ν (l.c. nu)   frequency   hertz (Hz)

ζ (l.c. zeta)   zeta potential   volts (V)

ξ (l.c. xi)   displacement from equilibrium   meters (m)

CHAPTER 2

Introduction to Electrostatics

CHARLES D. HENDRICKS, *Director*
Charged Particle Research Laboratory
Electrical Engineering Department
University of Illinois
Urbana, Illinois

2.1 DEFINITION

In this book we are dealing with the phenomena of electrostatics. Intuitively, most people have a rather well-defined opinion of electrostatic manifestations. However, a precise definition is less easy to invent. Since we are to discuss phenomena which we all understand as belonging to the class generally called electrostatics, it is by necessity that we attempt, as have others, a definition of electrostatics.

Electrostatics is that class of phenomena which is recognized by the presence of electrical charges, either stationary or moving, and the interaction of these charges, this interaction being solely by reason of the charges themselves and their position and not by reason of their motion.

Such a definition clearly rules out the possibility of magnetic and thermal interactions, which are phenomena produced by the motion of the charges themselves. The definition certainly does not exclude the possibility of studying electrostatics as a "mixed bag" of phenomena in which either magnetic effects or electrothermal effects accompany those which we would consider only electrostatic in nature. It is quite possible to find exceptions to the definition of any field of endeavor, since to include all possible phenomena in one field would encroach on many other fields and thus weaken the definition by making it too broad. Thus, under the definition of electrostatics as stated, we shall examine a number of subfields of electrostatics both on a phenomenological basis and on an analytical basis, insofar as is possible considering current theoretical development of the various areas.

8

2.2   TERMINOLOGY

Before it is possible to make a serious study of any subject, it is necessary to bring all parties concerned to a common understanding of the proposed symbolism, terminology, and system of units to be used. To some extent, our definition of electrostatics has roughly outlined the field of interest. If we class electrostatics as part of electrical engineering or of physics or both, and since we would like to use throughout the text a consistent set of units, we adopt the MKS system as being compatible with the usage of quantities such as volts, amperes, and ohms. The MKS system of units has as its fundamental quantities the meter as a unit length, the kilogram as a unit of mass, and the second as a unit of time. These units are sufficient to let us specify volumes, lengths, areas, forces, weights, and other mechanical quantities; but without some other units or defined quantities, electrical properties are not determined. It has been convenient in the past to specify the unit of charge or the unit of current as the fundamental electrical quantity. If the unit of charge is used in the MKS system it is the coulomb(C). Thus the system of units to be used in this book, is the MKSQ system. The symbol Q is used to indicate charge. It should be noted that a rationalized system of units is being used. The significance of this can be seen later in more detail, but we should note that the use of a rationalized system of units places a factor of 4π in the expressions for Coulomb's laws instead of in Maxwell's equations.

2.3   ELECTRONS, PROTONS, NEUTRONS, AND IONS

Basic to the understanding of electrostatics is a thorough background in what I would like to call *fundamental material concepts*. That is, an understanding of the basic materials of which everything is made.

Even at the risk of appearing too elementary, we begin with a discussion of the fundamental particles of matter. For our purposes these can be considered to be electrons, protons, and neutrons. There are others such as positrons, mesons, neutrinos, and perhaps even quarks; however, they appear and are important in fields such as high-energy physics and nuclear physics but not in our study of electrostatics. The electron is in many processes the only important particle because it is small and relatively loosely tied to atoms and molecules. It has a mass m_e of 9.1×10^{-31} kg and an electric charge q_e of -1.6×10^{-19} C. For our purposes, the electron is considered to be the smallest fundamental particle and the only one that has a negative charge.

The proton is another of the basic particles of which materials are composed. It has a positive charge q_p of 1.6×10^{-19} C and a mass M_p of 1.7×10^{-27} kg. The nuclei of all atoms are made up of protons, and hydrogen, the least complex atom, is made up of a proton and an electron. Thus, the hydrogen atom is electrically neutral and has a mass about the same as that of the proton (since the electron mass is only 1/1837 that of the proton). A convenient source of protons is a stream of hydrogen atoms from which the electrons have been removed. These particles are also called hydrogen ions. Neutrons are the third fundamental particle. We can consider the neutron to be identical to the proton except that it is electrically neutral. It has the same mass as the proton. A proton and a neutron combined form the nucleus of deuterium or heavy hydrogen, an important material in many nuclear reactors. A deuterium atom has one electron as does hydrogen.

The combination of two protons and two neutrons forms the nucleus of helium. We see immediately that for the helium atom to be neutral it must include two electrons. The removal of an electron from a helium atom would leave a helium ion.

A molecule is an electrically neutral combination of two or more atoms where the interatomic forces are strong. For example, a hydrogen molecule H_2 is made up of two atoms of hydrogen; a water molecule H_2O is two atoms of hydrogen and one of oxygen. If from any single molecule or atom we remove one or more electrons, we produce a positive ion. If extra electrons are added, a negative ion is produced. It should be noted that for many materials (e.g., chlorine, fluorine, and oxygen) it is easier to produce negative ions and for others (e.g., hydrogen, sodium and potassium) positive ions are more readily produced. Many other substances form positive or negative ions with about equal ease.

The term ion comes to us through its usage in electrochemistry, where it denotes the carriers of electrical current (moving charged particles) in conductive solutions. Here we make a distinction between ions and large charged particles–raindrops, for example. An ion is defined to be a charged particle, either positive or negative, of atomic or molecular size. The ions may exist either in a liquid or in a gas and occasionally in a solid (usually a transient situation).

2.4   ELECTRIC FIELDS

For many centuries it has been well known that if two isolated objects are charged electrically, there will be a force exerted on each by the other. A most common, simple example of this phenomenon is the effect of a hard rubber or plastic comb, charged by running it through someone's hair, on the hair or on a small bit of dust or paper. A plastic utensil can be easily charged by rubbing it with or on almost any material, including objects made of the same kind of plastic. Again, small bits of paper or dust particles are strongly affected by the charged object. Amber and glass rubbed with fur, silk, or wool have been used for centuries to demonstrate electrostatic effects.

To explain the influence of one charged object on another, the concept of *lines of force* has been developed. These lines of force or *electric field* lines are visualized as originating on positive charges and terminating on negative charges. The lines are thought of as elastic bands stretched between the positive and negative charges. The elastic lines also are thought of as repelling each other in a direction perpendicular to the line itself. The lines of force (or electric field lines) between a positive and a negative charge could be represented pictorially as in Fig. 2.1. By convention, the lines have a direction from the positive to the negative charge.

It should be noted that the visualization of field lines in this way can be a powerful tool to provide physical insight into electrostatic problems, but it can be misleading if carried too far.

Since we need quantitative answers as well as physical reasoning in most problems, we require definitions of electric field intensity and other quantities

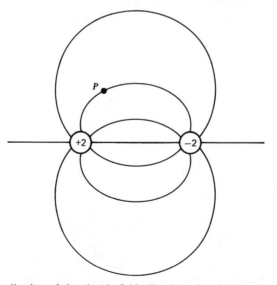

Figure 2.1. Visualization of the electric field "lines" in the vicinity of two charges, $+q$ and $-q$. The point P is a general point in the region at which we calculate the electric field intensity (magnitude and direction).

to be used in analysis. We take this opportunity to define the electric field intensity E as a logical extension of the concept of lines of force. If a minute positive charge q is placed in the vicinity of the two charges in Fig. 2.1, (say, at position P in the diagram) it will be repelled by the positive charge and attracted by the negative charge. The resulting force F will be along the line of force. If we divide the force F by the magnitude of our test charge q we will obtain the force per unit charge on our charge q. This is the electric field intensity E^* or

$$E = \frac{F}{q} \text{ N/C} \tag{1}$$

2.5   CHARGE TRANSPORT AND MOBILITY

Although we are here concerned with electrostatics, charges must be moved from one place to another. Such movement constitutes an electric current. In fact, let us define an electric current I through some surface S as the net number of coulombs of charge passing through that surface per second, as in Fig. 2.2. The current is measured in amperes. Often we need the current per unit area passing through the surface. This quantity, defined as the current density J, is another quantity with magnitude and direction and hence is a vector. Current density has the units of amperes per square meter.

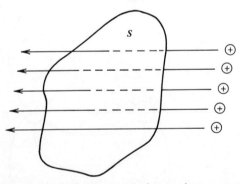

Figure 2.2. A surface S, through which charges are moving, used to compute the current density; N charges of q C each moving through the surface S per second constitute a current I, which is $I = Nq$ A, where 1 C/sec is 1 A.

* The informed reader will recognize that E has a direction associated with it as well as a magnitude. Such quantities, known as vector quantities, include force, electric field intensity, displacement, velocity, and many others. Quantities that only have a magnitude are called scalars. These are quantities such as temperature, mass, charge, and potential.

We consider in this section the transport of charge from one point in space to another and the ease with which the charge moves under the action of an electric field. The media through which the charge is to move is an important aspect in determining the charge behavior. We include solids, liquids, gases, and vacua as the various media to be considered.

Let us first consider motion of charge in a solid. In a good conductor such as most metals, the atoms of the material are quite rigidly held in place. Of course they are subject to vibrations about their average rest positions and, if the solid is heated sufficiently, the atoms may be given enough energy to overcome the binding forces that hold them immobilized. The atoms become relatively free to move around and we say that the material has "melted" or become a liquid. Addition of more heat may provide enough energy to put the material into a gaseous state. In the solid state, however, even though the atoms may be quite rigidly bound, the outer electrons of the atoms may be almost totally free to move about in the solid. When this is the case for a particular material, we class the material as a good conductor. Materials in which the electrons are more tightly bound or have difficulty moving through the material are designated poor conductors. If the electrons are not able to move at all or only with great difficulty, the material is designated a dielectric or an insulator.

A measure of the ease with which a current can be established in a material is called the conductivity of the material. Most materials obey Ohm's law,* which can be written in terms of electric field intensity E and current density J as

$$J = \sigma E \ \text{A/m}^2 \qquad (2)$$

where σ is the conductivity of the material in units of mhos per meter. Thus we see that if Ohm's law describes the electrical behavior of the material, the current density at a point in the material is directly proportional to the electric field intensity at that point and the conductivity σ is the constant of proportionality. The expression for Ohm's law may be written

$$E = \rho J \ \text{V/m} \qquad (3)$$

where ρ is the resistivity of the material and obviously $\rho = 1/\sigma$.

Table 2.1 lists several common materials and their conductivities. It should be clear from the table that the difference between a good conductor (e.g., copper) and a poor conductor or insulator (e.g., lucite or plexiglas) can be

---

* Ohm's law for small conductors such as wires can be written $I = V/R$ where V is the potential difference (V) between two points on the wire, R is the resistance between the two points (Ω), and I is the current that will flow (A).

Table 2.1. Several Common Materials and Their Conductivities

Material	Conductivity (mho/meter)
Aluminum	3.5×10^7
Copper	5.8×10^7
Iron	1.0×10^7
Lead	4.5×10^6
Mercury	1.0×10^6
Nichrome	1.0×10^6
Silver	6.1×10^7
Tungsten	1.8×10^7
Amber	2.0×10^{-15}
Celluloid	5.0×10^{-9}
Glass, plate	5.0×10^{-12}
Rubber, hard	3.0×10^{-15}
Ivory	5.0×10^{-7}
Mica	5.0×10^{-16}
Quartz, fused	2.0×10^{-17}
Sealing wax	1.2×10^{-14}
Shellac	1.0×10^{-14}
Sulfur	1.0×10^{-15}
Wood, very dry	3.0×10^{-9}
Lucite, plexiglass	1.0×10^{-13}
Polystyrene	1.0×10^{-17}

expressed by the conductivity. We can consider the current to be carried in almost all solids by electrons. The carriers in semiconductors can also be "holes" or vacancies where electrons have been removed, but for our purposes, we consider currents in solid materials to be only moving electrons.

2.6 ELECTRICAL CONDUCTION IN LIQUIDS

Mechanisms of conduction of electricity in liquids and gases are quite different from those in solids in many instances. In liquids such as liquid ammonia, mercury, and other molten metals, the conduction is still primarily by means of free electrons. However, in liquids such as salt water and other aqueous solutions, conduction occurs by means of positive and negative ions. Electrons entering the liquid from a metal electrode (cathode) attach immediately to a molecule or atom and form a negative ion or anion. At the other metallic electrode (anode), electrons are removed from atoms or molecules, thus forming positive ions or cations.

Dissociation of molecules in the body of the liquid may lead to the formation of positive and negative ions. If an electric field is present in the liquid, the anions will move toward the positive terminal or anode. The cations will move toward the negative terminal or cathode. Thus, because of the motion of these positive and negative particles, a current exists in the liquid. Since ions are much larger than electrons, they have much more difficulty getting through the liquid than do electrons in most good conductors. The average velocity acquired by a charged particle in a gas or a liquid is constant after its initial acceleration by a constant electric field. This can be expressed as

$$v = \mu E \text{ m/sec} \tag{4}$$

where v is the average velocity of the particle, μ is its mobility, and E is the applied electric field intensity. The units of mobility are meters squared per volt-second or velocity per unit field intensity.

Since current density can be written

$$J = nqv \text{ A/m}^2 \tag{5}$$

where n is the number of charges per unit volume with charge q traveling with a velocity v, we see that the current density is dependent on the number of carriers per unit volume and their velocity and charge. Since the velocity is dependent on the electric field intensity and the mobility as well as on the charge per particle, we see that current density is

$$J = nq\mu E \text{ A/m}^2 \tag{6}$$

If more than one carrier type is present, the current density is the sum of all carrier motion. For only two types, positive and negative, we can write

$$J = (n_+ q_+ \mu_+ + n_- q_- \mu_-)E \text{ A/m}^2 \tag{7}$$

for the current density. Thus the conductivity is

$$\sigma = (n_+ q_+ \mu_+ + n_- q_- \mu_-) \text{ mho/m} \tag{8}$$

If impurity particles are present in the liquid, they also may take part in the conduction process and can be treated as any other charged particle such as ions.

From the foregoing discussion, we see that low conductivity may be the result of either low mobility or low numbers of carriers present. Injection of charge carriers into a liquid insulator may increase its conductivity by many orders of magnitude. Such injection may take place at the tip of a sharp projection on an otherwise smooth electrode or from the sharp corner of an electrode. The injection, which can also be deliberate, can be by means of an ion or electron beam or by means of photoelectrons injected by pulsed or continuous beams of light.

2.7 ELECTRICAL CONDUCTION IN GASES AND IN VACUA

The fundamentals of conduction of electricity in gases follow very closely those of conduction in liquids. There are some differences, however, so we should briefly discuss gaseous conduction.

In all gaseous media there are a few ions and electrons present which are produced by cosmic rays, naturally occurring radioactive decay, and so on. Thus when an electric field is applied to a gas between two electrodes, some current will flow.

As the electric field is increased, the current increases until all the carriers that are naturally present are being removed as they occur. Further increase

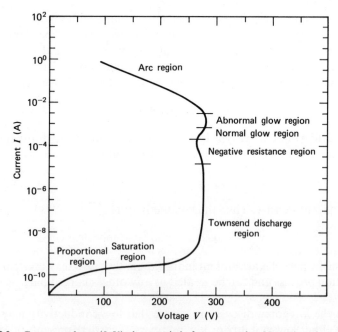

Figure 2.3. Current–voltage (I–V) characteristic for a gas tube. Note that the current and voltage scales are severely dependent on electrode geometry and on the gas composition.

of electric field will result in almost no increase in current until other effects begin to occur. If the electric field becomes large enough, electrons will be accelerated to velocities great enough to knock other electrons loose from neutral atoms or molecules when they collide. There is then an increase in the number of carriers present in the gaseous medium. Further increase of the electric field significantly increases the current flowing between the electrodes, but in a controlled manner. Still further increase of electric field results in current densities great enough to heat local regions of the electrodes to the point at which thermionic emission of electrons occurs. The current density rises almost uncontrollably, and we say an *electric arc* has developed. This is represented in Fig. 2.3. The various regions of the current *I* versus potential *V* curve are labeled with the commonly used terms.

The detailed behavior of such discharges depends on external circuit characteristics, gas pressures, and other parameters; therefore, the curve in Fig. 2.3 must be considered to be general at best. However, it illustrates the regions of interest.

2.8   CORONA

If any of the electrode surfaces have points, sharp corners, or other regions where the radius of curvature is very small, very high electric field intensities are possible in small regions. The high field strength gives rise to a discharge called corona. Corona is common when pointed conductors or fine wires are used as electrodes. If the pointed electrode is made positive, electrons in the vicinity of the point are accelerated toward the point and acquire sufficient energy between collisions to cause ionization of gas atoms and molecules. The process is self-sustaining, provided the source of high potential can maintain the field intensity. Recombination of electrons and positive ions in the corona region can give rise to high-energy photons which produce a few electrons in the region around the discharge. Electrons produced in the high field region will become involved in the corona discharge process. The electrons generated either by photoionization or other processes beyond some rather ill-defined boundary (beyond which the field is not high enough to accelerate the electrons to ionizing energies) do not aid in producing further ionization. Thus corona is usually seen in the near vicinity of sharp points and small radius of curvature regions on conductors at high potentials.

Sharp-pointed conductors which are maintained at high negative potentials will give rise to corona, but the processes may differ somewhat from those positive corona. The electrons may come from secondary emission or field emission or photoemission from the negative electrode.

If the spacing between two conductors is smaller than the distance from the pointed conductor to the outer edge of the corona, it is likely that a disruptive discharge will occur.

Corona discharges in air are usually reddish or violet and brushlike or branched in appearance. In addition to optical manifestations, corona gives rise to radiation of radio-frequency waves. This is quite often noticed as noise in receivers in the vicinity of high-voltage power lines or other high-voltage systems.

Corona discharge from aircraft in flight constitutes a rather severe nuisance. Elimination of the radio-frequency noise from such discharges is an active field in electrostatics on which hundreds of thousands of dollars are spent each year.

2.9   FIELD EMISSION

Electrons in a conductor move around much as though they were a gas at some temperature T. Some of the electrons have sufficiently high velocities that they can escape from the surface of the conductor. However, as soon as they leave the conducting surface, they produce an electric field which tends to pull them back to the surface. If a high-intensity external electric field is applied to the surface, the field can overcome the forces that return the electrons to the surface. The electrons are removed from the surface, and we say that they are the result of *field emission*. Large numbers of electrons can be obtained in very short times by this method. Pulsed systems using field emission of electrons are employed to generate short bursts of x-rays and to produce short flashes of light for high-speed photography.

2.10   THERMIONIC EMISSION AND SPACE CHARGE

If the temperature of the electrons in a material is sufficiently increased, some of the electrons will have energy to overcome the forces binding them to the material. They will still have some kinetic energy remaining after escaping from the surface. Thus it is possible to use a heated material as a source of electrons—an alternative well known to those of us who "grew up" during the vacuum tube era. If a slight electric field is applied to the surface to remove the electrons as they are emitted from the surface, an electric current can be produced. In a vacuum envelope a number of elements can be inserted to control the current by means of electric fields. This is the essence of vacuum tube amplifiers.

If no external fields are applied, the electrons emitted from a high-temperature material collect in a region slightly removed from the surface.

The cloud of electrons builds in density until as many electrons return to the surface as are emitted. The cloud or layer of electrons outside the surface is known as a space-charge layer. This is only one case of what we call a space charge. Space charge can be said to exist in any region in which the electric field that would be present if there were only a total vacuum in the region is modified by reason of the presence of large numbers of charged particles present.

In the previous sections we have seen a number of methods by which charge can be removed from a surface, as well as methods of moving charge through a solid, a liquid, or a gas; a number of necessary quantities and units have also been introduced. We should recognize at this point that the material is introductory and phenomenological for the most part. We continue in this fashion for a few more pages and then present some basic theory and analysis by which we can study in more detail many of the topics that have been mentioned in this introductory material.

2.11   CONVECTIVE CHARGE TRANSPORT

The motion of charges under the influence of electric fields has been mentioned. We should also point out that charge can be transported by moving material to which the charge is attached. A good example of this is found in the Van de Graaff high-voltage generator. In this type of generator, charge is sprayed on a moving belt which transports the charge mechanically to a collector where the charge is removed from the belt. Since large quantities of charge can be moved to an insulated conductor, large and smoothly rounded, very high voltages can be produced.

Suppose we have a moving belt, arranged, as in Fig. 2.4. At the bottom of the belt we spray on a negative charge. The belt is made to be a very good insulator, and thus it carries the negative charge on its surface up into the conducting dome. There the charge is removed from the belt by a brush or series of points near the belt. If we carry a surface charge density σ_s C/m^2 on the belt at a velocity v m/sec up into the dome and the belt is w m wide, a current

$$I = \sigma_s vw \text{ A} \tag{9}$$

is carried into the dome. In a time t sec there will be a charge

$$Q = It \text{ C} \tag{10}$$

placed on the dome. There will be a potential between the dome and the base of

$$V = \frac{Q}{C} \text{ V} \tag{11}$$

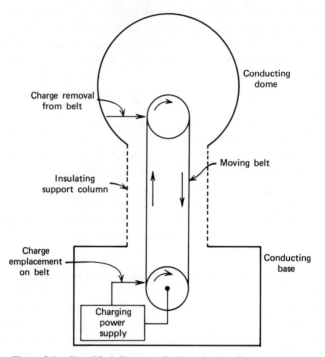

Figure 2.4. Simplified diagram of a Van de Graaff generator.

where C is a constant of proportionality called the capacitance between the dome and the base.

Now we must recognize that the charge can leak off down the belt (even though it carries charge up the column) and down the support column and by means of conduction through the air. If all these paths together have a resistance R Ω, the leakage current will be

$$I_L = \frac{V}{R} \text{ A} \tag{12}$$

We see that if the current carried up by the belt is greater than the leakage current down through the leakage paths, the charge on the dome will increase and the potential (voltage) from the dome to the base will increase. When the two currents are equal, a steady state will have been reached. The potential attainable can be calculated from Eqs. 9 and 12

$$V = \sigma_s vwR \text{ V} \tag{13}$$

The charge stored on the dome is

$$Q = C\sigma_s vwR \text{ C} \tag{14}$$

Let us assume some reasonable numbers for the quantities in Eqs. 13 and 14. Let $R = 10^{13}$ Ω, $w = 0.1$ m, $v = 10$ m/sec, $\sigma_s = 10^{-7}$ C/m^2, and $C = 10^{-10}$ F. With these parameters we can produce a potential difference V between the dome and the base of 10^6 V. There would be a charge Q of 10^{-4} C stored on the dome. As we see later, this corresponds to a stored energy of 50 J.

This short and rather unsophisticated treatment provides an idea of what can be done with quite simple points of view in electrostatics.

Many other examples of convective charge transport are found in electro-fluiddynamic generators, electrohydrodynamic experiments and, in fact, in almost all electrostatic machines. Chapter 8 covers these.

2.12   INDUCTION CHARGING

When an electric field is created in a region containing an insulated conductor or when a conductor is moved into a region where an electric field exists, charges are moved to new locations on the conductor. Figure 2.5 shows a simple example of this phenomenon. Electrodes A and B are connected to a source of high voltage so that an electric field is produced in the region between A and B. The cylindrical conductor C is placed in the space between A and B and immediately one end becomes positive and the other negative, as shown. The net charge is still zero. Now if a connection is made from C to ground, some of the electrons from C can run off to ground. Without moving any of the parts of the system, the connection to

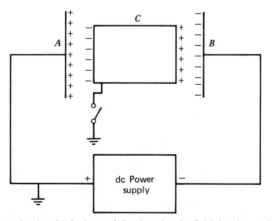

Figure 2.5.   A conducting body located in the electric field in the region between two charged plates.

ground from C is broken. This leaves C with fewer electrons than positive charges. If C moves out of the field produced by A and B, it will carry a net positive charge. We have charged C by "induction."

A similar situation appears in Fig. 2.6, where one surface of electrode A is removable. In this configuration the surface of B will be positively charged and the surface of A will be negatively charged. If the segment A' of A shown by the cross-hatching is removed by moving it slightly toward B to break the connection and then moving it out from between A and B, that segment

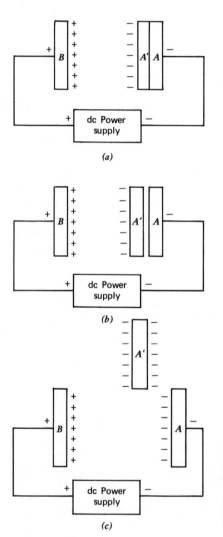

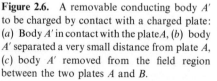

Figure 2.6. A removable conducting body A' to be charged by contact with a charged plate: (*a*) Body A' in contact with the plate A, (*b*) body A' separated a very small distance from plate A, (*c*) body A' removed from the field region between the two plates A and B.

will carry away a negative charge. If this process is repeated periodically by conducting plates mounted on a disc or belt, we have the beginnings of a generator.

It may be appropriate now to treat this problem in more detail, since it does form the basis of many charging systems. Let us consider two parallel plates X and Y, as in Fig. 2.7a. Let the area of the plates be A and

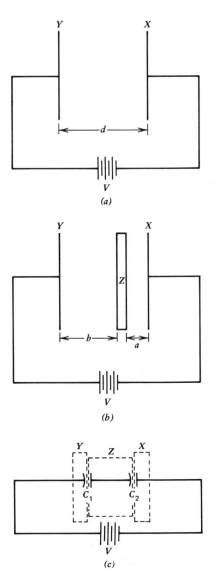

(a)

(b)

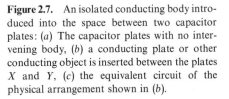

(c)

Figure 2.7. An isolated conducting body introduced into the space between two capacitor plates: (a) The capacitor plates with no intervening body, (b) a conducting plate or other conducting object is inserted between the plates X and Y, (c) the equivalent circuit of the physical arrangement shown in (b).

their separation be d, the space between the plates being filled with a medium whose permittivity is ε. The capacitance of this combination will be

$$C = \frac{\varepsilon A}{d} \text{ F}$$

If we now insert a plate (or other object) Z between X and Y (Fig. 2.7b), we will have created, in effect, two capacitors in series. The equivalent circuit is shown in Fig. 2.7c. Without any initial charge on plate Z, the potential will divide between the two capacitors inversely as the capacitances. There will be no free *net* charge on the plate Z. However, if plate Z is *momentarily* connected to one of the two other plates—X, for example—we have effectively shorted capacitor c_2 in the equivalent circuit of Fig. 2.7c. There will thus be no voltage applied to capacitor c_2 and the entire battery voltage will appear across c_1. This means that even after Z is no longer connected to X there will be no electric field present between Z and X as long as they are maintained in the same positions. There will be a net charge Q on Z where $Q = -c_1 V = -\varepsilon A/d\ V$ C. This charge would be carried away by Z if Z were now removed from between X and Y.

Conversely, if Z has a net charge Q before being placed between X and Y, the potential of Z with respect to X and Y can be adjusted by the magnitude of Q. The *special* case for which $Q = -\varepsilon A/b\ V$ C brings Z to the same potential as X. If $Q = +\varepsilon A/a\ V$ C, the potential of Z will be the same as that of Y.

This system of charging is sometimes called "contact" charging. In fact, the terms "contact" charging or "induction" charging mean different things to many people and almost always must be explained in discussions, to avoid confusion.

2.13 INSULATORS AND DIELECTRICS

In a broad sense we define an insulator as being a material having a very low bulk conductivity. A perfect insulator would, of course, have zero conductivity. A number of materials are very good insulators. Several of these are shown in Table 2.1. Quartz, along with some modern plastics, is probably the best. For many years quartz, sulfur, and sealing wax were used when exceptionally good insulating properties were needed.

One of the most difficult problems to overcome when very good insulators are needed is that of surface cleanliness. Regardless of how good the bulk properties are, if the surface of an insulator is dirty or damp, there will be high surface current leakage when voltages are applied to the insulator.

The shapes of many insulators are useful in keeping the surfaces uncontaminated but really do very little to decrease leakage currents otherwise. Deep slots cut into an insulator are effective in reducing leakage, however. This result is not always due merely to the increase in path length but often to the field configuration produced by the slots.

A word or two on the definition of an insulator may be helpful at this point. Let us consider a three-part thought problem. Assume that we have blocks of two materials which we will charge electrically. After being charged, these two blocks are set on a conducting surface which is connected to ground as in Fig. 2.8. After a very long time t, we measure the charge

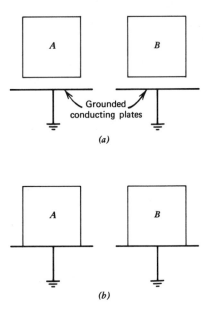

(a)

(b)

Figure 2.8. (a) Two blocks A and B, with different conductivities are given a charge while they are insulated from ground. (b) The blocks are placed on the grounded plates and allowed to discharge. The time for the charge on the blocks to decay to $1/e$ of the initial values is dependent on the ε/σ time constant of the material.

left on the two blocks and find that it is zero or at least below the sensitivity level of our instruments. If we are prohibited from measuring in a time shorter than t_1 we always find the charge is gone and we conclude that the two materials are both conductors.

Now let us perform the experiment again, this time making our measurements in a shorter time t_2. Now we find that the charge on block A is relatively unchanged from the initial value. The charge on block B is again zero as far as our instruments are concerned. Again, if we can never measure in a time shorter than t_2 and are not concerned with longer times, we conclude that block A is made of an insulator and block B is made of a conductor.

Now we perform the experiment in a very short time t_3. When we make our charge measurements we find that neither block A nor block B has lost any charge. Thus, if we are only concerned with times t_2 or less, both materials are considered to be insulators.

Now we must ask, What parameter is really of importance in the characterization of an insulator? From our thought experiment we must conclude that the quantity is the time constant for charge transfer. This time constant is

$$\tau = \frac{\varepsilon}{\sigma} \text{ sec} \tag{15}$$

where ε is the permittivity of the material.*

The quantity τ is sometimes known in the literature as the charge relaxation time constant. If a charge density ρ_0 is somehow put into a material and "released" at $t = 0$, the charge density at any later time t is given by

$$\rho = \rho_0\, e^{-\sigma t/\varepsilon}$$

where ε is the permittivity of the material, σ is the conductivity, and ρ is the charge density. This is similar to the RC time constant in a capacitive discharge circuit. It can be shown that the quantity τ is essentially the "RC" time constant for the material in bulk form. It is the time required for the charge density to fall to $1/e$ of its value, e being the base of natural logarithms and having an approximate value of 2.718. Thus $1/e = 0.368$, and the time constant or relaxation time is the time for the charge density to fall to about 37% of its original value.

The permittivity is a property of a material that indicates its ability to store energy when placed in an electric field. The energy per unit volume stored in a material in an electric field is

$$W = \tfrac{1}{2}\varepsilon E^2 \text{ J/m}^3 \tag{16}$$

The permittivity is also a property that is utilized in a capacitor. As we see in more detail later, if two conducting plates of area A are parallel and separated a distance d, they constitute what is called a parallel-plate capacitor. If a potential difference V is applied between the plates, a charge Q is found on each plate ($+Q$ on one and $-Q$ on the other). The magnitude of Q is proportional to V and may be expressed

$$Q = CV \text{ C} \tag{17}$$

---

* This expression is sometimes written $\tau = \varepsilon_r \varepsilon_0 \rho$ or $\tau = \varepsilon_r \varepsilon_0/\sigma$ where ε_0 is the permittivity of free space, ε_r is the relative permittivity or dielectric constant of the material, and ρ is the volume resistivity of the material.

The constant of proportionality C is called the capacitance of the system. For the parallel-plate configuration described, we can write

$$C = \frac{\varepsilon A}{d} \tag{18}$$

Thus a higher value of ε for a given potential difference V will lead to a higher magnitude of the charge Q stored. The quantity ε is the permittivity of the material between the plates, and for vacuum we assign the so-called free space value of ε_0, which is 8.85×10^{-12} F/m. Air has only a slightly different value. The permittivity ε of a material is expressed in terms of the permittivity of free space as

$$\varepsilon = \varepsilon_r \varepsilon_0 \text{ F/m} \tag{19}$$

where ε_r is the relative permittivity or is sometimes called the dielectric constant. The value of ε_r for most materials is found to be between about 2 and 10. For most gases ε_r is slightly greater than 1.0 and for water it is about 80. Some ferroelectric materials such as barium titanate and lead-zirconate-titanate (PZT) have relative permittivities as high as 1200. Some typical values of permittivity are presented in Table 2.2.

Table 2.2. Relative Permittivities (Dielectric Constant) and Dielectric Strengths of Some Common Insulating Materials

Material	Relative Permittivity	Dielectric Strength (V/m)
Amber	2.8	—
Bakelite	4.9	2.4×10^7
Cellulose acetate	3.8	1.0×10^7
Mica	5.4	1.0×10^8
Plexiglass (lucite)	3.4	4.0×10^7
Polystyrene	2.5	2.4×10^7
Porcelain	7.0	6.0×10^6
Titanium dioxide	90	6.0×10^6
Barium titanate	1200	5.0×10^6

A good dielectric should in general be a good insulator. This is not always the case, however. Water, which has a high dielectric constant, is not a very good insulator. The measure of quality of a good dielectric may then be its polarizability rather than its conductivity. If an electric field is applied to a dielectric, the mean position of bound electrons in the material will be shifted slightly. This is much as though the positive particles and the electrons were tied together by springs and the stronger the applied field, the more separation would occur between the average electron position and average positive particle position. When the field is removed, the electrons return to their original positions and the energy stored as a result of the displacement is returned to the field. If the dielectric has nonzero conductivity, an applied dc electric field will give rise to a flow of current in the material in addition to the average displacement of electrons from equilibrium positions. If we wait for long enough, the dielectric will appear as a conductor. That is, free charge will migrate to the extremities of the dielectric and the charge at the surfaces will completely neutralize the field inside the material. The current will then stop flowing.

Here we have again demonstrated the difference between a conductor and an insulator—the time scale of our measurement. If one material has a conductivity of 10^{-5} mho/m and another has a conductivity of 10^2 mho/m and both have the same dielectric constant, the time constant for charge transfer in the first will be 10^7 times that for the second. Thus if we apply an electric field to the first material, it may reach equilibrium in 10^{-6} sec. If the field is applied and measurements are not taken until several microseconds later, the material is to all intents a conductor. If we wait only 10^{-13} sec to make measurements on fields and charge configuration on the second material, it looks like a conductor. If our time scale of measurement is 10^{-10} sec, the first material appears to be an insulator; the second, a conductor. Formally, if the time constant for charge transfer τ_q is short compared with the interval between measurements of electrical quantities, the material will appear as a conductor. If τ_q is very long compared with the measurement interval, the material appears to be an insulator.

2.14 BREAKDOWN STRENGTH

Another insulator property of interest in electrostatics is the breakdown strength of the material. This is sometimes referred to as the dielectric strength. If an electric field is applied to an insulating material, an electrical stress is produced in the material. The field tries to move electrons one way and positive particles in the opposite direction. Because of cosmic rays, local natural radioactive decay, and thermal effects, there are always a few free electrons in the material for short times. A strong electric field can accelerate these free electrons before they attach to fixed molecules or atoms. If the electrons gain sufficient energy to dislodge other electrons, the process will "avalanche" and a spark or arc will result. The magnitude of the electric field intensity necessary to produce this breakdown process in a material is termed the breakdown field strength or dielectric strength. Table 2.2 lists several insulating materials and their breakdown field strengths.

CHAPTER 3

Mathematical Formulation of Electric Field Analysis

CHARLES D. HENDRICKS, *Director*
Charged Particle Research Laboratory
Electrical Engineering Department
University of Illinois
Urbana, Illinois

3.1 OVERVIEW

In this chapter we provide some of the mathematical formulation of electric field theory to enable the reader to analyze electrostatic field problems. This is necessary if he is to use the examples given throughout the book and extend the techniques to solve new problems.

We begin with Coulomb's law as our foundation. Although this law is in reality an experimental result, we treat it as a fundamental postulate. From the force exerted on one charge by another as expressed by Coulomb's law, we develop an expression for the electric field intensity at a field point owing to a point charge at some other location. The theory for point charges is then developed for extended distributions of charge. These may be volume, surface, or line charges.

Gauss's law is obtained from the expressions for electric field intensity, and the concept of electric flux density is discussed. The results of the classic ice pail experiment of Faraday are described, and their significance in field theory is pointed out.

To make it easier to solve field problems, the concept of potential is introduced. Examples are presented in which it is a distinct advantage to use the potential rather than to solve the field problem directly.

After an introduction to potential theory, the storage of energy in electric field systems arises rather naturally. This brings up the concept of capacitance

29

and charge accumulation. Polarization and induced free charge are discussed in their connection with capacitance and charge accumulation.

Fringing fields, guard rings, and shielding are also important in capacitance measurements and are treated at some length. The concepts of field lines and equipotential surface are introduced, along with field mapping and various analogs used in studying electric field systems.

The interaction of fields and material objects, forces, and motion make up the last part of the chapter. We see here the coupling of the electric field equations and Newton's second law. Also, we learn that energy can be converted from electrical to mechanical and from mechanical to electrical in electrostatic systems. These effects make possible electrostatic motors, generators, pumps, fans, and many other interesting and useful devices.

3.2 HISTORICAL BACKGROUND

Since early Greek times it has been known that when amber is rubbed it acquires a property that allows it to attract small bits of paper or other small objects. The Greek work for amber is $\eta\lambda\epsilon\kappa\tau\rho\nu$ from which we derive our words for electricity. Thales of Miletus knew of this property in the sixth or seventh century B.C. In the sixteenth century, William Gilbert investigated the electrification of many substances and concluded there were two types of materials which he termed "electrics" and "non-electrics." The "electrics," which he could charge by rubbing, were generally what we now call insulators. The "non-electrics" were materials he could not charge by rubbing; these included all the metals.

In the seventeenth century DuFay found it necessary to postulate two kinds of electricity, which he called "vitreous" and "resinous." Franklin, a few years later, renamed these "positive" and "negative" electrification— names that have persisted to this day! It was known that objects charged with the same type of charge repelled each other and objects charged with different types of charge attracted.

It remained for Coulomb in 1785 to show quantitatively that the force between two isolated, small charged bodies varies inversely with the square of the distance separating them and directly as the magnitude of the charges on the objects. To determine this experimental fact, Coulomb used the torsion balance. An appropriate starting point in a discussion of the theory of electrostatics is *Coulomb's law*. Although this is an experimentally determined relationship, we use it as though it were the result of a mathematical derivation, not subject to question. Stated in words, Coulomb's law is

The force exerted on one point charge by another is proportional to the magnitude of each of the charges, inversely proportional to the square of the distance between the charges, and has a direction along the line joining the two charges. It is an attractive force if the charges are opposite in sign and repulsive if the charges have the same sign.

In terms of an equation this can be written

$$F_{12} = k \frac{q_1 q_2}{r^2} \text{ N} \tag{1}$$

where F_{12} is the force on charge 1 owing to the presence of charge 2, or vice versa, q_1 and q_2 are the charge magnitudes, r is the distance between the charges as in Fig. 3.1, and k is a constant of proportionality that depends

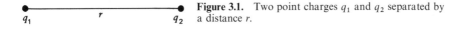

q_1 r q_2

Figure 3.1. Two point charges q_1 and q_2 separated by a distance r.

on the units we use for q and r. In the rationalized MKS system which we are using, k is usually written as $1/4\pi\varepsilon$. Now we have replaced k by another constant ε, which is the *permittivity* of the space in which the charges are located. The value of the permittivity is different for different materials. For empty space (perfect vacuum) we write $\varepsilon = \varepsilon_0 = 8.85\,\text{pF/m}$. The permittivities of various materials (see Table 2.2) are larger than the permittivity of empty space. It is convenient to write the permittivity of a material in terms of that of empty space (sometimes referred to as "free space"). Thus $\varepsilon = \varepsilon_r \varepsilon_0$ where ε_r is the relative permittivity or, as it is sometimes called, the dielectric constant. The dielectric constants of most common materials are between 1.0 and 80.0. Water has a dielectric constant of about 80. Some ferroelectric materials have much higher dielectric constants (e.g., barium titanate, with $\varepsilon = 1200$). However, most common insulators have dielectric constants of less than 10.

Let us now examine the ratio of the force on a charge q' under the influence of another charge q to the magnitude of the charge q'. We have from Coulomb's law

$$\frac{F}{q'} = \frac{qq'}{4\pi\varepsilon r^2/q'} \text{ N/C} \tag{2}$$

where r is again the distance between the two charges. If we let the magnitude of the charge q' become smaller and smaller, we see that the force F becomes smaller and smaller also. In fact, in the limit where q' becomes vanishingly small, the ratio of the force F to the charge q' becomes the value of the electric field intensity E at a distance r from the charge q. The expression for the electric field intensity E of a point charge q is given by

$$E = \frac{q}{4\pi\varepsilon r^2} \text{ N/C} \quad \text{or} \quad \text{V/m} \tag{3}$$

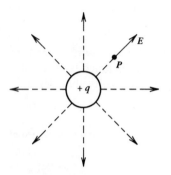

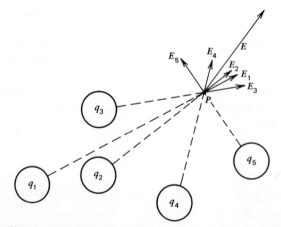

Figure 3.2. Visualization of the electric field in the vicinity of an isolated positive point charge q.

and the E field has a direction radially away from the charge q as in Fig. 3.2. We see immediately that the force on a charge Q in an electric field with intensity E is

$$F = QE \text{ N} \tag{4}$$

If there are a number N of point charges present as in Fig. 3.3, each will contribute to the electric field intensity E at some point P in space. To

Figure 3.3. The resultant field E as the sum of the fields from several point charges q_1, q_2, q_3, q_4, q_5.

determine the resultant field, the field from each charge is added vectorially to those from each of the other charges, since we must include direction as well as magnitude in combining the field intensities. This can be written very simply as

$$\mathbf{E} = \mathbf{E}_1 + \mathbf{E}_2 + \mathbf{E}_3 + \cdots + \mathbf{E}_n = \sum_{i=1}^{n} \mathbf{E}_i$$

where $\mathbf{E}_i$ is the electric field intensity at the point in question owing to the presence of the charge q_i. The $\mathbf{E}_i$ quantities are vectors and the summation must be done vectorially. The expression for $\mathbf{E}_i$ is given by

$$\mathbf{E}_i = \frac{q_i \mathbf{r}_i}{4\pi\varepsilon r_i^3} \tag{3'}$$

where $\mathbf{r}_i$ is the vector distance of the field point from the charge q_i as in Fig. 3.3. The force on a charge Q in magnitude *and* direction is then given as

$$\mathbf{F} = Q\mathbf{E} \tag{4'}$$

3.3 GAUSS'S LAW

If the charge configuration has an appropriate symmetry, Gauss's law provides an easy means for determination of field intensities. Let us examine a rather unsophisticated development of Gauss's law from the Coulomb's law expression for the electric field. We have

$$E = \frac{q}{4\pi\varepsilon r^2} \text{ V/m} \tag{5}$$

where the electric field direction is radially outward from the charge q.

Now let us define the Maxwell displacement D as

$$D = \varepsilon E \text{ C/m}^2 \tag{6}$$

Then for a point charge we can write

$$D = \frac{q}{4\pi r^2} \tag{7}$$

If the displacement D is integrated over a closed spherical surface with radius r, we have

$$\oint D_\perp \, dA = \int_{\theta=0}^{\pi} \int_{\theta=\phi}^{2\pi} \frac{q}{4\pi r^2} (r^2 \sin\theta \, d\theta \, d\phi) = q \text{ C} \tag{8}$$

where $D_\perp$ is the component of D perpendicular to the surface element $dA = r^2 \sin \theta \, d\phi$. Thus, for this simple case, the surface integral of the D vector over a closed surface is equal to the charge enclosed in that surface. This result can be generalized for extended charge distributions and irregular shapes and is known as Gauss's law. Formally stated, Gauss's law is as follows:

> If a closed surface S of any shape is constructed in a region in which an electric field is present, the surface integral of the normal component of D over the surface S is equal to the net free charge enclosed by the surface S.

Although Gauss's law enables us to find the fields produced by simple configurations of charge (e.g., spheres, cylinders, planes), the most significant use is in developing boundary conditions and establishing many theoretical aspects of field theory. Both these uses of Gauss's law are presented in later sections.

Let us now consider an example of a problem for which the fields can be determined by the use of Gauss's law.

Assume that we have a long circular cylinder of charge of radius R as in Fig. 3.4. Furthermore, let us assume that the charge is distributed uniformly

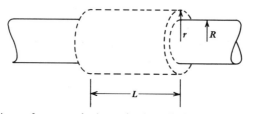

Figure 3.4.   Gaussian surface around a long, circular cylinder of charge.

throughout the volume of the cylinder and that there is a charge of λ C per unit length of the cylinder. Because of the symmetry of the charge distribution in both length and azimuth, the electric field will be radial and will be the same magnitude at every point on an imaginary cylindrical surface of radius r concentric with the charge cylinder. As the surface S over which we integrate the normal component of D we will, indeed, pick the cylindrical surface of radius r and length L, as in Fig. 3.4. We know from Gauss's law that the surface integral of D over the closed surface S must be equal to the charge enclosed by S.

There are essentially two problems here: one if we pick $r > R$, and another if we pick $r < R$. If $r > R$ then the charge enclosed by S is λL so that

$$\oint D \, dA = \lambda L \qquad r > R \text{ C} \tag{9}$$

But we know by symmetry that D is constant in magnitude and always normal to the surface everywhere on S. Thus in the integral on the left of the equation, D is a constant and can be taken out of the integral. It should be noted that the integrals over the ends of the closed cylindrical surface S contribute nothing because D is parallel to those surfaces and, hence, no flux of D passes through the ends. We have then

$$D \int_{z=0}^{L} \int_{\phi=0}^{2\pi} r \, d\phi \, dz = \lambda L \tag{10}$$

and after the integration we write

$$D = \frac{\lambda}{2\pi r} \tag{11}$$

Since $D = \varepsilon E$, we have for the electric field intensity

$$E = \frac{\lambda}{2\pi \varepsilon r} \tag{12}$$

Now let us examine the case where $r < R$. The left-hand side of the equation remains the same, but now the charge enclosed by the surface S is no longer λL. The charge enclosed is

$$q = \int_{\substack{\text{volume enclosed} \\ \text{by surface } S}} \rho \, dv \tag{13}$$

where the charge density ρ is given by

$$\rho = \frac{\lambda}{\pi R^2} \tag{14}$$

which is the charge per unit length of cylinder divided by the volume of a unit length of charge cylinder. Thus for our case where $r < R$, we have

$$q = \int_{r=0}^{r} \int_{\phi=0}^{2\pi} \int_{z=0}^{L} \frac{\lambda}{\pi R^2} r \, d\phi \, dr \, dz \tag{15}$$

which is

$$q = \frac{\lambda L r^2}{R^2} \tag{16}$$

and we have for the value of D at points inside the charge cylinder

$$D = \frac{\lambda}{2\pi R^2} r \tag{17}$$

and

$$E = \frac{\lambda}{2\pi \varepsilon R^2} r \tag{18}$$

Figure 3.5 plots the magnitude of E as a function of radius for this particular configuration.

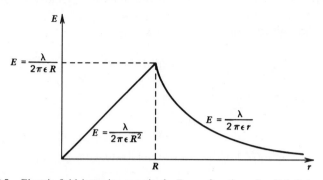

Figure 3.5. Electric field intensity magnitude E as a function of radial distance r from the axis of a long cylinder of charge whose outer radius is R.

3.4   FARADAY ICE PAIL EXPERIMENT

A classic experiment in electrostatics was performed by Michael Faraday in order to derive an exact law for induction. He lowered a charged object into a closed conducting (metal) container by means of an insulating thread (Fig. 3.6). To the outside of the container (ice pail) he connected an electroscope.* When the charged object was lowered into the bucket, the leaves on the electroscope diverged and when it was removed, the leaves collapsed. The maximum divergence occurred when the object was in the container with the metal cover in place.

If the experiment is repeated but the charged object is allowed to rest against the wall or bottom of the container before being removed, it is found that the leaves of the electroscope do not collapse when the object is removed. The leaves remain deflected by the same amount as with the object in the container.

The analysis of this experiment is quite interesting and very basic to electrostatics. We first note that the metal container, the lid, and the

---

* A device for determining the presence and relative magnitudes of charge or electric fields is called an electroscope. It is made up of a glass Erlenmeyer flask, a stopper, and a conducting rod with two attached gold foil "leaves" and a knob on top (Fig. 3.6). When a charge is placed on the knob, the gold leaves are similarly charged and repel one another. Thus they diverge. If the charge is removed the leaves collapse and hang straight down. It is possible to calibrate the leaves in terms of angular deflection per unit charge, but other more convenient instruments are now available for such measurements.

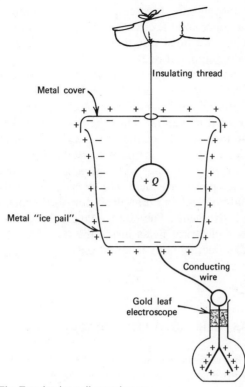

Figure 3.6. The Faraday ice pail experiment.

electroscope were uncharged (i.e., had a total net charge of zero). With the charged object in place but not touching the container, the net charge on the container and the electroscope was still zero. However, the positively charged object gave rise to field lines that acted on electrons in the container and caused a redistribution of charge on the container and the electroscope. Electrons were removed from the outside of the container and from the electroscope until there was a total negative charge on the inside container wall to equal the positive charge carried in by the charged object. Thus an equal positive charge was left on the outside surface of the container and the electroscope.

In this configuration, the external manifestations of the field are not such that the position of the object inside the container can be determined. It could even be touching the container and no charge would be observed on the electroscope. Gauss's law would tell us that the normal component of the displacement field flux (or D) for *any* closed surface must be equal to the charge enclosed by the surface. Consequently, if our Gaussian surface encloses the entire container and the electroscope, or if it is drawn to include only the charged object, we must obtain the same result for the surface integral of the normal component of D.

If the outside is grounded while the positively charged object is inside the container, the electroscope leaves will collapse. If then the charged object is removed (after the ground has been disconnected, the electroscope leaves will diverge to the original maximum. Thus the container is left with a net charge equal in magnitude to that of the charged object. However, it can be shown that the remaining charge is opposite in sign to that of the object.

In 1837 Faraday performed another very significant experiment which is related closely to the ice pail experiment. He insulated from the ground a large conducting metal box and had himself shut up in the metal box along with field measuring instruments. The box was then charged and Faraday detected no evidence of electrical fields inside the box. This demonstrated that the charge on a conducting body is found only on the outside surfaces and not on the inner surfaces.

3.5 ELECTROSTATIC POTENTIAL

A two-dimensional diagram of the magnitude of the electric field intensity in the vicinity of a point charge would look like Fig. 3.7. The "hill" would have cylindrical symmetry about the line down through the charge from the

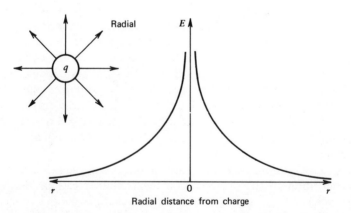

Figure 3.7. Electric field intensity in the vicinity of a point charge, indicating the magnitude as a function of radial distance r.

"peak" of the hill. A small positive charge q' placed at a distance r from the charge q would experience a force

$$F = \frac{qq'}{4\pi\varepsilon r^2} = q'E \text{ N} \tag{19}$$

in a direction away from q. To keep the charge q' at this point requires an external force to be exerted on q'. Now if q' is allowed to move to a position $r + dr$ from charge q, an amount of work $F\ dr$ is done on the external system. This is also $q'E\ dr$.

If we adopt the convention that this is energy leaving our electrical system and that incoming energy is positive, then the energy dw will have a negative sign. In terms of the energy w of our electrical system, we can write the small change in energy as

$$dw = -F\ dr = -q'E\ dr \text{ J} \tag{20}$$

Since only motion parallel to the direction of the force F (and consequently to E also) can result in energy changes, and since the direction of the force at a given point is constant regardless of the direction of motion of q', the system is *conservative*. That is, the total energy change by reason of a movement of q' along any path whatsoever is zero if q' is returned to its original position. This is equivalent to saying $-q'E\ dr$ is an exact differential. If we divide dw by q' we have an expression for an energy change per unit charge. We define this as a change in the electric potential dv. Thus

$$dv = -E\ dr \text{ V} \tag{21}$$

Since dv is an exact differential, we can write for the difference in electric potential at two points A and B

$$V_B - V_A = \int_A^B -E\ dr \tag{22}$$

The quantity $V_B - V_A$ depends only on the locations of the points A and B and not on any path taken by the charge q' to get from A to B. This is a property of the conservative nature of the electric field. The quantity $V_B - V_A$ is measured in volts and the units are joules per coulomb.

Let us examine the potential difference between two points p_B and p_A at different distances from a point charge q as in Fig. 3.8. The electric field intensity E of a point charge is $q/4\pi\varepsilon r^2$. Let the two points be located at r_A and r_B. Then we have

$$V_B - V_A = \int_{r_A}^r -\frac{q\ dr}{4\pi\varepsilon r^2} = \frac{q}{4\pi\varepsilon}\left[\frac{1}{r_B} - \frac{1}{r_A}\right] \tag{23}$$

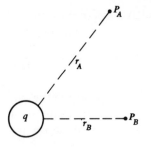

Figure 3.8. Two points P_A and P_B located at different distances r_A and r_B from a point charge q.

Since we can only define potential difference and never an absolute potential, it is convenient to have a reference potential which we designate as "zero potential." This is analogous to assigning zero altitude to sea level.

In the case of our point charge configuration, we let point A move to infinity and arbitrarily assign to V_A the value of zero. Then $V_B - V_A$ is the difference in potential between the point B and infinity and we say that the potential at point B is $V = V_B - V_A$ and $r_B = r$. Then

$$V = \frac{q}{4\pi\varepsilon r} \tag{24}$$

We should note that potential like temperature, has no direction. It is a scalar quantity and thus may be combined without regard to vector addition (which must be used in combining electric fields, distances, etc.). For the potential at a point owing to a number N of point charges $q_1, q_2, \ldots,$ at q_N points distant $r_1, r_2, r_3, \ldots, r_N$ from the field point, we can write

$$V = \sum_{j=1}^{N} \frac{q_j}{4\pi\varepsilon r_j} \tag{25}$$

If we have a continuous distribution of charge, the summation becomes an integral and we have for the potential at some point $p(x, y, z)$ in space

$$V(x, y, z) = \int_{\text{all charge space}} \frac{dq(x', y', z')}{4\pi\varepsilon r(x, x', y, y', z, z')} \tag{26}$$

where dq is an infinitesimal element of the charge distribution (which may be a function of position x', y', z') and r is the distance from p to the charge element dq. That is,

$$r = \sqrt{(x - x')^2 + (y - y')^2 + (z - z')^2} \tag{27}$$

The integral must be taken over all space, but usually it is limited because of limitations on the extent of the charge distribution. If, for example, we have a charge distribution in a spherical space, the limits of integration need only be such as to include the sphere in which the charge density ρ is nonzero.

Since it is easier to perform one nonvector integration to find the potential due to a charge distribution than it is to perform the three-dimensional vector integration, we should look for a method of obtaining the electric field intensity from the potential. If we refer to Eq. 21, we see that we had

$$dv = -E \, dr$$

If we generalize and instead of using dr we use $\cos \alpha \, ds$ where ds is a general linear displacement and α is the angle between the direction of E and the direction of the displacement ds, we have

$$dv = -E \cos \alpha \, ds \tag{28}$$

This leads us to consider the relationship between the space rate of change in v and the electric field intensity. We can, in fact, define the electric field intensity as the negative directional derivative of potential taken in the direction of the maximum rate of change of potential with distance. Thus

$$E = -\left(\frac{dv}{ds}\right)_{\max} \tag{29}$$

is the expression for E in terms of v. The electric field has as its direction the direction in which $(dv/ds)_{\max}$ is taken. Thus the electric field is the steepness of the potential slope at any point (in the direction of the greatest steepness).

3.6 ELECTROSTATIC POTENTIAL CALCULATIONS

As an example of the computation of the potential produced by a charge configuration, we consider the problem of two charges of equal magnitude but opposite sign separated by a distance d. Let the positive charge be placed at the origin of a coordinate system and the other on the z axis at $z = -d$, as in Fig. 3.9.

The point at which we wish to find the potential will be located at (x, y, z). We can write for the potential at (x, y, z)

$$V(x, y, z) = \frac{q}{4\pi\varepsilon\sqrt{x^2 + y^2 + (z)^2}} + \frac{-q}{4\pi\varepsilon\sqrt{x^2 + y^2 + (z + d)^2}} \tag{30}$$

or

$$V(R, z) = \frac{q}{4\pi\varepsilon}\left[\frac{1}{\sqrt{R^2 + (z)^2}} - \frac{1}{\sqrt{R^2 + (z + d)^2}}\right] \tag{31}$$

where, because of symmetry about the z-axis, we let

$$R^2 = x^2 + y^2 \tag{32}$$

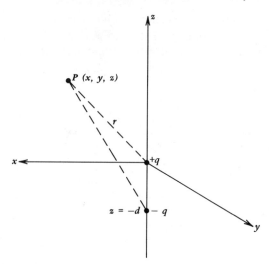

Figure 3.9. A linear dipole formed by two charges $+q$ and $-q$, located at $x = y = z = 0$ and $x = y = 0$ and $z = -d$, respectively.

This expression for V is exact and frequently is quite useful. However, in many problems we need the potential of this configuration when the charge separation d becomes very small compared with the distance of the field point from the dipole $r^2 = x^2 + y^2 + z^2$. To find this we obtain an approximation for V as d becomes small. We can write

$$V = \frac{q}{4\pi\varepsilon}\left(\frac{1}{r} - \frac{1}{r + d\cos\theta}\right) \tag{33}$$

and if we make the approximation for r large compared with d, the angles between the z-axis and the lines from the positive and from the negative charges are the same. Then we write

$$V = \frac{q}{4\pi\varepsilon}\left(\frac{d\cos\theta}{r^2 + r\,d\cos\theta}\right) = \frac{q\,d\cos}{4\pi\varepsilon r^2(1 + d\cos\theta/r)} \tag{34}$$

and recognize that since $d \ll r$, we have $d\cos\theta/r \ll 1$. Our final expression for V is

$$V = \frac{q\,d\cos\theta}{4\pi\varepsilon r^2} \tag{35}$$

which is the potential for a dipole oriented along the z-axis. A plot of the electric field lines and equipotential surfaces appears in Fig. 3.10.

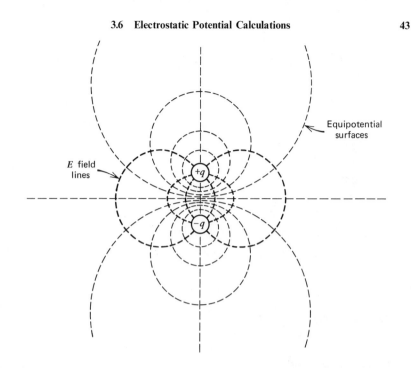

Figure 3.10.   The electric field lines and equipotential surfaces around two charges of equal magnitude but opposite sign separated a distance d.

The electric field intensity of a dipole can be found by taking the gradient of the potential. Alternatively, we can go back and solve the dipole problem again, this time using the expressions for the field intensities for two point charges and adding them vectorially to find the field intensity of the dipole. The same method of making the approximation for $d \ll r$ is used.

If we carefully plot the equipotential surfaces and the field lines for the dipole as in Fig. 3.10 we see that the equipotential surfaces and the field lines are perpendicular.

We should note at this point that with our definition of potential and the definition of a conductor (a material in which no electric field can exist) the potential at every point on a conductor must be the same. That is, a conductor is an equipotential surface. Also, every equipotential surface is equivalent to a thin conducting surface and can be replaced by such a surface. It is often useful in experiments in which particular potential or field configurations are desirable to be able to replace equipotential surfaces by conductors, and vice versa.

As an example of this technique, let us solve the problem of the point charge and conducting plane of Fig. 3.11. We would like the electric field

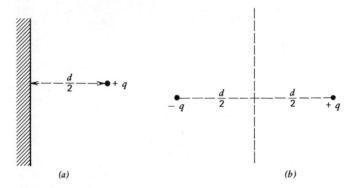

Figure 3.11. (*a*) A positive charge *q* near a conducting surface may be replaced by (*b*) two point charges. The negative charge at distance $d/2$ into the space where the conductor was located is the "image" of the charge *q*.

intensity and potential at a general point in space owing to the presence of the point charge q a distance $d/2$ from a conducting plate in the x–y plane. We have just seen an example (the dipole) in which a plane equidistant between the two charges is at zero potential (let $\theta = \pi/2$ in the expression for the dipole potential). If we replace the conductor in our present problem by a zero equipotential surface and add a negative charge $-q$ at a distance $d/2$ opposite q, we have the problem of two point charges, which is easily solved as before. Since we are only interested in the field intensity and potential distribution in the space on the "positive" side of the zero potential surface, we use only that part of the dipole problem for which $z \geq 0$ or for which $\theta \leq \pi/2$.

This example illustrates a technique widely used for solving problems involving charges near conducting surfaces. It is called *the method of images* because we replace the conducting surface by an equipotential surface and the "image" of the original charge.

3.7 CAPACITANCE AND CAPACITORS

Let us take two uncharged isolated conductors of any shape or size as in Fig. 3.12. The potential difference between the two conductors is initially zero. Now by some means, we carry a charge dq from one conductor to the other so that there is a potential difference dV between the two conductors. (If we transfer twice as much charge $2dq$ from one conductor to the other, we find that the potential difference between the conductors is twice what it was previously.) If we continue the process until we have transferred a

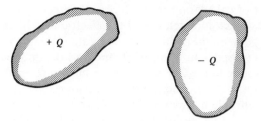

Figure 3.12. Two isolated conductors of any shape or size.

charge Q from one conductor to the other, there will be a potential difference V between the conductors. If we have kept a record of the values of the potential difference as we transferred the charge, it will be evident that the charge and potential difference are related linearly to one another. This may be expressed by the equation

$$Q = CV \tag{36}$$

where C is the constant of proportionality between Q and V. The constant C is called the capacitance of the system of two conductors in this particular arrangement. If the size or position of either conductor is changed or if the medium in which they are immersed is changed, the value of the constant will be different. It is seen that C has units of coulombs per volt. Since the capacitance is used so frequently, we give it a specific unit called the farad.

There are a number of simple configurations for which the capacitance is easily computed. Such calculations may seem trivial, but they give us a feeling for magnitudes of capacity for the configurations and also furnish basis from which to make approximations of capacitance for systems which are not amenable to exact (or even approximate) analytical calculation.

Let us first compute the capacitance of a pair of concentric spherical shells of radius a and b where $a < b$ (see Fig. 3.13).

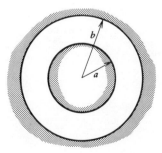

Figure 3.13. Concentric spherical shells of radius a and b, respectively, where $a < b$.

Our method of calculation is to place a charge $+Q$ on one conductor and $-Q$ on the other conductor, find the electric field intensity in terms of V in the space between the conductors, integrate E along a field line between the conductors to find V in terms of Q, and find the quotient Q/V which is, of course, the capacitance C. Let us place a charge Q on the inner sphere. We can use Gauss's law to find the E field because there is a high degree of symmetry in the problem. An imaginary spherical surface of radius r between the spheres is a suitable surface to us. We have then where $a < r < b$

$$\oint D \, dA = Q \tag{37}$$

and since D is constant in magnitude and perpendicular to the surface S, we can write

$$D \oint dA = Q \tag{38}$$

and we see that the integral is the area of the spherical surface S which is $4\pi r^2$. Since $D = \varepsilon E$, we have

$$E = \frac{Q}{4\pi\varepsilon r^2} \qquad a < r < b \tag{39}$$

Now we also know that the potential difference between the spheres is

$$V = V_A - V_B = \int_b^a - E \, dr = - \int_b^a \frac{Q \, dr}{4\pi\varepsilon r^2} \tag{40}$$

which is

$$V = \frac{Q}{4\pi E} \left(\frac{1}{a} - \frac{1}{b} \right) \tag{41}$$

The capacitance is Q/V, which is

$$C = \frac{4\pi\varepsilon ab}{b - a} \text{ F} \tag{42}$$

We should note that if the radius of the outer shell is permitted to get very large, $b \to \infty$, then

$$V = \frac{Q}{4\pi\varepsilon a} \tag{43}$$

and the capacitance of an isolated sphere is

$$C = 4\pi\varepsilon a \tag{44}$$

It is occasionally useful to have some idea of the capacity of specific objects, so let us take the value of a to be 0.01 m (a sphere about as large in diameter as a nickel), located in air. The value of ε for air is about the same as ε_0 which is 8.85 pF/m. Thus the capacitance of the small sphere is

$$C = 1.11 \times 10 \text{ pF} \tag{45}$$

The reader will recognize that even an irregularly shaped object of about the same size will have very nearly the same capacity as a sphere. This fact is quite useful in estimating the stray capacity of a diverse system or of an isolated body or piece of equipment.

Our next example is that of two parallel plates whose spacing is small compared with the lateral dimensions of the plates (Fig. 3.14). Let us assume

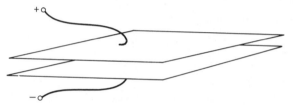

Figure 3.14. Two parallel conducting plates whose separation distance d is small compared with their lateral dimensions.

that the plates are very large in extent and calculate the capacity per unit area of the parallel system. Let us put a surface charge of σ (C/per unit area) on one plate and $-\sigma$ on the other plate. We can argue by symmetry and by Gauss's law that in this case the electric field is confined to the region of space between the plates, that little or no field exists outside this region, and that, consequently, all the charge is on the inner surfaces of the plates.

By using Gauss's law we can establish that the normal component of the displacement D is discontinuous at any surface by the amount of the surface charge density σ. Since the D field is zero in a conductor, we see that at the surface of a conductor $D = \sigma$. Thus in the region between the plates where the field is uniform in direction and magnitude we have

$$E = \frac{\sigma}{\varepsilon} \tag{46}$$

The potential difference between the plates is

$$V = \int_d^a -\frac{\sigma}{\varepsilon}\,dz = \frac{\sigma d}{\varepsilon} \tag{47}$$

The capacitance per unit area is the charge per unit area divided by the potential difference or

$$C = \frac{\sigma}{V} = \frac{\varepsilon}{d} \ \text{F/m}^2 \tag{48}$$

Let us suppose we have two parallel plates one square centimeter in area separated by one millimeter in air. The capacitance is

$$C = \frac{8.85 \times 10^{-12} \times 10^{-4}}{10^{-3}} \ \text{F} \tag{49}$$

which is approximately one picofarad ($1 \ \text{pF} = 10^{-12} \ \text{F}$). This also is a useful size to remember as a rough standard with which to estimate the capacity of other systems.

A third system whose capacitance is easily found is constructed of two concentric conducting circular cylinders of radii a and b where $a < b$. The capacitance per unit length is

$$C = \frac{2\pi\varepsilon}{\ln b/a} \tag{50}$$

This can be verified easily.

One of the early forms for a capacitor was the Leyden jar, which consisted of a glass jar silvered inside and outside up to about three-quarters of its height. Connection to the inner conductor was made by a length of metal chain hanging from a metal rod held in the neck of the jar by a cork or other stopper.

Occasionally the Leyden jar was constructed in the same manner except that instead of using a silver layer for the inner conductor, the jar was filled to about three-quarters its height with salt water. The jar was then set in a trough or other container with salt water to form the outer conductor up about three-quarters of the jar height.

A glass gallon jug about 8 in. in diameter and 12 in. tall can provide a reasonably good Leyden jar. We can consider the capacitor so formed to be a cylindrical capacitor in parallel with a circular plane parallel capacitor. Since capacitances in parallel are additive, we need only determine the two parts and add them. Assume that glass has a permittivity ε which is about five times that of air or vacuum, and let the thickness of the glass be about 2 mm. The jug has a radius of about 10 cm. Thus the area of the bottom of the jug is about 0.031 m^2. If we go back to our expression for a plane parallel capacitor, we see that the bottom of the jug has a capacitance of about 700 pF. The cylindrical part of the jug provides a capacitance of about 4000 pF. Thus the Leyden jar formed from the glass gallon jug provides a total capacitance of approximately 5000 pF.

3.8 ENERGY STORAGE

If we examine the force on a charge in an electric field, we see that we could (conceptually, at least) attach a cord to the charge and let it go, and the motion of the charge would raise a weight. We could thus extract energy from the electric field. This indicates that energy is stored in the electric field as a result of the position of charges and their magnitudes.

Let us examine the energy stored in a charged capacitor. Assume the system where capacitance C is uncharged and hence as zero potential difference between its component parts. Let us also assume that the system is conservative (no dissipative elements in the storage system) and that the stored energy w is a function of some position variable (e.g., plate separation x) and the charge Q. Since the capacitor is initially uncharged, and since no electrical forces are involved in putting it together mechanically, the change in energy dw as charge is moved from one plate to the other is

$$dw = V \, dQ \tag{51}$$

where V is the potential through which the charge dQ moves in being carried from one plate to the other. If we begin with the uncharged system and charge it to a potential V by transferring a charge Q, an external source must add w energy to the system where

$$w = \int_0^Q V \, dQ \tag{52}$$

But as before $V = Q/C$, which means that

$$w = \int_0^Q \frac{Q \, dQ}{C} = \frac{1}{2}\frac{Q^2}{C} \tag{53}$$

for the energy stored in the capacitor. Note that if the energy function is expressed in terms of Q and C, it is called the energy stored in the field. If we express the energy function in terms of potential V and C, it is called the coenergy. Thus we distinguish between the stored *energy*

$$w = \frac{1}{2}\frac{Q^2}{C} \tag{54}$$

and the stored *coenergy*

$$w' = \tfrac{1}{2}CV^2 \tag{55}$$

although they may be numerically equal. The reason for this becomes clear as we examine the forces in electrostatic systems.

3.9 FORCES IN ELECTROSTATIC SYSTEMS

Suppose we have a parallel-plate capacitor whose plates are separated by a distance z and whose area is A, as in Fig. 3.15. The capacitance is then

$$C = \frac{\varepsilon A}{x} \tag{56}$$

where ε is the permittivity of the material between the plates.

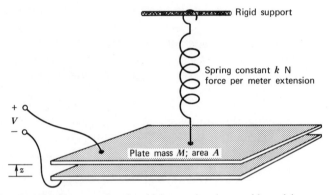

Figure 3.15. Parallel-plate capacitor in which one plate is movable and is suspended by a spring from a rigid support.

We would like to find the force on the plates. Let one plate (at $x = 0$) be fixed and let the other plate at $x = d$ be movable (in theory at least). If there is a force between the plates when the capacitor is charged, any motion of the plates should result in an energy change because of a change in C. If we move a plate a distance dx where there is a force f exerted on the plate (because of the fields present) there will be a change in potential V if the capacitor is isolate (open circuited) or a change in Q if V is maintained constant by means of an attached battery. Let us assume that a positive displacement (x increasing) results in an energy output $-dw = f_e l_x$ from the capacitor if the force f_e is in the $+x$ direction. If a charge dQ is added to the capacitor at potential V, there is an input of energy $dw = V\, dQ$. Thus for a positive displacement and an increasing charge, the energy input to the storage system is

$$dw = V\, dQ - f_e\, dx \tag{57}$$

We have tacitly assumed that w is a function of the independent variables Q and x so that we can also write

$$dw = \frac{\partial w}{\partial Q} \, dQ + \frac{\partial w}{\partial x} \, dx \tag{58}$$

We can equate the two expressions for dw and write

$$\left(\frac{\partial w}{\partial Q} - V \right) dQ + \left(\frac{\partial w}{\partial x} + f_e \right) dx = 0 \tag{59}$$

In terms of power balance, for the power into the system we can write

$$P = \frac{dw}{dt} = VI - f_e v = V \frac{dQ}{dt} - f_e \frac{dx}{dt} \tag{60}$$

This may be integrated with respect to time to give

$$dw = V \, dQ - f_e \, dx \tag{61}$$

Since Q and x are independent, the coefficients of dQ and dx must vanish independently so that

$$V = \frac{\partial w}{\partial x} \quad \text{and} \quad f_e = -\frac{\partial w}{\partial x} \tag{62}$$

This may seem to be a long way around to find f_e, but we know that we can obtain w without having f_e beforehand. For our parallel-plate capacitor,

$$w = \frac{1}{2} \frac{Q^2}{C} = \frac{1}{2} \frac{Q^2 x}{\varepsilon A} \tag{63}$$

so that

$$f_e = -\frac{\partial}{\partial x} \left(\frac{1}{2} \frac{Q^2 x}{\varepsilon A} \right) = -\frac{1}{2} \frac{Q^2}{\varepsilon A} \tag{64}$$

It is important to note that the force has a negative sign whereas Q^2, ε, and A are all positive quantities. This would indicate that the force tends to pull the plates together.

If, on the other hand, we wish to use the coenergy $1/2 \, CV^2$ in the calculation, we must change our equations. We note that

$$d(QV) = Q \, dV + V \, dQ \tag{65}$$

If we substitute for $V \, dQ$ we obtain

$$dw = d(QV) - f_e \, dx \tag{66}$$

and we let

$$dw' = d(QV) - dw \tag{67}$$

and

$$dw' = Q\,dV + f_e\,dx \tag{68}$$

We have now assumed that w' is a function of the independent variables V and x so that

$$dw' = \frac{\partial w'}{\partial v}\,dv + \frac{\partial w'}{\partial x}\,dx \tag{69}$$

and as before we find

$$Q = \frac{\partial w'}{\partial v} \tag{70}$$

Now w' is written for the charged capacitor as

$$w' = \frac{1}{2}\,CV^2 = \frac{1}{2}\frac{\varepsilon A}{x}\,V^2 \tag{71}$$

and

$$f_e = \frac{\partial}{\partial x}\left(\frac{1}{2}\frac{\varepsilon A}{x}\,V^2\right) = -\frac{1}{2}\frac{\varepsilon A V^2}{x^2} \tag{72}$$

which can be written

$$f_e = -\frac{1}{2}\frac{Q^2}{\varepsilon A} \tag{73}$$

We should call attention to the presence or absence of the negative sign in the calculations, depending on whether energy or coenergy is used in the calculation.

Now let us discuss the foregoing section without equations.

If we have two oppositely charged objects such as the plates of a capacitor (or any other objects near each other) there will be a force that tends to pull the objects together. In the case of capacitor plates where the separation is small, the force can be very large. Charged sheets of paper or Saran wrap behave similarly.

If we wish to separate our capacitor plates while they are charged, we must exert a force to pull them apart. When they move, even a small distance, the force times the distance they move is the energy we must expend (or work we must do) to separate them. This energy must go somewhere! And indeed, it shows up as a change in the energy stored in the capacitor or in the field configuration around other nearby objects. That change in energy divided by the distance the object was moved gives us the force exerted to produce the motion. Thus by calculating the energy change that *would* take place *if* we moved the object a very small distance, we can calculate the force without ever actually measuring it directly. This is a very powerful theoretical *and* practical tool in electrostatic studies.

3.10   FORCES AND MOTION

Whenever we have part of a system that is free to move under the action of an electric field, we must bring the laws of mechanical motion into the computations of the system behavior. Newton's second law is written

$$F = Ma \tag{74}$$

where F is the force exerted on a mass M, and a is the resulting acceleration of the mass.

In our case the F involved is composed of the forces exerted by electric fields, springs, friction, and gravity. In many systems friction is minimal, gravity can be ignored except for finding equilibrium positions, and we need only work with spring or suspension forces and forces of electrical origin. When fluids are involved there are sometimes drag forces of viscous origin to contend with.

As a simple example of the coupling of electric field forces and forces external to the electrical energy storage system, let us consider a plane parallel capacitor system as shown in Fig. 3.15. The upper plate is of mass M and is suspended by means of a spring whose elastic constant is k. The lower plate is fixed and the separation of the two plates is z. In the uncharged state, we can calculate the frequency of small oscillation of the upper plate for small displacements from equilibrium, and we also can find the equilibrium position of the plate. The plate hanging on the spring will extend the spring until $-Mg$, the force downward owing to gravity, is balanced by $k(z_0 - z)$, the force upward owing to the stretched spring, where $z = z_0$ is the position of the mass when the spring is unstretched and z is measured upward from the bottom plate. There is no motion of the mass, and therefore we can write

$$Ma = \sum F_j = 0 \tag{75}$$

where the F_j are all the forces acting on the mass. Thus

$$k(z_0 - z) - Mg = 0 \tag{76}$$

and

$$z = z_0 - \frac{Mg}{k} = z_1 \tag{77}$$

is the equilibrium position of the mass (i.e., the upper plate).

If we now displace the plate slightly from its equilibrium position and release it, it will oscillate about the equilibrium position. The equation of motion is

$$M \frac{d^2 z}{dt^2} = +k(z_0 - z) - Mg \tag{78}$$

Since $kz_0 - Mg = kz_1$, the equation of motion becomes

$$M \frac{d^2 z}{dt^2} = k(z_1 - z) \tag{79}$$

which can be written

$$\frac{M d^2 \xi}{dt^2} = -k\xi \tag{80}$$

where ξ is the displacement from z_1. That is, $\xi = z - z_0$. This equation has solutions of the form

$$\xi = A \sin \omega t + B \cos \omega t \tag{81}$$

where $\omega = k/M$. The initial conditions for our problem were a displacement ξ_0 of the mass and no velocity initially. Thus

$$\xi = \xi_0 \cos \sqrt{k/m}\, t \tag{82}$$

is the solution to the problem of the motion of the upper capacitor plate with no applied potential.

Now if we apply a potential between the plates we have another force with which to contend, namely,

$$M \frac{d^2 z}{dt^2} = k(z_0 - z) - Mg - \frac{1}{2} \frac{\varepsilon A V^2}{z^2} \tag{83}$$

If there is no motion, there will be another equilibrium position in which the plates are closer together unless the applied voltage is high enough to overcome the spring tension—in that case, the plates will be pulled together by electrical forces. At equilibrium, $d^2 z/dt^2$ will be zero, and we can solve for the value of z which is the equilibrium position. Let us call this z_2, writing

$$k(z_0 - z_2) - Mg - \frac{1}{2} \frac{A V^2}{z_2{}^2} = 0 \tag{84}$$

For low enough values of V, there will be a real positive value of z_2 as a solution of this equation which will prove to be a stable position for the plate. The equation of motion can be rewritten in terms of small excursions ξ from equilibrium, and we have

$$M \frac{d^2 \xi}{dt^2} = -K\xi \tag{85}$$

where

$$K = k - \frac{\varepsilon A V^2}{Z_a^{\ 3}} \tag{86}$$

The solutions for ξ are the same as before except for a different value of ω which is now

$$\omega = \sqrt{(k - \varepsilon A V^2/Z_2^{\ 3})/m} \tag{87}$$

Thus, with an applied voltage V, the frequency of oscillation is reduced compared with the frequency with no voltage applied. When V is large enough to reduce ω to zero, the upper plate does not oscillate but rather, when displaced slightly, travels directly to the lower plate.

Let us now examine another problem of interest in which we have coupling between electrical forces and Newton's second law. In this problem we place a charge on a macroscopic particle and let the particle interact with an electric field. If the particle is forced to move against the electric field by an external force (a gas stream, momentum of the particle, or other means) then the particle will supply energy to the electrical system and we have a generator. If the particle is accelerated by the electric field and given kinetic energy, then we have a motor, a pump, or a means of obtaining high-velocity particles.

We begin by writing Newton's second law once again and including the field forces. This time we do not include gravitational terms, since they usually prove to be small or unimportant to the problem except as a perturbation. For a charged particle of mass M and carrying charge Q in an electric field E we can write

$$M \frac{d^2 x}{dt^2} = QE \tag{88}$$

where we assume for this problem that E is in the x direction and constant. Then if the particle is at rest initially at position $x = 0$ we have

$$v = \frac{QEt}{M} \tag{89}$$

and

$$x = \frac{QEt^2}{2M} \tag{90}$$

for the velocity and position as a function of time.

If the particle is in a viscous medium such as a liquid or a gas, we must include drag terms in the equations. For example, we might include a term kv for drag if Stokes's law is applicable where

$$k = 6\pi\eta r \tag{91}$$

In this expression η is the viscosity of the medium and r is the radius of the particle if it is a sphere. If the particle is irregularly shaped, we assume an equivalent sphere in order to make the analysis easier. The equation of motion then becomes

$$m\frac{dv}{dt} = QE6\pi\eta rv \tag{92}$$

This is admittedly a very simple case, but it illustrates the effect of drag on the particle motion.

CHAPTER 4

Charging Macroscopic Particles

CHARLES D. HENDRICKS, *Director*
Charged Particle Research Laboratory
Electrical Engineering Department
University of Illinois
Urbana, Illinois

4.1 OVERVIEW

In nature, in the laboratory, and in many industrial processes we are faced with problems involving electrically charged macroscopic particles. In atmospheric clouds (precipitation or dust) the particles are well known to be rather highly charged. In fact, almost every small particle found floating in the atmosphere is almost certain to be electrically charged. In industry we have many situations where charged particles are at least a nuisance and may, indeed, be a serious safety problem. For example, flour dust that may be charged can be a potential source of disastrous explosions. In pneumatic particulate material transfer systems, the particles invariably become charged and cause difficulty by adhering to walls of pipes or containers, or they produce electric fields high enough to initiate discharges that may ignite inflammable materials or gases.

Situations or environments in which charged particles may be present either in a beneficial or in a detrimental sense include: flour mills, smoke and other particulate waste discharges, pneumatic transfer systems, radioactive dust and fallout, pesticide and herbicide dispersal systems, paint spray systems, electrofluiddynamic power generators, electric propulsion systems, particle deposition processes (flocking, sandpaper, printing, etc.), dispersed particulates in liquids systems, and many others.

This chapter presents a discussion of the means by which particulate charging can occur or be accomplished. The fundamental phenomena are explored, and means of enhancing or reducing charging are presented. The physical processes of primary importance are discussed, and their relative importance in complex processes is estimated. Strong emphasis is placed on physical reasoning rather than on detailed mathematical analysis.

4.2 PARTICLE CHARGING

The phenomenon of particle charging can be described in an almost trivial manner as any process by which a particle,* initially uncharged, can be caused to give away or accept one or more positive or negative elemental charges (protons, positive or negative ions, or electrons). The particles may be in essentially any medium including liquids, gases (atmospheric or other), vacuum, or solids. In our discussion of some of the processes of particle charging, we do not necessarily cover every possible aspect of particle charging, although representations of the processes likely to be met in most industrial and research work are included.

The division of the discussion into sections according to the use or field of occurrence of the particles should not conceal the related nature of many of the charging phenomena.

4.3 LABORATORY–GENERATED PARTICLES

In several fields of laboratory research it is convenient (or, in some cases, necessary) to have charged particles for experimental purposes. These fields include studies of microparticle impacts, cloud physics, air pollution, precipitators, paint spraying, printing, electrostatic imaging, and many others. Laboratory methods of charging particles may be quite different from those encountered in, for example, an industrial process involving dusts or smokes. Thus we discuss some laboratory techniques for charging which may or may not be the same as those occurring naturally in the nonlaboratory processes.

4.4 IMPACT STUDIES

Shelton, Hendricks, and Wuerker[1] initiated studies of hypervelocity impacts by macroscopic particles accelerated electrostatically. In this work it was necessary to place as much charge as possible on small particles

---

* When we speak of particles in this sense we mean primarily macroscopic entities rather than atoms, molecules, or ions. When these types of "particles" are discussed, the distinction is noted.

because the velocity of the accelerated particles was dependent on the square root of the charge. That is,

$$v = \left(\frac{2qV_{acc}}{m}\right)^{1/2} \text{ m/sec} \tag{1}$$

where v is the particle velocity, q is the particle charge, V_{acc} is the electric potential through which the particle is accelerated, and m is the particle mass. For example, an iron particle whose diameter is one micron ($1 \mu = 10^{-6}$ m) and which has a charge of 10^{-14} C can be accelerated to a velocity of about 700 m/sec by a potential of 100 kV. To obtain such velocities with small particles and to eliminate drag effects, the particles should be accelerated under vacuum. A prime requisite, of course, is a source of charged particles.

If we examine a simple parallel-plate capacitor, we see that a surface charge is present on the inner surface of each of the capacitor plates. The surface charge density ρ_s (charge per unit area) is

$$\rho_s = D = \varepsilon E \text{ C/m}^2 \tag{2}$$

where D and E are the normal components of the displacement field and electric field intensity at the surfaces, respectively, and ε is the permittivity of the insulating medium adjacent to the surface. If we now remove a small portion of the surface of one of the plates without disturbing the field configuration, the small portion removed will carry with it a charge

$$Q = \rho_s A \text{ C} \tag{3}$$

where A is the area of the small piece of plate removed.

Now, in what is a less than ideal case, we can place a small particle in contact with one of the surfaces as in Fig. 4.1. If the particle is small, the field configuration in the capacitor is relatively unaffected except very near the particle. The particle in the figure will acquire a positive charge and experience a force away from the plate. If the upper plate has a hole above the particle as shown, the particle will leave the lower plate, be accelerated upward, and leave through the hole as a charge particle. Cho[2]

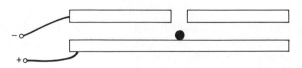

Figure 4.1. A particle resting on the bottom plate of a horizontal parallel-plate capacitor. The applied voltage will result in an electric field that will charge the particle. The hole in the upper plate allows the extraction of charged particles from between the plates.

has shown that a spherical particle on a flat plate will acquire a charge given by

$$Q = 1.65 \times 4\pi\varepsilon R^2 E_0 \text{ C} \qquad (4)$$

where R is the particle radius and E_0 is the electric field intensity between the plates in the absence of the particle.

If the charging by this method is accomplished in air, we can assume that $E = 1.65E_0$ (which is the average field intensity on the sphere in contact with the flat plate) must be less than 30 kV/cm (the breakdown strength of air). If we take that value of E as a limiting case, then the maximum charge the particle can be given can be estimated to be approximately

$$Q = 3.3 \times 10^{-4} \times R^2 \text{ C} \qquad (5)$$

where R is in meters. A particle whose radius is 1 μ will acquire a charge of $Q = 3.3 \times 10^{-16} \text{ C}$.

If we perform the charging process under vacuum, the maximum field strength before breakdown is about 10^9 V/m (the field intensity at which field emission of electrons occurs). Thus under vacuum the maximum particle charge by this method can be estimated to be

$$Q = 0.11 \times R^2 \text{ C} \qquad (6)$$

A 1-μ radius particle, therefore, can be given a charge of $Q = 0.11 \times 10^{-12} \text{ C}$.

The foregoing discussion has assumed conducting plates and a conducting particle. If the particle is a very good insulator, only the surface of the particle directly in contact with the charge plate will acquire a charge. Thus the total charge on such a particle may be much less than it would be for a conducting particle. There is a value of conductivity below which the particle appears to be an insulator and above which the particle acts as a conductor. That value, of course, depends on the time the particle spends in contact with the plate. If we let the time of contact be τ_m (a "mechanical" time constant) and let the time constant for charge transfer be $\tau_E = \varepsilon/\sigma$ (σ is the electrical conductivity of the material) then for $\tau_m > \tau_E$ the particle behaves as a conductor and for $\tau_m < \tau_E$ the particle behaves as an insulator.

When the particle is brought into contact with the surface, it begins to charge; if left there for a time that is long compared with τ_E, it will charge to the value given by Cho. If the particle is removed before charging is complete (as would be the case with an insulating material), we can write for the charge Q in terms of the maximum possible Q_∞ and the time t for which the particle is in contact with the surface

$$Q = Q_\infty \left(1 - \exp \frac{-t}{\tau_E}\right) \text{ C} \qquad (7)$$

If we examine Eq. 4 we see that the charge put on the particle is dependent on the surrounding applied electric field. Any increase of the field will give an increase in particle charge. A method of accomplishing particle charging in high fields obtained in a quasi-spherical geometry is described by Shelton et al. (SHW).[1] The principle of the method remains that of contact charging (sometimes referred to as induction charging). In the SHW method the small particle is caused to contact a sphere of conducting material (e.g., tungsten) maintained at a high potential with respect to surrounding surfaces. A small spherical particle in contact with a larger sphere will acquire a charge Q given by

$$Q = \frac{2\pi^3 \varepsilon_0 R r^2 V}{3(r + R)^2} \, C \tag{8}$$

where R is the large sphere radius, r is the small particle radius, and V is the potential of both spheres. The configuration appears in Fig. 4.2. Particles

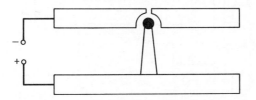

Figure 4.2.   Schematic diagram of a device to study charging of particles to very high charge levels. Particles introduced between the plates bounce back and forth between the plates until, on a purely random basis, one bounces between the ball and the curved region of the hole in the upper plate. The particle then exits through the hole with very high charge.

introduced onto the lower plate will acquire a positive charge and be accelerated toward the upper plate. When the particles impact on the upper plate an exchange of charge occurs and the particles acquire a negative charge and are accelerated downward. For a given particle, this process continues until it is either lost to the right or left of the plates or until the particle enters the region around the sphere on the post. A particle that strikes the curved surface of the cavity above the sphere obtains a negative charge and is accelerated toward the surface of the sphere. Upon striking the sphere, the particle becomes positively charged with Q given by Eq. 8. The combination of particle momentum and electrical forces make it highly probable that the particle will be accelerated through the hole and will leave the system with a velocity given by Eq. 1.

It has been shown by SHW[1] that the method described here can be used to obtain single charged particles or a beam of particles by utilizing the

device schematized in Fig. 4.3. Until particles are needed, the potential at
A is maintained at the same as that of B (which may or may not be at
ground). When we want positive charged particles to be emitted from the
aperture, we place a positive potential on A with respect to B. The particles
in the cup on the lower plate become charged and bounce about in the
space between the lower two plates. Some of the particles travel through the
holes in the plate attached to A. The particles then bounce back and forth
between the upper plates and the action is as described in the previous
paragraph.

The discussion on charging by contact methods has been extensive because
of the widespread use of these methods in many processes.

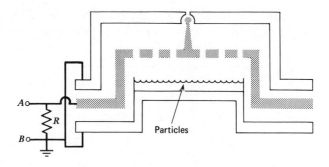

Figure 4.3.   An operating device for charging particles as described by Shelton, Hendricks, and
Wuerker.[1]

4.5   CORONA CHARGING

If a particle is caused to travel through a region of ionized gas in which
the charge is predominantly of one sign or the other, the particle acquires
a net nonzero charge.

Let us examine the situation that would occur if a spherical particle with
a radius of 0.1 mm were put into a gaseous region in which there were
only neutral molecules and negative ions of oxygen at a temperature of
27°C (300°K).

Phenomenologically we see that the particle would become charged by the
negative ions impinging on the particle surface until the electric fields
around the particle became large enough to decelerate the ions and turn
them away. This, of course, would depend on the velocity with which
the ions approached the particle, the ion mass, the ion charge, the radius

of the particle, and the particle charge. An ion in a gas at a temperature $T°K$ will have a most probable speed given approximately by

$$v_i = 1.3 \times 10^2 \times \left(\frac{T}{m_i}\right)^{1/2} \text{ m/sec} \tag{9}$$

If we now consider a bit of electrostatic theory, we see that if an isolated ion with charge q_i and mass m_i is placed at rest at the surface of a charged sphere of radius R and charge Q and released, the ion will be accelerated away from the sphere if the sign of their charge is the same. After traveling a large distance from the sphere, the ion will attain a velocity of

$$v_i = \left(\frac{2q_i Q}{m_i 4\pi\varepsilon_0 R}\right)^{1/2} \text{ m/sec} \tag{10}$$

Similarly, if the ion starts from a great distance from the sphere with a velocity v_i toward the sphere, it will stop just as it reaches the particle surface at a radius R if the particle charge is

$$Q = \frac{1/2 m_i v_i^2}{q_i} 4\pi\varepsilon_0 R \text{ C} \tag{11}$$

If we now substitute for v_i from Eq. 9, we find the particle charge is

$$Q = 3.38 \times 10^4 \times \frac{\pi\varepsilon_0 T R}{q_i} \tag{12}$$

We must recognize that many ions have velocities much greater than the most probable velocity given by Eq. 9. Since there is a Maxwellian distribution of velocities in a gas, there should be some ions with very, very high velocities. We would expect a particle in such a situation to charge to a potential high enough to cause field emission from the particle or to cause local breakdown of the gas. In reality, the situation is more complex because in any ionized gas there are ions of both signs (or positive ions and electrons and negative ions) present, even though ions of one sign may predominate. In the latter case, the particle will charge to the sign of the most predominant ions but will only charge to the value that will lead to an equal number of impacts per unit time of ions of both signs. The particle will be charged to a value depending on the temperatures of the ions and electrons in the gas and on their masses.

It should be pointed out here that the detailed behavior of macroscopic particles in ionized gases is not entirely understood, and research in this area is still going on.

4.6   CONTACT CHARGING AND TRIBOELECTRIC EFFECTS

It has long been recognized that if two dissimilar metals are placed in contact (Fig. 4.4) there will be a potential difference between the metals. This is because the electrons find it easier to go from metal A to B than from B to A. The potential difference between the metals is called the contact potential. If the two metals are carefully separated and kept isolated from other systems, we see that A will be charged positively and B will have a negative charge. The potential difference between the metals A and B will be at *most* 3 to 4 V and for the most common materials is a few tenths of a volt to about 2 V. Thus, even though no externally applied field is present, a metal particle that comes into contact with another metal surface may leave the surface carrying a charge.

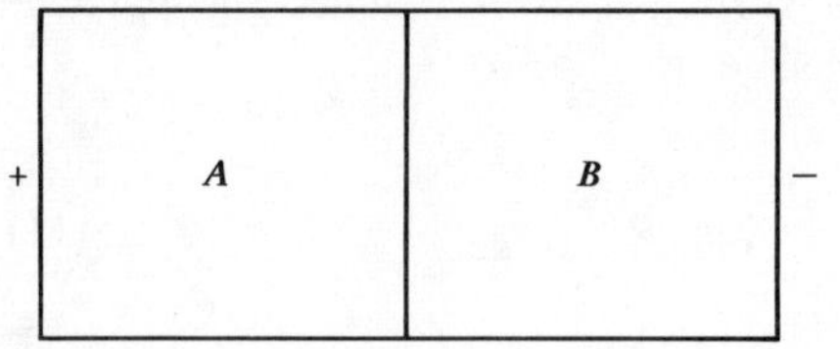

Figure 4.4. Two material blocks in contact. The metal in block A has a lower work function than the metal of block B. Thus electrons move more easily from block A to block B than vice versa. The result is that block A becomes positive and block B becomes negative.

If the two materials A and B are nonconductors, they may still become charged. The triboelectric effect or charging by moving surface interaction is a significant means of charging particles, deliberately or otherwise. However, it is very difficult to separate the phenomena of contact charging from triboelectric (or charging by frictional effects only). It has been stated by Loeb ". . . in most cases, the frictional effect is not important, charge transfer requiring chiefly intimate contact. True triboelectrification can result for two identical substances on asymmetric rubbing but it is rare."[3]

Although ideally, as mentioned by Loeb, triboelectric charging refers to charge transfer by friction effects *only*, common usage lumps contact charging and frictional charging of insulating surfaces together. As an example, if quantities of small particles of two dissimilar insulating materials are shaken together in a container, it is quite likely that particles of one material would become positively charged and particles of the other material would become negatively charged. Ideally we should recognize that the primary mechanism of charge transfer has been contact charging and almost none as the result of friction alone (triboelectric charging). However, it is normally said that the particles became charged triboelectrically. And, indeed, the rubbing together of the particle surfaces contributes greatly to the charging process. Thus we speak of the combination of frictional charging plus the "rubbing"-assisted contact charging as triboelectric charging.

Since most so-called triboelectrification is the result of mechanical exposure of clean surfaces which are charged by coming into contact with each other, let us examine in more detail the contact charging process.

An energy-level diagram for the electrons in a metal indicates the energy states of the electrons in the material with respect to a reference level outside the metal (the "vacuum" energy level). A typical energy-level diagram for a metal is illustrated very simply in Fig. 4.5. This diagram indicates that it is necessary to give the electrons in a metal an amount of energy equal to ϕ the "work function"* in order for them to escape from the metal. This can be done by heating them (thermionic emission), by introducing a strong electric field (field emission), by impact by other electrons (secondary emission), or by other means.

If two dissimilar metals are placed in contact, there will be a transfer of electrons from one metal to the other until the potentials of the metals are such that the Fermi levels are aligned. This is shown in Fig. 4.6 for tungsten and cesium, whose work functions are 4.5 and 1.9 V, respectively. Electrons have been transferred from the cesium to the tungsten until the Fermi levels are aligned, and we see that the potential difference between

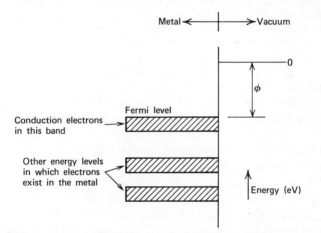

Figure 4.5. Typical energy-level diagram for a metal. The value of ϕ (V) is called the work function of the material. The work function is usually given in volts and the "energy" is in electron-volts (eV), which is the charge on an electron times the potential difference in volts: 1eV is 1.6×10^{-19} J.

* The work function of a metal is that energy which must be given an electron in the metal to get it just free of the metal with no kinetic energy. It is the potential difference between the Fermi level and the vacuum energy level.

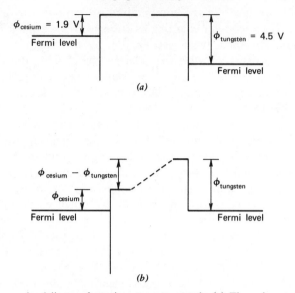

Figure 4.6. Energy-level diagram for cesium–tungsten couple: (*a*) The cesium and the tungsten are separate and thus the "vacuum" energy levels must match if both are at the same potential. (*b*) The two materials have been brought into contact and electrons have transferred from the cesium to the tungsten until the Fermi levels are equal (on the energy-level diagram). Thus the tungsten has become negative and the cesium has become positive. The potential we observe externally is the difference in the work functions, which is 2.6 V in this case.

the two metals is the difference in the work functions $\phi_{Cs} - \phi_{W}$. The tungsten is negative, of course.

The same basic type of behavior can be postulated for insulating surfaces in contact, although the details are not at all well understood. This is mostly a result of the difficulty of completely characterizing insulating surfaces. The energy levels are not well defined in many materials, particularly those which are amorphous. It is very likely that only electrons very near the contacting surfaces take part in the charging of highly insulating materials. If two materials A and B which have conductivities and permittivities σ_A, σ_B, ε_A, and ε_B, respectively, are brought into contact, we would expect the charging of those surfaces away from the contact area to depend on the time constants $\tau_A = \varepsilon_A/\sigma_A$ and $\tau_B = \varepsilon_B/\sigma_B$. These are the time constants for charge transfer* in the materials, and they give a reasonable idea of the time necessary for remote portions of the contacting bodies to learn electrically in terms of charge density that contact has been made.

---

* This time constant is also known as the "charge relaxation time constant."

This method of particle charging is used in a number of applications. One in particular is that of the cascade development process used in electrostatic imaging. A small charged particle is needed to develop the latent image. To obtain the particles, toner particles are mixed with larger carrier particles of a different material. The small toner particles are charged by triboelectric or contact charging by the carrier particles and are actually held to the surface of the carrier by the electrostatic forces. In the development process the toner particles are removed from the carrier, leaving the carrier with one sign (e.g., positive) while the toner particles are oppositely charged (negative in this case). Careful choice of both carrier and toner material is necessary to ensure proper charging of the toner. Materials are available for charging of either sign and even some control of charge magnitude is possible.

Many researchers working in the field of triboelectricity have set up a "triboelectric series" similar to the "electrochemical potential series" for the metals. There is some agreement on the locations of some materials, but most of the series are very dissimilar, even when the same materials are used. The data on individual materials used are apparently not reproducible. This is undoubtedly owing to the complexity of the surfaces of the materials themselves and also to the treatment of the surfaces in the many processes necessary for manufacture and handling. A typical triboelectric series, presented in Table 4.1, serves only as an example.

Table 4.1.   A Typical Triboelectric Series[a]

Rabbit's fur	Cotton
Lucite	Wood
Bakelite	Amber
Cellulose acetate	Resins
Glass	Metals
Quartz	Polystyrene
Mica	Polyethylene
Wool	Teflon
Cat's fur	Cellulose nitrate
Silk	

[a] It is certain that this series order will only be reproducible in rare instances. Conditions such as cleanliness and humidity affect the series drastically. The materials at the top of the list are positive with respect to those lower in the list.

4.7 SPRAY CHARGING

Lenard and many others have observed and studied the charging of drops that arise during mechanical disruption of liquids. This has been called "balloelectric charging" by some research workers.

In the absence of externally applied fields, we would expect the sum of the charge on all drops produced from an uncharged source to be zero. This is only true for very special circumstances, however. The reason is the existence of many electric field sources in any spraying apparatus.

Natanson[4] described the balloelectric effect as the phenomenon of the formation of charged drops in a cloud or fog produced by an atomized spray of a liquid. He concluded, after much research, that the electrification of drops during mechanical atomization of the liquid occurs because the drops carry with them excess charge of one or the other sign and that these charges form in the volume of the drop as a result of fluctuations in the ion distribution in the liquid. In his experimental work, Natanson employed an air atomization system and related the droplet charge to the ion density in the liquid from which the drops were formed and to the expected statistical fluctuations of ion density in the liquid. Lenard, on the other hand, examined charging of liquid drops from waterfalls and shower baths.

Lenard explained his results on the basis of a double layer of charge at the surface of the liquid. He has been criticized for his lack of theoretical competence, but he was a very able experimentalist and did some excellent work in the field of spray electrification.

The two points of view do not seem to be in conflict when it is recognized that the ion concentration fluctuation theories of Natanson and others lead to a statistically symmetric charging of sprayed particles, whereas the double-layer theory leads to a net charge of the sprayed particles.

In the double-layer theory, a layer of oriented dipoles is found at the liquid–gas interface. Depending on the liquid, the dipoles may be oriented with negative poles outward and positive ends inward. The inward ends of the dipoles attract and in some cases bind very tightly some of the negative ions present in the liquid. The ions of positive sign in the liquid are less strongly held to their positive counterparts, and they therefore move with more ease randomly in the liquid.

Thus, on the average, the double layer and its attached negative ions constitutes a charged layer which, when sprayed or otherwise disrupted mechanically into small droplets, will lead to net negative charges on the droplets. The positive charge remains on the container, leaves with larger particles, or is compensated by a negative charge conducted into the liquid through a contact to ground or other reservoir of charge. The presence or addition of impurities or salts of various kinds and in various amounts can alter quite drastically the spray charging process. Small quantities of salts in distilled water, for example, increase the symmetrical ionization (Natanson's ion density fluctuation theory) and decrease the net charge effect. Larger quantities may even reverse the sign of the net charge imparted to the droplet cloud.

When liquids such as transformer oil and butyl esters are disrupted to form droplets whose radii are in the 0.01 to 1.0 μ range, the particle charge may be from 0 to approximately 500 electron charges. On the other hand, when water is sprayed by means of an air blast, we might expect total currents of 10^{-6} A with flow rates of 1 kg/sec. This would correspond to an average specific charge of 10^{-6} C/kg.

It should be pointed out here as it has been in cases discussed earlier that time constants of various types play a very important role in any of these phenomena. Those with which we are concerned in the spray electrification processes are the time constant for charge transfer in the liquid and the time constant for mass transfer or the mass spray rate. If we are mechanically removing charge (by removing the liquid mass to which the charge is attached) more rapidly than charge can be transferred through the liquid to replace that removed, the charge per unit mass removed will begin to fall. For a highly insulating liquid, we may reach the point where a net charge is not possible, and we go from an asymmetrical charging situation to a symmetric situation where each drop may be slightly charged, although the net charge (averaged over all drops sprayed) will be zero.

We should also reiterate the caution that any local electric fields at the liquid surface will affect the results drastically and will lead to incorrect interpretation of any data obtained unless their presence is taken into account.

4.8 INDUCTION CHARGING OF SPRAYED LIQUIDS

It was mentioned in the discussion of spray electrification of liquids that the true results may be masked if external electric fields are present. This is because we may have an induced surface charge on the liquid. Figure 4.7 shows schematically how such a situation may occur. If a liquid has other than zero conductivity, an applied electric field will cause a current to flow in the liquid. The negative charge in Fig. 4.7 will accumulate at the liquid surface until a zero field results in the liquid. This condition is necessary for the current flow to be zero in the liquid. The time constant for this charge transfer to take place is $\tau = \varepsilon/\sigma$ where ε is the permittivity of the liquid and σ is its conductivity.

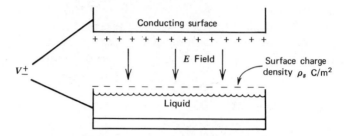

Figure 4.7. Two conducting surfaces with a layer of liquid on the lower surface. In this case the liquid will attain a negative surface charge as a result of the potential applied between the two plates.

If we now arrange to remove the upper surface of the liquid (or any part of it) while the external field is maintained, the liquid will carry with it some negative charge. Thus droplets formed from the liquid surface in the presence of the field as shown will have a net negative charge.

Now let us carry this a step further and increase the electric field significantly. We should note again that an electric field exerts a force on a charge and the force is given by

$$F = qE \tag{13}$$

The surface charge density is ρ_s C/m². Thus the force per square meter on the surface will be

$$F_s = \rho_s E \tag{14}$$

and the force will be trying to pull the charge out of the surface. A sufficiently strong field will bring about disruption of the liquid surface, and electric field spraying of the liquid will result. Drops of liquid will be formed and they will be very highly charged. It should be stated explicitly that no hydrodynamic or other mechanical forces are used to cause the liquid to spray in the situation just described. The disruptive force is entirely electrical.

To estimate very roughly the lower limit of charge per drop to be expected, let us assume a drop radius r. The cross-sectional area of such a drop will be $A = \pi r^2$ m². Now let us assume that the charge on the drop came from a surface area equal to the projected area of the drop on the surface (πr^2). Since the surface charge density is ρ_s C/m², the drop charge will be

$$q = \pi r^2 \rho_s \ \text{C} \tag{15}$$

The mass of such a drop will be

$$m = \tfrac{4}{3}\pi r^3 \rho_m \ \text{kg} \tag{16}$$

where ρ_m is the liquid mass density. The specific charge of the particle, which is found by dividing Eq. 15 by Eq. 16, is

$$\frac{q}{m} = \frac{3}{4}\frac{\rho_s}{r\rho_m} \text{ C/kg} \tag{17}$$

The surface charge density ρ_s is present as a result of the applied electric field intensity E; by the use of Gauss's law (see Chapter 3), this quantity can be directly determined to be

$$\rho_s = -\varepsilon_g E \text{ C/m}^2 \tag{18}$$

where ε_g is the permittivity of the gas above the surface of the liquid and the negative sign is present because of the direction of the E field (into the liquid). The specific charge is approximately

$$\frac{q}{m} = -\frac{3\varepsilon_g E}{4\rho_m r} \text{ C/kg} \tag{19}$$

and the charge on the drop is

$$q = -\pi r^2 \varepsilon_g E \text{ C} \tag{20}$$

Recall that we have made a rather arbitrary assumption, namely, that the charge on the drop came from the area of the surface which is equal to the droplet cross section. However, the results indicated in Eqs. 19 and 20 are not too much at variance with experimental results. In electrostatic spraying experiments with Octoil,* Hendricks[5] found that droplets with a 1-μ radius had specific charges from 0.5 to 0.3 C/kg when the applied electric field intensity was about 10^9 V/m. If the values $\rho_m = 980$ kg/m^3 for Octoil and values $\varepsilon_g = 8.85 \times 10^{-12}$ F/m, $E = 10^9$ V/m, and $r = 1.0$ μ are put into Eq. 19, we obtain a specific charge of

$$\frac{q}{m} = 0.765 \text{ C/kg} \tag{21}$$

Thus our approximate development has led to a value of droplet specific charge which lies in the range of experimentally obtained values.

A further modification of our "spraying system" can provide other interesting results. Let us examine the system of Fig. 4.8. In this system we will spray a liquid through a capillary or an orifice to form a jet that breaks up some distance downstream from the orifice.

The cylinder surrounding the jet is arranged so that the breakup of the jet occurs inside the cylinder. Now consider the situation if a potential

---

* A common name given to a diffusion pump oil which has the chemical name diethylhexyl phthalate.

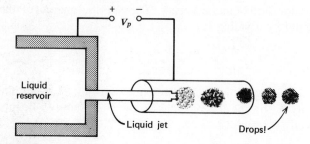

Figure 4.8. A liquid jet issuing from a reservoir and traveling through a concentric charging electrode. A potential is applied between the jet and the charging electrode. The jet need not be concentric with the charging electrode (which need not even be a cylinder, nor even surround the jet); however, the analysis is much easier if the nice geometry is maintained.

V_p is applied between the cylinder and the jet by way of the connection to the reservoir. Assume the liquid of the jet to have an electrical conductivity of about that of water (10^{-6} mho/m or greater). With such a high conductivity, the jet will appear essentially as a center conductor in a coaxial capacitor. There will be a negative charge per unit length of λ on the jet which is

$$\lambda = \frac{CV_p}{L} \text{ C/m} \tag{22}$$

where C is the capacity of the cylinder–jet combination, L is the unbroken length of the jet inside the cylinder, and V_p is the applied voltage. The capacity C is given approximately by

$$C = \frac{2\pi\varepsilon_g L}{\ln b/a} \text{ F} \tag{23}$$

where b is the inner radius of the cylinder and a is the radius of the jet. If the drops into which the jet is breaking have radii r, they must come from a length of jet which is

$$x = \frac{4/3\pi r^3}{\pi a^2} = \frac{4}{3}\frac{r^3}{a^2} \text{ m} \tag{24}$$

Since the jet breaks up into drops while it is in the cylinder and hence has a charge on the jet, each drop will carry away a quantity of charge given by the product of Eqs. 22 and 24.

$$q = x\lambda = \frac{4r^3 CV_p}{3a^2 L} \text{ C} \tag{25}$$

Upon substitution of Eq. 23 for C in Eq. 25, we obtain for the drop charge

$$q = \frac{8\pi V_p \varepsilon_g r^3}{3a^2 \ln b/a} C \tag{26}$$

For a jet radius a of 25 μ, the drop radii will be about 50 μ. If we assume $V_p = 100$ V, $\varepsilon_g = 8.85$ pF/m, and an inner radius b of 1/6 in. (625 μ), we will find a drop charge of approximately $q = 4.6 \times 10^{-13}$ C.

Now we note that by changing the potential V_p we can change the charge on the drops. If the potential V_p is normally zero and is pulsed negatively only during the formation time of one drop and brought back to zero, that drop can be made positive and all foregoing drops will have zero charge.

With the potential V_p back to zero and the positive drop in front of the end of the jet, a negative charge will be induced on the next drop. Thus to have one positive drop and all following drops be uncharged, V_p must be

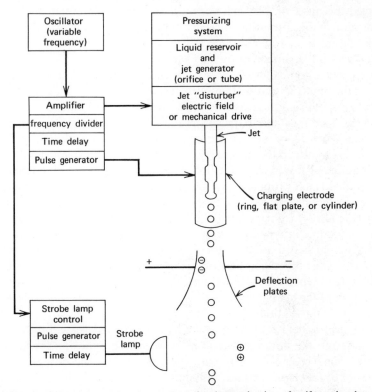

Figure 4.9. Schematization of droplet generator for the production of uniform droplets. Two droplets are selectively charged positively, two droplets charged negatively, and the remaining droplets are uncharged.

made negative for one drop formation time, be made much less negative (but not zero) for the next drop formation time (to balance the effect of the positive foregoing drop), and brought to zero from then on.

A suitable, programmed signal V_p can give rise to well-controlled charges of either sign or zero on the drops coming from the liquid jet. If a small periodic mechanical wave is put on the jet to cause uniform breakup of the jet, the control of both drop charge and size can be accomplished. This has been reported by Hendricks and others in his group at the University of Illinois.[6-9] Figure 4.9 shows some of their apparatus and some uniform particles.

4.9   CHARGING OF PARTICLES IN LIQUIDS

If a solid particle is put in an insulating liquid, it most likely will become charged because of the existence of a contact potential by selective adsorption of ions from the liquid, by inducing or causing local dissociation of the liquid or its own surface molecules, or because of the adsorption of dipoles on the surface of the particle.

If a particle somehow acquires a charge or a dipole layer on its surface, ions of the opposite sign will be attracted to the vicinity of the particle but will be present in a diffuse layer around the particle. The diffuse layer extends some distance from the particle surface into the liquid. The initial layer which is close to or on the particle surface is a very thin layer and is bound very tightly to the surface. The second layer is much less tightly bound and is quite well mixed with other ions of the opposite sign. This phenomenon of a double layer, proposed by Helmholtz, has been studied extensively.

It was suggested in 1879 by H. von Helmholtz that an electrical double layer (Fig. 4.10a) is formed at the surface between two phases of material. At a solid–liquid interface the layer is formed as has already been mentioned. The potential distribution in the vicinity of the surface is represented in Fig. 4.10b. The potential at the dotted line with respect to the liquid is known as the zeta potential. The dotted line is taken as the plane representing the separation of the charge layer fixed to the surface and the diffuse layer that extends into the liquid.

If an electric field is applied to the liquid, the particles will move in one direction and the liquid will tend to move in the opposite direction. The tightly bound layer will move with the particle and the diffuse layer will move oppositely with the sheared liquid being located at the dotted surface. Figure 4.11a shows the situation that would obtain if a free particle were put between two plates, one positive and one negative. The particle would become a dipole as a result of polarization in the electric field but if the electric

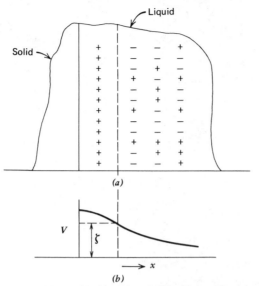

4.10. A liquid in contact with a solid: (*a*) This solid–liquid combination results in a strong binding of the positive ends of some liquid molecules to the surface of the solid. Beyond the bound charge is a diffuse region of charge where the charge density gradually becomes that of the ambient liquid. (*b*) The potential distribution in the region of the solid surface that is in contact with the liquid.

field were uniform, there would be no net force on the particle. On the other hand, if the particle were put into a liquid between the positive and negative plates, the particle could become a dipole as before but in addition could have a net charge as a result of the process just described. Figure 4.11*b* indicates schematically that the particle may be forced by the electric field in one direction (to the left if the particle becomes charged positively in the figure) and the liquid may be forced in the opposite direction in the near vicinity of the particle. This phenomenon of particle migration in an electric field is known as electrophoresis.

If the double layer is considered to be equivalent to the two plates of a capacitor, the zeta or Helmholtz potential can be found as

$$\zeta = \frac{4\pi\eta\mu}{\varepsilon_r} \, \text{V} \tag{27}$$

where η is the fluid viscosity, ε_r is the dielectric constant (relative permittivity) of the liquid, and μ is the electrophoretic mobility* of the

* The velocity of the particle will reach a limiting value v where $v = \mu E$. The electrophoretic mobility is defined as the velocity of the particles under the influence of a field of one volt per meter.

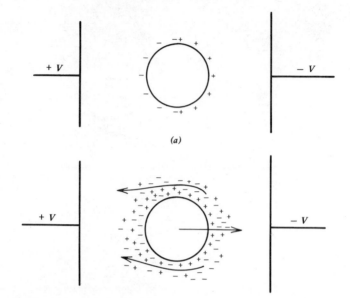

Figure 4.11. (a) An uncharged sphere between two electrodes maintained at plus and minus V V as shown. There is no ambient material between the plates other than the sphere. (b) The sphere has been placed between the plates and a liquid added to the system. The net charge on the sphere is no longer zero but is somewhat positive because of the interaction of the solid with the liquid. The electric field between the plates provides a force on the positive particle toward the negative plate. Since the liquid is slightly negative in the region of the sphere, there will be a force on the liquid toward the positive plate.

particle. By measuring the mobilities of the particles and the viscosity and the dielectric constant of the liquid, it is possible to determine zeta potential. The values range from about 0.2 V to zero. As a rule, the values range from 0.02 to about 0.05 V. Both positive and negative values are found. Aqueous colloidal dispersions of metallic hydroxides and hydrated oxides and basic dyestuffs usually contain positively charged particles (positive zeta potentials), whereas the aqueous sols of metals, sulfur, acidic hydroxides, and acidic dyestuffs usually carry negative charges. The sign of the particle charge may even depend on the method of preparation of the colloidal suspension. Similarly, dispersions of various materials in nonaqueous liquids may take either positive or negative charges depending on the liquid, the particle material, and the method of preparation.

A condensed discussion of the Helmholtz double layer and the zeta potential is to be found in Taylor and Glasstone's *Treatise of Physical Chemistry*,[10] along with a discussion of the Debye–Hückel theory of the double layer. (See also Chapter 5.)

4.10   CHARGING BY FREEZING

A number of observers have postulated a mechanism for charging particles during the process of freezing. The best known is that of Workman and Reynolds,[11] who found large potential differences between the ice and water phases during the freezing of water. A probe was located in contact with the ice phase and another in contact with the water. As the freezing process occurred, potential differences between ice and water of up to 230 V were observed. This effect has been postulated as being a charging mechanism in thunderstorms.[12] If, during the freezing of a water drop, some liquid is blown away, the remaining solid ice particle will carry a net charge. Such a mechanism could lead to the charge separation in thunderstorms.

B. J. Mason[13] has postulated that as a water drop freezes, small "splinters" are ejected from the freezing drop and the splinters are charged oppositely to the remaining solid ice particle. This mechanism could lead to thunderstorm electrification. Recent results of Cheng[14] seem to support to some extent the suggestions of Mason. The typical particle charges obtained by Mason in his experiment were of the order of 3×10^{-14} C for drops of 1.0-mm diameter freezing at $-5°C$. The drops ejected about 20 splinters during the freezing process. The ejection of splinters occurs when a shell of ice is formed on the drop. Further freezing occurs interior to the shell and generates stresses in the ice shell which can cause fracturing and the formation of splinters from the interior of the freezing drops. The ejection of charge along with the splinters can be related to the Workman–Reynolds charge separation mechanism or, alternatively, it can be explained by positing the formation of a double layer, as the Helmholtz double-layer theory suggests for other solid–liquid boundaries. Differing rates of diffusion of H^+ ions and OH^- ions through ice crystals form the basis for yet another possible explanation.[15]

4.11   ION AND ELECTRON BEAM CHARGING

In a number of experiments it has been convenient to charge particles by bombarding them with ions, electrons, or x-rays. Let us now examine these in two parts. First we lump the ion and electron beam effects together under the assumption that the particle charges by collection of ions or electrons. Second, we lump together x-ray beam charging and photoelectric charging, since in each case constitutes an electron ejection process.

Consider these beams as charged particle beams in which the particles carry a charge q and have a mass m. For simplicity we assume that the

Charging Macroscopic Particles

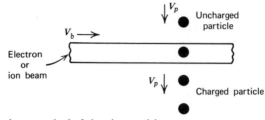

Figure 4.12. The beam method of charging particles.

particles have a velocity that is small compared with the velocity of light. Figure 4.12 illustrates an experiment in which macroscopic particles may be charged by an ion or electron beam. First let us assume that the diameter of the particle is small compared with the diameter of the electron or ion beam through which it must pass. Let the beam current density be J A/m², the particle radius be r m, and the beam radius be R m. If the particle falls through the diameter of the beam with a velocity v m/sec, it will be in the beam a time

$$t = \frac{R}{v} \text{ sec} \tag{28}$$

During that time (assuming an ideal case) there will be a current of

$$I = \pi r^2 J \text{ A} \tag{29}$$

impinging on the particle. The total charge transfered to the particle surface will be

$$Q = \frac{r^2 J R}{v} \text{ C} \tag{30}$$

Now let us put in some typical numbers for the various parameters: $r = 10^{-5}$ m, $R = 10^{-3}$ m, $J = -0.1$ A/m², and $v = 1.0$ m/sec. This gives a charge of $Q = -3.14 \times 10^{-14}$ C for the particle after it traverses the electron beam.*

The charging of the particle in the beam depends on the energy of the electrons or ions. This is a result of the particle charging process. Initially the particle has no charge on its surface; hence the incoming electrons find no retarding fields to keep them from the surface. As the particle becomes charged, the electrons are acted on by the field produced by the electrons

---

* Note that a current density $J = -0.1$ A/m² was assumed. This would be an electron or negative ion beam and would produce a negative particle charge. A positive ion beam would have a positive value for J and would give a positive charge on the particle.

on the particle. If the beam electrons have an energy that is high compared with the potential at the surface of the particle when it is fully charged, they will reach the surface of the particle during the entire time period the particle is in the beam. On the other hand, if the potential of the particle becomes comparable to the beam energy while it is still in the beam, the electrons in the beam will be repelled and charging will cease or be markedly reduced. A similar situation holds in the case of ion beams used for charging particles.

A complication that may arise in the case of electron beam charging is that of secondary emission of electrons from the particle surface. Incoming electrons may have sufficient energy to eject secondary electrons from the surface. These electrons would be accelerated away from the particle by local electric fields around the particle. Secondary emission would reduce the rate of charging but would not decrease it to zero except in rare circumstances where beam energy was very high (or at some value giving a maximum in the secondary emission yield*).

Beam charging is further discussed in Chapter 13.

4.12   CHARGING BY THERMIONIC EMISSION

In some instances particles are raised to temperatures high enough to permit thermionic emission of electrons to take place. Such temperatures occur in the gases from furnaces, in high-temperature chemical reactors, and in a number of other environments. The phenomenon of thermionic emission depends on an increase of velocity of the electrons inside a heated material. When the component of electron velocity in the direction perpendicular to the material surface becomes great enough to overcome the image forces (see Chapter 3) acting on the electron as it leaves the surface, the electron escapes. This effect is relatively well understood and is employed in hot cathode vacuum tubes.

Since electrons behave somewhat like a gas in a conducting material, we would expect a few electrons to have enough energy to escape the surface even at room temperatures (say, 20°C). This is actually the case, but the number per unit area per unit time is very small. As the temperature is increased, the number escaping increases rapidly. The equation we use to obtain the current density from the surface (charge per unit area per unit time), called Richardson's equation, is

$$ J = A_0 T^2 \exp \frac{-e\phi}{kT} \ \text{A/m}^2 \tag{31} $$

* The secondary electron emission yield is the number of secondaries per incoming primary electron.

where A_0 is a universal constant* equal to 1.2×10^6 A/(m^2)($^\circ$K^2), T is the temperature in degrees Kelvin, ϕ is the work function of the material (V), e is the electron charge (C), and k is the Boltzmann constant (J/$^\circ$C).

To find the current density at 727°C (1000°K) from an iron sphere, we would substitute into the Richardson equation $T = 1000$, $A = 0.26 \times 10^6$, $e = 1.6 \times 10^{-19}$, $\phi = 4.48$, and $k = 1.38 \times 10^{-23}$. This gives a current density J of 8.2×10^{-12} A/m^2. At a temperature of 1500°C (slightly below the melting point) the iron particle will emit electrons sufficient to provide a current density of about $J = 0.15$ A/m^2. It is obvious that the current density is a strong function of temperature.

If a particle were negatively charged, a high temperature would lead to neutralization very quickly. On the other hand, it is possible to visualize situations in which a cloud of electrons could exist near a hot particle. If some of these electrons were removed by an external field or by a near encounter with another surface, the particle could end up with a positive charge.

Although these situations are not usual at ordinary temperatures, conditions are found in which thermionic emission in the charging and discharging process must be considered.

4.13 PHOTOELECTRIC CHARGING

If light falls on the surface of a particle, the light quanta can impart sufficient energy to electrons in the surface to eject them from the particle. The Einstein photoelectric equation

$$E = h\nu - e\phi \text{ J} \tag{32}$$

tells us that the energy E of an electron ejected from the surface by a photon is the difference between the incoming photon energy $h\nu$ and the work function energy $e\phi$, where h is Planck's constant (6.63×10^{-34} J-sec), ν is the frequency of the light, and e and ϕ are the electron charge and the work function of the material, respectively.

Metallic surfaces are notably poor photoemitters, but they do yield some photoelectrons. There are certain exceptions (e.g., zinc, which was used in many experimental studies); however, "good" photoemissive surfaces are very complex. To quote Sproull, "Practical photoemitting surfaces are so complicated that the reasons for the large [electron] yields are not well understood."[16]

In the visual range of light wavelengths we expect very few photoelectrons from most surfaces. However, as we go into the ultraviolet and x-ray

---

* The theoretical value of A_0 is given. Experimental values for the constant are much lower than the theoretical value.

regions, the yield becomes greater and we have a more efficient electron ejection process. This is partly owing to the greater penetration of the waves into the surface (lower reflection coefficients) and partly to the greater interaction with bound electrons. Free electrons are not easily ejected from a surface because of the relative impossibility of satisfying both energy and momentum conservation conditions. These conditions are easily satisfied in the case of a bound electron.

Experimental studies on the charging of macroscopic particles by x-rays were undertaken by Hendricks and Shelton in 1959 and the results indicated, as expected, that a particle placed in an intense beam of x-rays became charged positively. The method was studied for use in charging particles to be electrostatically accelerated to meteoric velocities. Other methods (contact charging) proved more efficient, and the x-ray photoelectric charging system was dropped.

There may be significant photoemissive charging of particles in naturally occurring dispersions in the upper atmosphere. This is more likely to occur at higher altitudes, since much ultraviolet and soft x-ray radiation is filtered out by the atmosphere and does not reach low altitudes with high intensity.

4.14 RADIOACTIVE DECAY

A little-used method of particle charging is that of radioactive decay in the particle material. If a particle is composed of a material in which there is present a beta emitter, for example, an isolated particle will be charged to the level at which local breakdown of the surrounding atmosphere takes place or, alternatively, until field emission of positive ions occurs from particle asperities. For every β^- particle which is emitted, the particle effectively acquires a positive charge of 1.6×10^{-19} C.

As an example, let us assume we have a particle of bismuth having a 10-μ radius. The bismuth will be assumed to be 99% $_{83}\text{Bi}^{209}$ which is a stable isotope and 1% $_{83}\text{Bi}^{210}$ which is a β^- emitter with a half-life of 5.0 days. Each atom of bismuth has a mass of approximately 3.57×10^{-27} kg. The particle mass, assuming a spherical particle, will be

$$m = \frac{4}{3} \pi r^3 \rho_m \tag{33}$$

and, since bismuth has a density of 9800 kg/m^3, the 10-μ particle will have a mass of $m = 4.10^{-14}$ kg. Thus in this particle there will be 1.15×10^{13} bismuth atoms. Of these there will be 1.15×10^{11} unstable $_{83}\text{Bi}^{210}$ atoms. If our particle is set free and its charge is monitored, we will see a β^-

emitted about every 7.5 μsec. In one second, the bismuth particle will acquire a charge 2.13×10^{-14} C.

This example indicates the process and the charge calculations. Since $_{83}Bi^{210}$ is radium E and is rather difficult to find, this example is not particularly useful. However, other available β emitters are more easily obtained and can be incorporated chemically into material from which particles are to be made. This method of charging provides a means by which particles may be charged to high levels at a known charging rate.

Materials which emit β^+ or α particles are also available and will provide charging of the opposite sign to those materials which emit β^- particles.*

4.15   FIELD EMISSION CHARGING

If an uncharged particle (liquid or solid) enters a region in which there is an electric field, the particle will polarize. That is, there will be a transfer of charge within the particle so that one end of the particle becomes negative and the other positive. If we assume the particle material has a conductivity σ which is nonzero, the time necessary for charge transfer to occur and equilibrium to be established will be related to the time constant $\tau = \varepsilon/\sigma$. As has been discussed in Chapter 3, the surface charge density at any surface location will be at a time t after application of the field

$$\rho_s = \rho_{s0}\left(1 - \exp\frac{-t}{\tau}\right) \tag{34}$$

where ρ_{s0} is the surface charge density that will be reached after an infinite time. In a time $t = 4\tau$ we see that $\rho_s \approx 0.98\,\rho_{s0}$ and in a time $t = 10\tau$, $\rho_s \approx 0.99995\rho_{s0}$.†

The particle in any electric field thus becomes a dipole. If the field intensity E is high enough, electrons may be extracted by the high field from the negative end (electron field emission) and/or positive ions may be extracted from the positive end (positive ion) field emission. If electrons leave, the particle becomes positively charged and experiences a force in the direction of the electric field.

A very high field intensity‡ is necessary to cause field emission of either electrons or ions from a solid. Field strengths high enough to cause field

---

* A β^- particle is an electron with -1.6×10^{-19} C charge. The β^+ is a positive particle having the same mass as an electron $(9 \times 10^{-31}$ kg) and a charge of $+1.6 \times 10^{-19}$ C. The α particle is composed of two neutrons and two protons and thus has a charge of $+3.2 \times 10^{-19}$ C and a mass of about 6.8×10^{-27} kg—the same as a helium nucleus.

† We assume $(1 - e^{-5}) = 1$, since it is really about 0.99 and in most electrostatic experiments an error of much more than 1% is tolerable.

‡ Approximately 10^{10} V/m field strength is required for electron emission and 10^{11} V/m for positive ion emission.

emission are almost impossible to attain in gaseous environments because electrical breakdown of the gas occurs at lower field strengths. However, on a very small scale, under vacuum, or in an insulating liquid environment, it is possible to achieve the necessary values of the E field to cause field emission.

Let us examine a situation in which a particle is placed in a nonuniform electric field such as that found in the vicinity of a sharp point. The geometry will be that of Fig. 4.13.

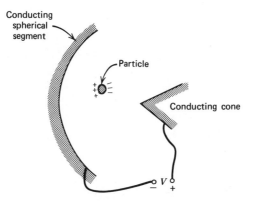

Figure 4.13. Particle charging by field emission or charge exchange at a sharp point held at a high potential. The particle becomes an induced dipole that moves toward the region of higher field intensity (in the direction of increasing field gradient). When the dipole becomes close to or touches the point, charge is transferred to or from the particle to the tip and the particle becomes positively charged. It then experiences a force *away* from the tip.

The particle will be polarized and become a dipole whose equivalent *dipole moment** will be M. Since the orientation of the dipole is important, we note that the induced dipole always is along the field lines and that the positive end is in the direction of the electric field intensity E. It is easy to see that in a uniform field the charge at each end of the dipole experiences the same force but in opposite directions. Thus the dipole experiences no net force. However, in a nonuniform field such as that of Fig. 4.13, the positive end of the dipole will be in a slightly weaker field than the negative end. The entire particle will experience a net force toward the stronger region of the field. The particle will then move (if it is free) in the direction of the *field gradient* and almost along the field lines. As the particle approaches the positive electrode (as in Fig. 4.13), the field intensity

* The dipole moment is the magnitude of the charge q on either end of the dipole times the separation d of the charges.

becomes very high, and field emission of electrons can occur from the particle surface next to the electrode. The particle then becomes positively charged and the direction of the total force on the particle can change 180°. Acceleration of the particle away from the electrode occurs and the result is a positively charged particle.

This is not a usual method for charging particles, but it does occur frequently when particles are in insulating liquids or when the particles are liquid drops. The emission of charge then may be either electrons or ions or tiny droplets carrying a net charge.

We should note here that in a relatively good vacuum, the molecules of the residual gas may serve as the particles. The molecules become dipoles in the E field and near a sharp point are accelerated to the point where they become positively charged. When they leave the point they leave more frequently from microscopic asperities such as grain boundaries and certain crystal planes and protruding surface atoms. The ions will be accelerated to the negative spherical segment, which may be coated with a phosphor. The phosphor emits light on impact of an ion and is brightest where the largest number of ions per second strike. This entire process is the essence of Müller's field ion microscope.[17]

4.16   CHARGING BY MECHANICAL FRACTURE OF MATERIALS

It has long been known that the cleavage of mica and other crystalline materials leaves the fresh surfaces charged. If the crystals are cleaved or split under vacuum, electrons may literally be sprayed out of the cleaved region. Electron energies of hundreds of kilovolts were found by Krotova and others.[18] The mechanism proposed for the generation of the electrons is that of emission from ruptured, strained, or otherwise mechanically broken bonds within solids. Such emission also occurs when adhesively joined materials are separated. The separated pieces of material are highly charged and in some cases even emit electrons as long as 1.5 hr after the surfaces are separated.

It has been suggested that the existence of a charge double layer at the separation surface is the location of the charges. When the double layer is disturbed or separated, high electric field intensities occur and field emission and acceleration of electrons can take place in the high field region.

We can visualize the same process occurring during a pulverization process or a process where small particles are broken or torn from a surface. It would be difficult outside the laboratory to distinguish charging by mechanoelectric effects (fracture or rupture of materials) from charging by triboelectric effects. Indeed, they may well involve some of the same processes in many cases.

REFERENCES

1. H. Shelton, C. D. Hendricks, and R. F. Wuerker, "Electrostatic acceleration of microparticles to Hypervelocities," *J. Appl. Phys.*, **31**, 1243 (July 1960).

2. A. Y. H. Cho, "Contact charging of micron-sized particles in intense electric fields," *J. Appl. Phys.*, **35**, 2561 (September 1964).

3. L. B. Loeb, *Static Electrification*, Spinger, Berlin, 1958, p. 166.

4. G. L. Natanson, "The problem of balloelectric phenomena," *Dokl. Akad. Nauk, SSSR*, **73**, No. 5, 975–978 (1950).

5. C. D. Hendricks, "Charged droplet experiments," *J. Colloid Sci.*, **17**, 249–259 (1962).

6. C. D. Hendricks, and J. M. Schneider, *Rev. Sci. Instr.*, **35**, 1349 (1964).

7. J. M. Schneider, N. R. Lindblad, and C. D. Hendricks, *J. Colloid Sci.*, **20**, 610 (1965).

8. C. D. Hendricks, and J. B. Y. Tsui, *Rev. Sci. Instr.*, **39**, 1088 (1968).

9. ——— and T. Erin, *Rev. Sci. Instr.*, **39**, 1296 (1968).

10. H. S. Taylor, and S. Glasstone, *Treatise of Physical Chemistry*, 3rd ed., Vol. 2, D. Van Nostrand, New York, 1951, p. 628.

11. E. J. Workman and S. E. Reynolds, *Phys. Rev.*, **78**, 254 (1950); *Phys. Rev.*, **74**, 709 (1948).

12. ———, "Thunderstorm electricity," Proceedings of the 3rd International Conference on Atmospheric and Space Electricity, Montreux, Switzerland, May 1963.

13. B. J. Mason, "Charge generation in thunderstorms," Proceedings of the 3rd International Conference on Atmospheric and Space Electricity, Montreux, Switzerland, May 1963.

14. R. J. Cheng, *Science*, **170**, 1395–1396 (1970).

15. B. J. Mason, *Clouds, Rain and Rainmaking*, Cambridge University Press, Cambridge, England, 1962 Chapter 7, p. 132.

16. R. E. Sproull, *Modern Physics*, Wiley, New York, 1952, p. 385.

17. E. W. Müller, *Advances in Electronics and Electron Physics*, Vol. 13, Academic Press, New York, 1960, p. 83.

18. A. M. Polyakov and N. A. Krotova, *Dokl. Akad. Nauk SSSR*, **151**, No. 1, 130–133 (1963).

CHAPTER 5

Static Electrification of Dielectrics and at Materials' Interfaces

ION I. INCULET, *Professor, Chairman*

Electrical Engineering Group
The University of Western Ontario
London, Ontario, Canada

5.1 OVERVIEW

At the basis of all engineering applications of electrostatics we find matter in an electrified state. *Electrified matter* may be defined as possessing a net charge of one type, an electric dipole moment, or both* At the atomic level, the so-called neutral matter contains in effect equal amounts of positive and negative charges in the form of protons and electrons. The total electric charges present in matter of microscopic size are very large (e.g., in 1 g of water the total negative charge balanced by an equal positive charge is about 53,000 C.)

Under the influence of an applied electric field, electrification of certain matter may occur internally throughout its volume. The phenomenon is called polarization, and the materials are *generally* known as dielectrics. *In practice* the "dielectrics" are all the electrical insulation materials.

Matter also becomes electrified at an interface which may be solid-to-solid, solid-to-liquid, solid-to-gas, and so on.

An interface is the site of natural transfer of electric charges. Depending on the materials in contact, the surface condition, and external factors, such as

---

* Two electric charges q equal and opposite in sign, separated by a distance x form a dipole of moment qx.

temperature and electric fields, an electric charge crosses the interface; some, none, or all of its remains there after the separation. The result may be that both surfaces become electrified. Although the electrification at interfaces is fundamental to modern electronics engineering, the electric charges that remain on the surface after separation are in the domain of the applied electrostatics engineer. In the same domain are the surface charges produced by other means such as: (a) corona currents (presented in detail in Chapter 9), (b) thermionic emission, and (c) atmospheric ions.

Dielectrics and surface electrification are inseparable. For example, a dielectric ("perfect" insulator) covering a conductor maintained at a constant positive electric potential will become polarized internally and the dielectric surface will at first appear positively electrified. Depending on the surface charge and the ion content of the ambient air, it may take as long as several hours for the negative ambient ions to completely neutralize the apparent surface charge. Once the high-voltage connection to the conductor is interrupted and the dielectric depolarizes, its surface will appear as negatively charged.

The importance of dielectrics in the electrical technology is brought into focus when we consider that the life of a capacitor, an electric motor, or a relay is determined primarily by the life of its insulation or the dielectric materials used and not by the life of its conducting or ferromagnetic materials (e.g., copper, aluminum, or silicon steel). In spite of this, the technology and the understanding of the polarization of dielectric materials have not paralleled the rapid developments in the field of semiconductor materials that require similar theoretical and practical analysis.

The static electrification of dielectrics—or polarization—under the influence of an electric field takes place at the atomic or molecular level and there are three types:

1. *Electronic (induced) polarization*: An atom considered as a nucleus of positive charge surrounded by a spherical cloud of negative charge equal in magnitude will not produce a constant field outside the sphere. When the atom is placed in an external electric field, however, the electron cloud and the positive nucleus shift in opposite directions, and an electric dipole or "induced polarization" appears.

2. *Ionic polarization* appears only in ionic dielectrics. For example, a hydrogen chloride molecule has a positive hydrogen ion and a negative chlorine ion. Under the influence of an electric field, the two ions are pulled farther apart, producing a stronger dipole or ionic polarization.

3. *Orientation polarization* is produced in dielectrics with molecules having a *permanent* dipole, such as the hydrogen chloride molecule just described. Before the electric field is applied, the molecules are randomly oriented. Under the influence of the electric field all dipoles that are not

lined up with the field experience a torque which tends to orient them in the direction of the field.

The effects of these three types of polarization can be seen on a molecular scale at the dielectric surfaces that are traversed by the external electric field. The last dipoles in the chains will expose their negative or positive charge, depending on whether the electric field is directed toward or away from the surface, and the entire surfaces appear electrified. For good insulators, in ambient air, the surface charges are gradually neutralized by atmospheric ions. However, when the electric field in the dielectric is produced by the conductive plates of a capacitor containing the dielectric, the dielectric surface charges are instantaneously neutralized by electric charges on the capacitor plates. If voltage is maintained constant, the ratio of the capacitor charge when a dielectric is present between its plates to the charge of the same capacitor with vacuum between its plates is called relative permittivity.

The permittivity of a dielectric is a direct indication of its polarizability. Table 5.1 gives the relative permittivities of several media. Not all high-permittivity materials are insulators. Water is an example. It is interesting to observe the considerably larger values for polar liquids, such as nitro-

Table 5.1   Some Representative
Values for Relative Permittivities

Medium	Value
Vacuum	1.0
Air	1.00059
Teflon	2.0
Benzene	2.3
Shellac	3.5
Glass	4.0–10.0
Ether	4.4
Mica	6.0
Alumina	10.0
Acetone	21.0
Ethyl alcohol	24.0
Tantalum oxide	28.0
Methanol	32.0
Nitrobenzene	36.0
Glycerine	56.0
Water	80.4

benzene. In the polar liquids the induced and the rotational polarizations play a very important role, and they are responsible for the large value of the permittivity.

The published "internal electrification" or permittivities of dielectrics are reliable figures which may be used by designers of electrical equipment operating in normal ambients. By contrast, the electrification at interfaces and the charges remaining after the surfaces in contact are separated, are very often completely unpredictable in normal ambients. Normal variation in temperature and humidity of the ambient air, and surface contaminants of the size of a molecular layer may change the work function and the surface state of a metal or semiconductor sufficiently to alter not only the amount of charge transfer but also the direction.

To achieve perfectly reproducible static electrification at interfaces, it is necessary to reproduce the surface conditions first. From the moment a prepared surface is exposed to the normal air ambient, the surface conditions are no longer known. It is for this reason that investigators of static electrification at interfaces have gone to great pains to clean the surfaces prior to contact and to carry out experiments under vacuum.

To illustrate, it is sufficient to look at some past experiments with nickel–borosilicate interfaces. In a rolling friction experiment published in 1954, Peterson[20] found that nickel charged positively. Thirteen years later, practically simultaneous experiments by Davies[29] and Inculet and Wituschek[30] showed that nickel charges negatively when in contact with borosilicate glass. The applied electrostatics engineer can hardly rely on any published figures to predict static electrification in our normal ambient air with its temperature and humidity variation and its many contaminants such as gases and particulates.

Considering the technological applications and implications and the different combinations of solid, liquid, and gas interfaces, the static electrification of solid-to-solid, solid-to-liquid, and liquid-to-liquid are the most important. Table 5.2 shows the major fields of application.

Table 5.2 Static Electrification

Solid-to-Solid	Solid-to-Liquid	Liquid-to-Liquid
Textile industry	Petroleum industry	Petroleum industry
Mineral beneficiation	Electrostatic painting technology	
Printing technology		

5.2 STATIC ELECTRIFICATION (POLARIZATION) OF DIELECTRICS UNDER CONSTANT ELECTRIC FIELDS

5.2.1 Macroscopic Phenomena

The mathematical expression of the electrification of a dielectric material may be developed in a simple way by assuming that the dielectric fills the original free space between two parallel plates made out of conductive material. The potential difference is maintained constant and the dielectric is assumed to be a perfect insulator. Let

A = area of the plates		m^2
x = distance between plates		m
V = electric potential difference applied to the plates		V
E_i = electric field intensity in the dielectric		V/m
E_{free} = electric field intensity in free space		V/m
D_i = electric displacement (or flux density) in the dielectric		Cm^{-2}
D_{free} = electric displacement (or flux density) in free space		Cm^{-2}
Q = electric charge accumulated on the plates when the space between them is filled with dielectric		C
Q_{free} = electric charge accumulated on the plates in free space (with the dielectric removed)*		C
ε_r = dielectric constant or relative permittivity†		(dimensionless)
ε_0' = permittivity of free space		8.85 pF/m

The following relations exist between the foregoing quantities:

$$Q_{free} = \varepsilon_0 \frac{V}{x} A \tag{1}$$

$$Q = \varepsilon_r \varepsilon_0 \frac{V}{x} A \tag{2}$$

$$Q = \varepsilon_r Q_{free} \tag{3}$$

* Q_{free} is very close to the charge value in air (within $\simeq 0.6\%$); for practical purposes in the following demonstrations, we can substitute free space with air.

† For static electrification conditions ε_r, the dielectric constant is considered under constant voltage conditions. For varying electric fields a "complex permittivity" must be defined which takes into account energy losses in the dielectric as well as the dependence on the frequency.

As explained qualitatively in the overview, the introduction of the dielectric between the two plates calls for an additional charge on the plates. This additional charge is called "bound charge" and is obtained from eqs. 1 to 3.

The bound charge

$$Q - Q_{\text{free}} = Q - \frac{Q}{\varepsilon_r} \tag{4}$$

is neutralized by the ends of the last dipoles in the dielectric's chains ending on the conductive plates.

According to Gauss's theorem (see Chapter 3), and neglecting any leakage fluxes linking the backs and sides of the plates, we can write the following equations:

The electric-flux density, or displacement, when the plates are in free space is produced by the charge Q_{free}

$$D_{\text{free}} = \frac{Q_{\text{free}}}{A} \tag{5}$$

The electric flux density when the plates have the dielectric in between is produced by the charge Q

$$D_i = \frac{Q}{A} \tag{6}$$

The additional flux density produced by the dielectric electrification is called **polarization:**

$$P = D_i - D_{\text{free}} = \frac{Q - Q_{\text{free}}}{A} \tag{7}$$

Because in the chosen example the applied potential difference V is the same, with or without the dielectric, we have

$$E_i = E_{\text{free}} = \frac{V}{x} \tag{8}$$

hence

$$D_i = \varepsilon_0 E_i + P \tag{9}$$

or the electric flux density in the dielectric is equal to the electric flux produced by the charge Q_{free} plus the dielectric polarization.

Equation 7 may also be written as

$$P = \varepsilon_0 (\varepsilon_r - 1) E_i \tag{10}$$

$$\varepsilon_r - 1 = \chi = \text{susceptibility of the dielectric} \tag{11}$$

which may also be defined as the ratio, of bound charge to free charge; hence

$$P = \varepsilon_0 \chi E_i \qquad (12)$$

The preceding mathematical equations and definitions are most useful because they allow us to relate the macroscopic static electrification of a dielectric, or its polarization, to measurable electrical or geometrical quantities. In the general case, both D and P are best treated as vectors.

Exercise

It has been explained that the bound electric charges of a capacitor neutralize the charges of the dielectric dipoles ends immediately adjacent to the conductive surface. How do the charges and electric field intensities change when a layer of dielectric of thickness d_d is applied on the inside surfaces of a capacitor, leaving a free space region of thickness d_{free}? (The dielectric is assumed to be of zero conductivity.)

Assuming two conductive plates of area A in Fig. 5.1 where E_{free} and E_i are the respective electric field intensities in the free space and dielectric regions, V the applied potential difference, and Q the charge on the plates, we have

$$V = E_i d_d + E_{\text{free}} d_{\text{free}} + E_i d_d \qquad (13)$$

The electric flux density produced by the charges developed on the two plates is the same in the dielectric and the free space regions.

$$D_i = D_{\text{free}} = \frac{Q}{A} \qquad (14)$$

From Eqs. 9 and 10 we see that

$$E_i = E_{\text{free}} - \frac{P}{\varepsilon_0} \qquad (15)$$

$$E_i = \frac{E_{\text{free}}}{\varepsilon_r} \qquad (16)$$

From Eqs. 13, 14, and 16 we have

$$Q = A\varepsilon_0 \frac{V}{d_d/\varepsilon_r + d_{\text{free}} + d_d/\varepsilon_r} \qquad (17)$$

$$E_{\text{free}} = \frac{V}{d_d/\varepsilon_r + d_{\text{free}} + d_d/\varepsilon_r} \qquad (18)$$

$$E_i = \frac{V}{\varepsilon_r(d_d/\varepsilon_r + d_{\text{free}} + d_d/\varepsilon_r)} \qquad (19)$$

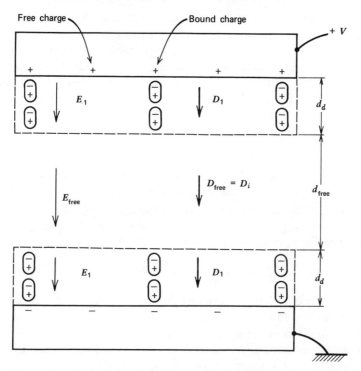

Figure 5.1. Combination of dielectrics and free space in an electric field.

The results of this exercise lead to two conclusions of general application:

1. When a plane dielectric slab is placed in a uniform electric field E_{free} of a free space, the electric field intensity E_i in the dielectric diminishes. Its value may be calculated by Eqs. 15 or 16; P/ε_0, the term subtracted from E_{free}, is called the "depolarizing field."

2. Depending on the relative permittivity of the dielectric, the value of the total charge produced on the two plates of a capacitor will have a minimum: $Q_{\text{min}} = A\varepsilon_0 V/(\text{distance between plates})$ and a maximum: $Q_{\text{max}} = A\varepsilon_0 V/d_{\text{free}}$. For very strong polarization of the dielectric (ε_r large), it is as if the dielectric layers disappear and the distance between the plates is reduced by the thickness of the dielectric layers.

5.2.2 Molecular Phenomena

A fundamental analysis or any attempts to explain the static electrification (polarization) of dielectrics leads to the molecular level. The key factor permitting the link between the macroscopic and molecular phenomena is the polarization P.

At the macroscopic level, the polarization P was defined by Eq. 7 as the bound charge per unit area of a plane capacitor, and Eq. 12 gave the mathematical expression in terms of susceptibility and electric field.

At the molecular level, let us consider the case when each atom represents a dipole. Assume that a rectangular parallelepiped of base A and height h contains n such dipoles and that they are oriented by an external field perpendicular to A. Let

$$\mu = qx \tag{20}$$

where qx is the dipole moment, and let

n_A = number of dipoles in the top layer of thickness x
n_v = number of dipoles per unit volume

The layer of dipoles along the base or the top of the parallelepiped will expose a total charge qn_A, and the surface charge density will be $q_A = qn_A/A$. If we replace $n_A = xn/h$ and q from Eq. 20 we have

$$q_A = \mu \frac{n}{Ah} \quad \text{or} \quad q_A = \mu n_v \tag{21}$$

Since the surface charge density at the macroscopic level is equal to the polarization P, it follows that *the polarization at the molecular level is the number of dipole moments per unit volume.*

The dipole moment μ is assumed to be proportional to the local field strength E' ($\mu = \alpha E'$) where α is polarizability. Hence

$$P = q_A = \alpha n_v E' \tag{22}$$

All three types of polarization—electronic, ionic, and orientational—may exist individually, or they may coexist in gaseous, liquid, and solid dielectrics. In all cases, in order for the polarization to be measurable at the macroscopic level, the dielectric must be placed in an external electric field and generally, the stronger the electric field, the stronger the polarization. The electric field applied to an individual molecule or atom which is far away from any other mass may be assumed to be the only force producing the polarization.

In an approximate way, such is the case with dilute gases. The molecules are sufficiently far apart that the electric field produced by a polarized molecule does not influence the polarization of its neighbors. However, when atoms or molecules are gases under high pressure, or when they form liquids or solids, the distances from one to another approach molecular dimensions and the influence of the polarization of one atom on the polarization of the neighboring atoms can no longer be neglected. The local polarizing force, or effective field, inside liquid or solid dielectrics must be considered to be the resultant of the externally applied field and of the neighboring dipoles' fields.

The calculation of the resultant field, designated as the "local," "internal," or "effective" field, generally becomes very involved.

In a simple case of polarization, let us assume an atom or molecule located in the interior of a symmetrical crystal lattice which is exposed to a uniform external field E_{free}. The lattice may be approximated by two regions:

I. A small free space spherical cavity surrounding the atom under consideration.

II. The remainder of the dielectric containing the spherical cavity.

The inside walls of the cavity will contain the ends of the various dipole chains present in the dielectric. The method of dividing the dielectric into two regions, introduced by Mossotti in 1850, has been widely used ever since in the mathematical calculation of the local field. In the original assumption, after the various electric fields are calculated, the free space of the spherical cavity is filled back with the dielectric material. For certain symmetrical crystal lattices, it has been demonstrated that the electric field produced by the dipoles of the dielectric sphere in the center of the sphere is almost zero. Let

E' = local field (resultant)
E_{free} = applied external field
E_{dep} = depolarizing field*
E_I = electric field produced by the dipoles of the dielectric region I after the spherical cavity is filled with the dielectric
E_{II} = electric field produced by the dipoles of region II, lining the cavity

We have then

$$E' = E_{free} + E_{dep} + E_I + E_{II} \tag{24}$$

The two electric field intensity components E_I and E_{II}, being produced locally by internal dipoles, must not be considered as linked directly to the surface charges of the dielectric.

Mossotti assumed that E_I, the electric field produced by the dipoles of the dielectric in the spherical region, was zero. (In a very approximate way, we could say that the spherical region I is in fact the space between the polarized atom under consideration and its neighbors. In such a case, E_I is zero because there are no dipoles in region I. However, the "spherical cavity" idea becomes a difficult approximation in a cubic lattice crystal.)

* Refer to exercise, p. 92, and Eq. 15. The depolarizing field E_{dep} is defined as

$$E_{dep} = -\frac{P}{\varepsilon_0} \tag{23}$$

Hence, $E_i = E_{free} + E_{dep}$.

The calculation of the electric field produced inside the cavity by the dipole ends lining the walls (E_{II}), based on the polarization P of the dielectric under the influence of the applied electric field, gives

$$E_{II} = \frac{P}{3\varepsilon_0} \quad \text{(called Lorenz field)} \tag{25}$$

From Eq. 12 we have

$$E_{II} = \frac{\chi}{3} E_i \tag{26}$$

By substituting the values of the dielectric field E_I and E_{II}, Eq. 24 becomes

$$E' = E_{\text{free}} + E_{\text{dep}} + \frac{P}{3\varepsilon_0} = E_i + \frac{\chi E_i}{3} = \frac{E_i}{3}[3 + \chi]$$

Substituting χ from Eq. 11 yields

$$E' = \frac{E_i}{3}[\varepsilon_r + 2] \tag{27}$$

and combining Eqs. 10, 22, and 27, we obtain

$$\frac{n_v \alpha}{3\varepsilon_0} = \frac{\varepsilon_r - 1}{\varepsilon_r + 2} \tag{28}$$

Equation 28 is called the Clausius–Mossotti equation. It permits the calculation of the molecular polarizability α from the macroscopic relative permittivity experimental values. For gases at atmospheric pressure, the relative permittivity is close to that of vacuum, and the Clausius–Mossotti equation becomes

$$\alpha = \frac{\varepsilon_0(\varepsilon_r - 1)}{n_v} = \frac{\varepsilon_0 \chi}{n_v} \tag{29}$$

At a certain pressure and temperature, n_v may be derived from the Loschmidt number, which is $= 2.687 \times 10^{25}$ molecules/m^3 at 0°C and 1 atm (e.g., for argon $\varepsilon_r = 1.000517$ at 20°C and 1 atm and $\alpha = 1.83 \times 10^{-40}$ Fm^2.

The polarizability α is the resultant of the three types of polarization (electronic, ionic, and orientational) and hence may be expressed as

$$\alpha = \alpha_{\text{electronic}} + \alpha_{\text{ionic}} + \alpha_{\text{orientational}} \tag{30}$$

The references listed at the end of the chapter can be consulted for an in-depth study of the attempts that have been made to measure the individual effects of each type of polarization. Of particular interest to the electrostatics engineer is the investigation of polar dielectrics and hence the ionic and orientational polarization.

When the individual molecule of a dielectric is formed such that an infinite plane can separate the centers of electric charges $\pm q$ of opposite polarity, the molecules are said to be polar. If the distance between the centers of charge q is x, each molecule has a dipole moment qx. All hetero-nuclear diatomic molecules are polar. Two atoms of different elements generally have different attractions for electrons. It is normal to expect that the common electron cloud will be shifted toward the atom that attracts it most, thus forming a dipole (e.g., in hydrogen chloride the electrons spend more time near the chlorine atom). The unit for the dipole moment of a polar molecule has been named in honor of the Dutch physicist, P. Debye: 1 debye (D) $= 3.33 \times 10^{-30}$ Cm.

For polyatomic molecules, the resultant dipole moment depends greatly on their geometry. The bonds between each two different atoms have a dipole moment, but the molecule as a whole may not. As an example, the geometry of the carbon dioxide molecule may be simply represented as

$$O^- = C^{2+} = O^-$$

Nowhere can an infinite plane be drawn to separate two equal and opposite *total* charges. The dipole moment of carbon dioxide is zero. By contrast, the geometry of the nitrogen dioxide molecule is

$$
\begin{array}{c}
N^{2+} \\
\diagup \ \diagdown \\
O^- \ \ O^-
\end{array}
$$

A horizontal plane can separate the two oxygen atoms from the nitrogen atom. The dipole moment of nitrogen dioxide is approximately 0.4 D.

When polar dielectrics are placed in an electric field, all the molecular dipoles are subjected to a torque tending to align their axes in the direction of the field. In gases and liquids the orientation of the dipoles takes place easily. Thermal motion hinders the orientation and tends to maintain the random distribution. It can be observed that as the temperature increases, the orientational polarization decreases.

For dilute gases in electric fields encountered experimentally (in the order of 10^4 kV/m) the orientational polarizability in Eq. 30 may be derived as

$$\alpha_{\text{orientational}} = \frac{\mu^2}{3kT} \tag{31}$$

where μ is the dipole moment, k is Boltzmann's constant, and T is the absolute temperature.

For liquids and solids there is no general mathematical formula for the orientational polarizability. In most cases, at the temperature point at which a liquid changes into a solid, a sudden decrease in the orientational polarizability occurs.

Recently, highly polar liquids have been the object of intensive studies. The reader is referred to the work of N. J. Felici on nitrobenzene.[5]

For many years, polar liquid dielectrics have been neglected on account of their unsatisfactory resistivity or high dielectric loss. The findings on nitrobenzene have brought the polar liquids back into focus. Deionizing nitrobenzene by means of electrodialytic membranes, the resistivity increased by orders of magnitude (from 10^7 to 10^{10} Ω-cm). At the same time it was also found that strong electric fields could be applied without subsequent reduction of the electrical resistivity.

It appears that there are some bright horizons in this field. Dielectrics in particular are discussed in References 1 to 7.

5.3   STATIC ELECTRIFICATION AT INTERFACES

5.3.1   General

Among the possible types of interfaces formed by two different materials in solid, liquid, or gaseous state, the solid-to-solid, solid-to-liquid, and liquid-to-liquid are of considerable current interest in applied electrostatics. Under the broad definition of electrification, all interfaces become electrified. Solid-state physics and modern electrochemistry provide the needed theoretical background for some understanding of the complex phenomena present in the electrification of the three types of interfaces mentioned previously. To the applied electrostatics engineer, however, understanding of electrification alone is not sufficient. He must be able to predict the electric charges left on the surfaces after separation or, indirectly, the charge backflow.

It is perhaps the present lack of knowledge of the surface charge *backflow* laws that compounds the problem of predicting the static electrification. When dealing with surfaces of materials which are generally encountered in practice—not, for example, the (010) plane of a perfect single crystal—the best we can do with the present status of knowledge is to predict the polarity of the charge left after separation. Such a prediction requires very careful preparation of the surfaces and could only be based on an average of several experiments. It would be impossible to cover in the space allocated the vast body of knowledge dealing with the topic. For an up-to-date collection of many recent experiments in static electrification, the reader is particularly directed to References 8 to 10. In what follows, some data are presented to help the engineers or manufacturers in the optimization of their particular application.

For this purpose, the following topics are considered:

Solids	Surface preparation
	Types of contact
	Charge backflow
	Synopsis of some experiments for the period 1937 to 1971
	Electric fields and ambient temperatures
Liquids	Helmholtz double layer
	Flow of liquids in pipes
	Charge relaxation
Molecular Phenomena	Types of bonds and crystals
	Work function
	Surface states
	Electronegativity

5.3.2 Solids

5.3.2.1 *Surface Preparation*

The word "preparation" is often preferred to "cleaning." It is virtually impossible to produce a truly clean surface, free from any contamination. A monolayer of adsorbed gases is sufficient to change the work function of the surface, completely altering the electrification (see definition of work function, Section 3.3.4). Adsorption refers to surface only and is distinct from absorption, which involves penetration (e.g., absorption of a gas into a solid or liquid). Adsorption is influenced by the pressure of the gas up to a limit represented by a complete monolayer of gas molecules. Such layers are difficult to remove because they often form bonds with the atoms of the materials (oxygen forms oxides, hydrogen forms hydrides, etc.).

The adsorbed molecules are very mobile, and the monolayer is often compared with a two-dimensional gas or liquid. The phenomenon is not limited to solids. Liquid surfaces also adsorb monolayers of other molecules.

Particulate contamination of surfaces is another variable to consider. Very fine dust settled on a surface is difficult to remove. For light pressures, the dust could play an important part in the interface electrification.

Determining the means of preparation of a surface by standard mechanical, chemical, or electrical procedures would be a good starting point for comparing notes and accumulating a pool of data on which to build the static electrification laws of materials used in practical applications. A review of the literature shows a great variety of surface preparations. No two methods are identical, and in many cases it is doubtful whether an investigator could reproduce a certain set of experiments based solely on the published data.

Among the recent techniques, baking in high vacuum below 10^{-5} torr, nitrogen or argon atmospheres, and surface preparation by means of electric sputtering have yielded fairly meaningful results to some experimenters.

5.3.2.2   *Type of Contact*

Another area where standardization would be very beneficial is in the type of contact. W. R. Harper[8] pinpointed the problem and proposed the following terminology:

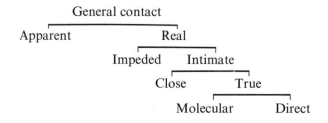

Such a terminology must be supplemented by specification of the nature of the contact, including: rolling, sliding, or simple contacting; shape of surfaces; dimensions with specific manufacturing tolerances; speed of contacting; duration; speed of separation; masses, and moments of inertia involved in the motion.

The standardization of the surface preparation techniques and the types of contact, coupled with a complete description of the materials involved, the test procedures, and the working ambients will go a long way toward achieving reproducibility.

5.3.2.3   *Charge Backflow*

The charge backflow is an integral part of the final static electrification of surfaces. Being able to control the electric charge backflow, in many cases, it would represent the control of the majority of the charge in static electrification. Although a considerable number of studies and theories have been advanced for the electrification of interfaces prior to separation, the control of the charge backflow has been a neglected field. Based on the knowledge to date, we can make only a few general statements of interest to applied electrostatics:

1. When an electrified interface is separated in its two component surfaces, the amount of charge that flows back may be an order of magnitude or more larger than the charge remaining.

2. The charge backflow depends on several parameters, including: the materials in contact, the surface preparation, the geometry of the contact,

the speed of separation, the temperature, the ambient gas, and the external fields. Special consideration must be given to the gases adsorbed on the surface or present in the separation gap.

3. It appears that when all the foregoing variables are fixed, the charge backflow can be expressed as a fairly reproducible percentage of the original electrification.

4. For an understanding of the molecular phenomena involved, further studies of the density of surface states on the various materials and of the tunneling of charge seem to be essential.

5.3.2.4 *Synopsis of Some Experiments in Solid-to-Solid Electrification*

There are two questions that must be answered in any application of the solid-to-solid electrification phenomena:

1. What is the polarity of the charge?
2. How much charge is left after separation?

With the help of solid-state physics and through carefully controlled experiments performed under vacuum, considerable progress has been made toward the answer to the first question. It has been shown by Davies[29] and Inculet[30] that when two contacting solids of different work function are separated, the charge left on the surface with greater work function is of negative polarity. The surface of smaller work function becomes positively charged.

Regarding the amount of charge left after separation, there are no fast rules. The reader is directed to Table 5.3, a synopsis of some experiments carried out over a period of 35 years, giving the materials in contact, the type of contact, the polarity of the charge, the ambient, the surface preparation details, and the variables studied.

It is of interest to note that as time passed, the various researchers used increased control of the surface preparation and experimental ambients. W. Greason's work (reported last in the synopsis as Ref. 37) includes electric sputtering as the last step in surface preparation of the metals, as well as 10^{-9} torr vacuum ambient.

5.3.3.5 *Electric Fields and Ambient Temperature*

The recent developments in the electrostatic beneficiation or separation of mineral ores in fluidized beds have prompted intensive studies on the influence of externally applied electric fields on the interface electrification of various materials. Generally the electric field applied to the fluid bed is unidirectional. However, the collisions of the particles of two different materials A and B will, at times, be such as the electric field crosses the interface from A to B and at other times from B to A. The two cases were

Table 5.3. Some Contact Electrification Studies (1937–1972)[a,b]

Materials	Type of Contact	Charging Polarities	Ambient	Surface Cleaning Preparation	Variables Studied	Year	Ref.
Quartz-nickel	Rub	X	Vacuum, 60 cm—10^{-4} mm Hg	M	T, P	1937	11
Quartz-Ni, Sn, Bi, Zn, Al, Cu, Pd	Rub	Quartz (+) with all metals except above 350°C, when it became (−)	Air and vacuum to 10^{-4} mm Hg	M	T, P	1937	12
Nickel-quartz, NaCl	Rub	Quartz (−), NaCl (+) with nickel	Air, O_2, N_2	C	P	1944	13
Ebonite, sand–unknown metal	Rub	Sand and ebonite (−)	Air	X	E	1949	14
Steel-Rubber	Roll	Steel (+)	Air	C	T	1946	15
Quartz–quartz	Rub	Quartz particles sliding over a stressed quartz surface charged (−)	Air	X	F	1949	16
Cr–Co, Ni, steel, Ag, Cu, Rh, Pt, Au	Point	Cr charged (+) with all other metals	Air	X	W	1951	17
Au, Cr–silica, polyethylene, amber, polystyrene, nylon, perspex, teflon, silicone	Point	Charging characteristics of silica depended on surface cleaning, others unknown	Air	C, M	—	1953	18
Hg–nylon, alkathene, mica, cellophane, Keratin, filter paper	Rub	X	Air, slight vacuum	C, M	E	1953	19

Materials	Method	Notes	Atmosphere			Year	No.
Glass–Ni	Roll	Glass (−)	Vacuum: 10^{-6} torr and dry air	X	V, E, H	1954	20
Quartz, glass–ni	Roll	Quartz and glass (−)	Vacuum: 10^{-7} torr	C, M	V, P	1954	21
Pt, Au, Al, Mg, Fe, stainless steel, viscose, nylon, cotton, acetate, orlon, Dacron, polyethylene, PVC, Dynel, PFTE, wool, silk, lucite, Velon	Rub	Polarities depended on whether metal surface was scraped	Air	C, M	F, V, H	1955	22
Ni, Pt, Cu–quartz, Al$_2$O$_3$, MgO, NaCl, KCl, KBr, KI	Roll	Quartz charged (−) with all metals	Vacuum: 10^{-8} torr	C	V	1956	23
Pt, Au, Ag, W, Cr, Ni, Mo, Fe, V, Mg, Al, Cu, Zn, Sn, various semiconductors	Point	X	Air	X	W	1957	24
Cellulose nitrate, ethyl cellulose, cellulose acetate casein, ebonite Perspex, polystyrene, Tufnol, PTFE, polythene–Cu, Zn, brass, Fe, Bi, Al	Point	Triboelectric series given	Air	C	F	1957	25
Nylon, polyethylene–Ta PTFE, polyethylene, polystyrene, amber, PMMA, mica	Rub	X	Air	C	V, F	1958	26
	Rub	Triboelectric series given	Air	C	—	1958	27

Table 5.3. Some Contact Electrification Studies (*Continued*)

Materials	Type of Contact	Charging Polarities	Ambient	Surface Cleaning Preparation	Variables Studied	Year	Ref.
Polyethylene–polyethylene	Impulsive	X	Air	C	T	1961	28
Pt, Au, Pd, Al, Cd–polythene, glass	Rub	$\phi_{glass} = 4.3$ $\phi_{polythene} = 4.7$	Vacuum: 10^{-5} torr	C, M	W	1967	29
Zr, Sn, Ag, Cu, Au, Ni, Pt-glass	Rub	$\phi_{glass} = 4.6$	Vacuum: 10^{-7} torr	Z, C	—	1967	30
Cu, Al, Ni, Mg, Cd, Zn, Sb, W, Mo, Ag, Au, Pt–nylon	Rub	$\phi_{nylon} = 4.2$	Vacuum: 10^{-3} torr	C	W	1967	31
PVC, polyimide, PTFE, Polycarbonate, PET, polystyrene, nylon 66–Cd, Au, Zr, Pt, Al	Rub	$\phi_{PVC} = 4.85$ $\phi_{polyimide} = 4.36$ $\phi_{polycarbonate} = 4.26$ $\phi_{PTFE} = 4.26$ $\phi_{PET} = 4.25$ $\phi_{polystyrene} = 4.22$ $\phi_{nylon\,66} = 4.08$	Vacuum: 10^{-6} torr	C, M	W	1969	32
Ni, Au-glass	Rub	Ni (−), Au (+)	Vacuum: 10^{-7} torr and air	Z, C	E	1970	33
Ni, Au-glass, quartz	Point	Ni (−), Au (−)	Vacuum: 10^{-9} torr	Z, C	E, T	1971	34
Fe-copolymerization of methyl methacrylate (MMA) and styrene	Rub	Charges of both polarities present	Air	C	% MMA	1971	35

| Cu–PTFE | Rub | | Air | C | V, F | 1971 | 36 |
| Au, Ni, Pt, Al–nylon 6, nylon MD, polystyrene, acetal, polyethylene, polypropylene, glass, quartz, PVC, Teflon, Lexan, acrylic | Point | X Effective dielectric work functions to be calculated for all insulators | Vacuum: 10^{-9} torr | C, Z | T, E | 1972 | 37 |

---

[a] Key to surface cleaning preparations:

Z = sputtering (electrical)

C = chemical

M = mechanical

X = unknown.

[b] Variables studied:

T   temperature

P   pressure of gaseous ambient

H   humidity

V   velocity of rubbing

F   force at contact

E   electric field

W   measured contact potential difference of metals

% MMA % methyl methacrylate in copolymer

ϕ = work function (eV).

Table 5.4. Charges on Various Metals for Single Contact under Vacuum, 10^{-11} C

Insulator	T (°C)	Aluminum, 10^{-11} C Electric Field (kV/m)			T (°C)	Gold, 10^{-11} C Electric Field (kV/m)			T (°C)	Nickel, 10^{-11} C Electric Field (kV/m)			T (°C)	Platinum, 10^{-11} C Electric Field (kV/m)		
		−300	0	300		−300	0	300		−300	0	300		−300	0	300
Polypropylene SR	70	0.590	0.606	0.622	70	0.168	0.177	0.186	70	0.091	0.150	0.209	75	0.083	0.080	0.078
	25	0.738	0.845	0.953	30	0.432	0.401	0.36	30	0.157	0.190	0.222	25	0.060	0.072	0.085
	−25	0.883	0.883	0.883	−5	0.029	0.035	0.039	−10	0.119	0.215	0.316	−10	0.059	0.055	0.052
Polypropylene NSR	65	0.684	0.629	0.574	75	0.279	0.251	0.223	75	0.182	0.200	0.217	75	0.083	0.108	0.132
	25	0.483	0.691	0.898	30	0.171	0.158	0.144	25	0.130	0.172	0.215	25	0.065	0.069	0.072
	−25	0.864	0.871	0.878	−15	−0.136	−0.145	−0.153	−10	0.130	0.197	0.264	−10	0.044	0.050	0.055
Acrylic	75	1.40	1.58	1.76	70	−0.918	−0.722	−0.525	80	−1.31	−1.30	−1.27	80	−0.226	−0.286	−0.346
	30	1.28	1.46	1.67	30	−0.902	−0.820	−0.739	30	−0.970	−1.00	−1.04	30	−0.545	−0.473	−0.493
	−20	1.23	1.31	1.39	−5	−0.837	−0.799	−0.762	−5	−1.31	−1.21	−1.12	−5	−0.534	−0.542	−0.549
Lexan	70	0.757	0.669	0.580	75	−0.506	−0.421	−0.335	80	−1.02	−1.12	−1.22	80	−0.067	−0.137	−0.207
	30	−0.121	−0.448	−0.775	30	−0.718	−0.691	−0.663	30	−0.755	−0.717	−0.678	30	−0.113	−0.102	−0.092
	−20	−0.401	−0.477	−0.533	−25	−0.444	−0.554	−0.663	−5	−1.28	−1.52	−1.76	−5	−0.226	−0.256	−0.285
Polyethylene LDNSR	60	0.629	0.597	0.565	70	0.086	0.072	0.057	70	0.882	1.12	1.37	60	0.164	0.140	0.116
	25	0.426	0.481	0.536	30	0.074	0.066	0.058	30	0.814	0.797	0.781	30	0.058	0.060	0.062
	−15	0.331	0.533	0.736	−5	0.085	0.038	−0.008	−10	0.070	0.156	0.242	−10	0.028	0.036	0.043
Polyethylene LDSR	60	0.354	0.450	0.546	70	−0.338	−0.271	−0.204	65	0.987	0.978	0.970	60	0.065	0.070	0.075
	25	0.298	0.371	0.444	30	0.203	0.115	0.267	30	0.620	0.612	0.604	30	+0.049	+0.037	0.026
	−15	0.226	0.472	0.718	−5	−0.115	−0.129	−0.144	−10	−0.011	0.082	0.177	−10	−0.009	−0.013	−0.017
Polyethylene HDNSR	65	0.224	0.417	0.608	75	0.169	0.178	0.188	65	0.628	0.775	0.922	60	0.082	0.089	0.096
	30	0.261	0.291	0.321	30	0.114	0.142	0.170	30	0.554	0.661	0.767	30	0.049	0.058	0.067
	−20	0.175	0.253	0.331	−5	0.037	0.069	0.101	−10	0.013	0.145	0.276	−10	0.025	0.033	0.042
Polyethylene HDSR	70	0.460	0.608	0.757	75	0.139	0.198	0.258	65	0.855	1.16	1.47	80	0.126	0.119	0.111
	30	0.409	0.467	0.526	30	0.156	0.145	0.133	30	0.913	1.05	1.18	30	0.087	0.099	0.111
	−20	0.246	0.293	0.341	−5	0.054	0.056	0.058	−10	0.031	0.197	0.363	−10	0.057	0.072	0.086

Material															
PTFE	80	0.239	0.307	0.375	70	0.171	0.231	0.291	65	0.251	0.321	0.392	70	0.106	0.121	0.135
	30	0.367	0.372	0.378	25	0.154	0.150	0.147	30	0.202	0.303	0.405	25	0.063	0.092	0.121
	−30	0.181	0.210	0.239	−15	0.181	0.180	0.180	−5	0.142	0.249	0.356	−10	0.055	0.071	0.087
Acetal	80	2.31	0.116	−2.08	75	0.954	−0.079	−1.11	80	7.71	0.381	−6.95	80	3.09	0	−3.1
	30	−0.264	−0.227	−0.190	25	−0.192	−0.176	−0.160	30	−0.174	−0.189	−0.204	30	−0.076	−0.080	−0.085
	−20	−0.454	−0.546	−0.638	−5	−0.260	−0.239	−0.217	−20	−0.482	−0.446	−0.411	−5	−0.116	−0.108	−0.101
PVC	55	1.55	1.31	1.08	100	35.2	0.709	−21.0	70	0.459	0.392	0.325	75	0.324	0.368	0.413
	30	1.32	0.979	0.633	25	−0.136	−0.166	−0.196	30	0.374	0.408	0.441	25	0.111	0.117	0.124
	−30	0.852	0.760	0.668	−15	−0.142	−0.202	−0.261	−5	0.493	0.582	0.672	−10	−0.128	−0.097	−0.067
Polystyrene	75	3.22	3.38	3.53	70	0.055	0.073	0.091	70	−0.813	−0.903	−0.992	80	0.061	0.030	0
	25	2.16	2.42	2.68	25	−0.346	−0.401	−0.470	30	−0.438	−0.511	−0.583	30	−0.150	−0.121	−0.091
	−20	1.35	1.23	1.09	−5	−0.224	−0.268	−0.313	−25	−1.17	−0.886	−0.597	−5	−0.196	−0.170	−0.144
Nylon 1203	50	0.091	−1.53	−3.16	75	21.9	−0.542	−23.1	80	40.3	−1.26	42.8	80	16.5	−0.355	−17.2
	25	−0.342	−0.364	−0.385	25	−1.1	−1.33	−1.56	30	−1.1	−1.28	−1.47	30	−0.606	−0.618	−0.630
	−30	−0.492	−0.766	−1.04	−5	−1.43	−1.43	−1.43	−20	−2.28	−2.64	−3.00	−5	−0.828	−0.752	−0.677
Nylon MD1211	60	14.1	2.03	−10.0	75	19.7	−0.859	−21.5	80	40.2	−1.19	−42.5	80	6.08	−0.358	−6.8
	25	−0.325	−0.415	−0.505	25	−0.592	−0.621	−0.645	30	−0.572	−0.819	−1.07	30	−0.365	−0.588	−0.812
	−25	−0.556	−0.712	−0.868	−5	−0.738	−0.760	−0.782	−20	−0.930	−0.88	−0.831	−5	−0.359	−0.412	−0.465
Pyrex 7740	100	7.58	0.874	−5.84	125	24.4	−0.310	−25.1	100	30.0	−0.724	−28.5	90	9.40	−0.554	−8.28
	25	1.81	1.93	2.05	20	−0.258	−0.209	−0.161	30	−0.023	−0.137	−0.047	30	−0.142	−0.153	−0.163
	−20	1.12	0.631	0.139	−5	−0.463	−0.416	−0.369	−15	−0.108	−0.103	−0.098	−5	−0.125	−0.139	−0.153
Quartz	90	1.53	0.878	0.223	125	−0.629	−0.525	−0.421	100	−0.077	−0.086	−0.095	100	−0.171	−0.208	−0.243
	25	1.19	1.22	1.25	30	−0.388	−0.363	−0.338	30	−0.089	−0.110	−0.131	30	−0.163	−0.141	−0.120
	−20	0.874	0.361	−0.151	−5	−0.387	−0.416	−0.445	−15	−0.123	−0.125	−0.126	−5	−0.096	−0.10	−0.105

arbitrarily designated $(+)$ when the field was directed from metal to the other material and $(-)$ for the reverse case. Some electrification studies indicate a lack of symmetry in the characteristic curve.[33]

Also a considerable interest is the influence of temperature. Over a relatively small range of temperatures $(-40–+100°C)$ we find pronounced changes in the electrification of certain interfaces under the influence of electric fields.

Table 5.5.   Charge on Various Metals, 10^{-11} C, for Single Contact[a] in Air at 25 °C and 20% Relative Humidity

Insulator	Aluminum	Gold	Nickel	Platinum
Polypropylene SR	0.164	0.151	0.188	0.120
Polypropylene NSR	0.179	0.173	0.273	0.116
Acrylic	−0.257	−0.156	−0.194	−0.320
Lexan	0.113	−0.420	0.205	−0.151
Polyethylene LDNSR	0.338	0.465	0.372	0.154
Polyethylene LDSR	0.271	0.339	0.322	0.157
Polyethylene HDNSR	0.282	0.325	0.317	0.164
Polyethylene HDSR	0.243	0.264	0.285	0.158
PTFE	0.169	0.200	0.198	0.152
Acetal	−0.157	−0.142	−0.286	−0.140
PVC	0.345	0.383	0.330	0.314
Polystyrene	0.237	0.133	0.189	0.232
Nylon 1203	−0.356	−0.331	−0.238	−0.238
Nylon MD1211	−0.301	−0.385	−0.265	−0.263
Pyrex 7740	0.182	−0.198	−0.148	−0.703
Quartz	−0.249	−0.259	0.173	−0.116

[a] Average of 30 single contacts.

Tables 5.4 and 5.5 give results reported by W. Greason[37] for various fields and temperatures under vacuum as well as some corresponding results in air. All measurements were for point contacts between a spherical surface (the metal) and a plane (the other solid). For complete details of the experimental setup the reader is directed to the original paper. For each metal–insulator combination at each temperature in Table 5.4, an average of 40 data points were taken with voltage varying between ±10 to 15 kV. The best straight line was fitted to the points. These readings are the intercept of the Q axis (i.e., 0 V), the Q value at -10 kV, and the Q reading at $+10$ kV, respectively.

5.3.3 Liquids

Compared with solid interfaces, the knowledge of the electrification of liquid-to-solid or liquid-to-liquid interfaces is not further advanced. The published empirical formulas developed from certain experimental conditions are not always reliable, unless the experimental conditions can be accurately duplicated.

Among the many applications, there are three types of liquid interfaces applications which deserve a systematic investigation:

1. Flow of liquids in pipes.
2. Filtration of liquids by means of solids.
3. Settling of a dispersed liquid in another.

Any such investigation should be based on standard materials, techniques, and ambients and must take into account such factors as: (1) the thickness of the Helmholtz double layer, (2) the electrical resistivity of the liquid, (3) the mobility of the ions, (4) the viscosity of the liquid, (5) the relative velocity, and (6) the temperature.

For a detailed review of the topic the reader is directed to References 8 to 10.

5.3.3.1 *Helmholtz Double Layer*

The interface between normally encountered solids and liquids is generally characterized by the appearance of the so-called Helmholtz double layer. Figure 5.2 represents, in a very simplified and idealized form, the structure of the interface between a liquid (in this case chosen to be polar) and a metal.

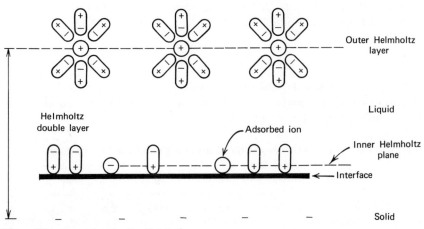

Figure 5.2. A metal–polar liquid interface.

Some positive metal ions are assumed to be released into the liquid near the interface. (If negative ions in a different case are released, the following polarities will be reversed in all respects). The polar nature of the liquid will cause some of its molecules to rotate and align themselves in the direction of the electric fields around the positive ions and near the interface. It is also assumed that some negative ions present in the liquid are adsorbed at the metal surface.

The layer of positive ions at the level surrounded by polar molecules forms the outer Helmholtz layer. The outer Helmholtz layer, together with the layer of opposite charge on the metal, form the Helmholtz double layer. As already mentioned, this is an idealized plane representation; in fact, the layer of positive ions must be visualized as a somewhat diffuse layer. The locus of the negative ions adsorbed by the metal surface is sometimes called the inner Helmholtz plane.

The importance of the Helmholtz double layer in solid-to-solid electrification is evident. The fast removal of the outer layer leaves behind a charged surface on the solid, whereas the liquid carries with it an equal amount of opposite charge.

5.3.3.2   Flow of Liquids in Conducting Pipes

The electrification of liquids in pipes has been analyzed by many investigators and is perhaps the furthest advanced area. Yet it still relies on very empirical formulas. All formulas that have been developed focus on the streaming current, or the amount of net charge per unit time entrained by the liquid as it flows through the pipe. Among the many types of interfaces and liquids, we consider the common occurrence of a high-resistivity liquid in turbulent flow. Designating the pipe diameter by d and the linear velocity of the liquid by v, most investigators seem to agree that the streaming current for an infinite pipe length (i_∞) and a certain metal liquid interface, can be expressed as

$$i_\infty = A v^\alpha d^\beta \qquad (32)$$

where A is a constant depending on various physical constants of the liquid, degree of ionization, and so on, and α and β are exponents which vary from investigator to investigator (Table 5.6).

Table 5.6.   Variations in α and β According to Several Investigators

Investigator(s)	α	β	Year	Ref.
Koszman and Gavis	1.88	0.88	1962	39
Schön	1.8–2.0	1.8–2.0	1965	40
Gibson and Lloyd	2.4	1.6	1970	42

5.3.3.3   *Notes of General Interest*

1. It has been observed[41] that when the liquid flow changes from laminar to turbulent, the streaming current increases markedly.

2. The main effects of contaminants in a liquid are reflected in its resistivity. The other physical constants such as viscosity or permittivity (which also affect the streaming current) do not change markedly for contaminated liquids.

3. Equation 32 does not apply to laminar flow or to flow in nonconducting pipes.

5.3.3.4   *Charge Relaxation*

The relaxation of electric charge from a stationary liquid is generally described by the equation

$$Q = Q_0 e^{-t/\tau} \tag{33}$$

where Q = total charge at time t
$\quad Q_0$ = initial charge
$\quad \tau$ = time constant $= \varepsilon_r \varepsilon_0 \rho$
$\quad \rho$ = resistivity of the liquid

It is important to consider that ρ is not generally constant. It has been pointed out by Klinkenberg[8] that since some of the ions in the double layer are being carried away at a faster rate than they are regenerated, the resistivity of the liquid in that region must change.

5.3.4   Molecular Interface Phenomena

For a better understanding of the interface electrification phenomena, we must investigate what happens at the molecular level. It is believed that types of bonds, work function, surface states, and electronegativity are some of the important tools with which applied electrostatics engineers must become familiar in any systematic attempt to improve or control the interface electrification.

5.3.4.1   *Types of Bonds*

a.   VAN DER WAALS FORCES.   Van der Waals forces are explained by considering two identical atoms separated by a distance that is relatively large when compared with the radii of the atoms. If the electrons' charge of the atoms were in the form of a fixed cloud surrounding the nucleus, there could not be any interaction. Because of the electrons' motions, fluctuating dipoles appear on the two atoms. Any instantaneous dipole moment on one atom induces a dipole moment on the second one. The result of this interaction is an attraction force between the two atoms called Van der Waals force or the "fluctuating dipole interaction."

b.   IONIC OR ELECTROVALENT BOND.   Generally an ionic bond involves the formation of two oppositely charged ions through the transfer of electrons from a metal atom to a nonmetal one, followed by the mutual attraction of the ions. The reasons for the electron transfer is found in the tendency for all atoms to approach the stable electronic configuration of the nearest rare gas which has no further combining power.

Let us consider sodium chloride (NaCl) as an example: Na has one electron more than neon and Cl has one electron less than argon. When Na gives an electron to Cl, the two ions that are formed, Na^+ and Cl^-, will have rare gas electron cloud configurations. Because of their opposite charges, electrostatic forces appear to form the generally strong ionic bond.

c.   COVALENT BOND.   Bonds based on the sharing of electrons to achieve rare gas electronic cloud configurations are called covalent. When two atoms of nonmetals need additional electrons to reach a rare gas configuration, they cannot transfer electrons as in the case of an ionic bond. If the two electronic clouds are such that by sharing some electrons, both atoms achieve rare gas configurations, a covalent bond emerges.

Water is an example of a covalent bond. The electrons of the two hydrogen atoms are shared with the oxygen atom to form a rare gas configuration of eight electrons around the oxygen nucleus. At the same time, two of the oxygen electrons are shared by the hydrogen atoms to form a stable helium configuration of two electrons around the hydrogen nuclei.

d.   METALLIC BOND.   The quantum-mechanics theory explaining the metallic bond is rather complex. It would require considerable background to lead to the model that is envisaged to represent this type of bond, believed to comprise a "free electron gas" surrounding the individual atoms. The reader is referred to References 43 to 47.

5.3.4.2   *The Work Function*

The work function is defined as the energy required to remove one electron from the Fermi level out of the solid to infinity. The Fermi energy level for a metal may be considered to be the energy of the free electron gas at zero absolute temperature.

The work functions of most metals have been known for a long time, but it is interesting to note that work functions of several insulators were measured only in the past three or four years through static electrification techniques.[32]

5.3.4.3   *Electronegativity*

In broad terms, electronegativity is the electron-attracting capacity of atoms. It is sometimes defined as the average of the ionization energy and electron affinity.

The ionization energy is the energy necessary to remove one electron from an atom (the atom becomes a positive ion). The electron affinity is the energy released when a neutral atom acquires an electron (the atom becomes a negative ion).

5.3.4.4 *Surface States*

The symmetry surrounding an atom in a crystal lattice ends abruptly at the surface. In addition to surface imperfections, the surface often becomes contaminated with foreign atoms, adsorbed gases, or both. The result is that the energy band configurations that apply to the interior of the material are no longer valid for the surface, where large densities of localized quantum states may appear.

The density of the impurities and imperfections may be so high that separate energy bands are formed. The energy levels of these new bands may lie in the forbidden gap. Although the crystal may be an insulator, the surface states could generate "impurity conduction."

At the 1971 Static Electrification Conference in London, H. Krupp,[10] reviewing the static electrification of solids, justifiably placed major emphasis on the responsibility of the surface states for static electrification.

REFERENCES

1. A. von Hippel, *Dielectrics and Waves*, Wiley, New York, 1952.
2. ———, *Dielectric Materials and Applications*, Wiley, New York, 1954.
3. J. B. Birks, *Progress in Dielectrics*, Academic Press, New York, 1963.
4. J. C. Anderson, *Dielectrics*, Chapman and Hall, London, 1964.
5. N. J. Felici, *Direct Current*, pp. 24–32 (1967).
6. W. R. Harper, *Contact and Frictional Electrification*, Oxford University Press, London, 1967.
7. L. B. Loeb, *Static Electrification*, Springer, Berlin, 1958.
8. "Static Electrification," Proceedings of the London Static Electrification Conference, Institute of Physics and Physical Society, London, 1967.
9. "Advances in Static Electricity," Proceedings of the Vienna 1st International Conference on Static Electricity, Auxilia S.A., Brussels, 1970.
10. "Static Electrification 1971," Proceedings of the 3rd Conference on Static Electrification, The Institute of Physics, London, 1971.
11. P. A. Mainstone, *Phil. Mag.*, **23**, 702–708 (1937).
12. ———, *Phil. Mag.*, **23**, 620–629, (1937).
13. D. Debeau, *Phys. Rev.*, **66**, 9–16 (1944).
14. E. W. B. Gill and G. F. Alfrey, *Nature*, **163**, 172 (1949).
15. R. S. Havenhill, H. C. O'Brien, and J. J. Rankin, *J. Appl. Phys.*, **17**, 338–346 (1946).
16. J. W. Peterson, *Phys. Rev.*, **76**, 1882 (1949).
17. W. R. Harper, *Proc. Roy. Soc. (London)*, **A205**, 83–103 (1951).

18. ———, *Proc. Roy. Soc. (London)*, **A218**, 111–121 (1953).

19. J. A. Medley, *Brit. J. Appl. Phys.*, Suppl. 2, p. S29 (1953).

20. J. W. Peterson, *J. Appl. Phys.*, **25**, 501–504 (1954).

21. ———, *J. Appl. Phys.*, **25**, 907–915 (1954).

22. S. P. Hersh and D. J. Montgomery, *Text. Res. J.*, **24**, 279–295 (1955).

23. P. E. Wagner, *J. Appl. Phys.*, **27**, 1300–1310 (1956).

24. M. M. Bredov and I. Z. Kshemianskaia, *Sov. Phys.-Tech. Phys.*, **2**, 844–850 (1957).

25. G. S. Rose and S. G. Ward, *Brit. J. Appl. Phys.*, **8**, 121–126 (1957).

26. R. G. Cunningham and D. J. Montgomery, *Text. Res. J.*, **28**, 971–979 (1958).

27. E. Fukada and J. F. Fowler, *Nature*, **181**, 693–694 (1958).

28. A. H. Bowles, *Proc. Phys. Soc.*, **231**, 388 (1961).

29. D. K. Davies, IPPS Conference Series 4, *Static Electrification*, Institute of Physics and the Physical Society, London, 1967, pp. 29–36.

30. I. I. Inculet and E. P. Wituschek, IPPS Conference Series 4, *Static Electrification*, Institute of Physics and the Physical Society, London, 1967, pp. 37–43.

31. R. G. C. Arridge, *Brit. J. Appl. Phys.*, **18**, 1311–1316 (1967).

32. D. K. Davies, *Brit. J. Appl. Phys. (J. Phys. D)*, **2**, 1533–1537 (1969).

33. I. I. Inculet, *J. Colloid Interf. Sci.*, **32**, 395–400 (1970).

34. ——— and W. D. Greason, IP Conference Series 11, *Static Electrification*, Institute of Physics, London, 1971, pp. 23–32.

35. N. Murasaki, N. Kono, M. Matsui, and H. Mada, IP Conference Series 11, *Static Electrification*, Institue of Physics, London, 1971, pp. 44–51.

36. A. Wahlin and G. Backstrom, IP Conference Series 11, *Static Electrification*, Institute of Physics, London, 1971, pp. 52–58.

37. W. D. Greason, Ph.D. thesis, University of Western Ontario, London, "Effects of electric fields and temperature on the electrification of metals in contact with insulators and semiconductors in vacuum," 1972.

38. A. Klinkenberg, *Electrostatics in the Petroleum Industry*, Elsevier, New York, 1958.

39. I. Koszman and F. Gavis, *Chem. Eng. Sci.*, **17**, 1013–1023 (1962).

40. G. Schön, *Handbuch der Raumexplosionen*, H. H. Freytag Verlag Chemie GmbH (Weinheim, Bergstr.), 1965

41. J. C. Gibbings and E. T. Hignett, *J. Electroanalyt. Chem.*, **16**, 139 (1968).

42. N. Gibson and F. C. Lloyd, *Brit. J. Appl. Phys. (J. Phys. D.)*, **3**, 563 (1970).

43. C. Kittel, *Introduction to Solid State Physics*, Wiley, New York, 1967.

44. J. O'M. Bockris and A. K. N. Reddy, *Modern Electrochemistry*, MacDonald, London, 1970.

45. J. E. Spice, *Chemical Binding and Structure*, Pergamon Press, London, 1966.

46. A. Nussbaum, *Electromagnetic and Quantum Properties of Materials*, Prentice-Hall, Englewood Cliffs, N.J., 1966.

47. C. A. Wert and R. M. Thomson, *Physics of Solids*, McGraw-Hill, New York, 1970.

CHAPTER 6

Long-Lasting Electrization and Electrets

OLEG D. JEFIMENKO
Department of Physics
West Virginia University
Morgantown, West Virginia

6.1 OVERVIEW

Magnets—bodies possessing long-lasting magnetization—have been known since the dawn of civilization, and a multitude of immensely useful applications have been found for them. It is natural to expect that if there existed bodies possessing long-lasting electrization,* these bodies also would be most useful.

A search for methods of obtaining such bodies started more than 200 years ago. But only relatively recently, with the advent of the so-called electret, has it met with evident success. Electrets, if properly handled, retain their electrization for many years. They can be manufactured from a variety of materials. Numerous applications have been proposed for them, and some of these have been already adopted by industry.

A peculiar and unfortunate circumstance adversely affecting contemporary investigations of long-lasting electrization is that almost all generally available publications on the subject completely ignore pertinent studies performed prior to 1919, the year when the first modern electret was made. †

---

* We use the term "electrization" to describe a state of a dielectric body manifested by the presence of an electric field outside (or inside) the body not attributable to sources external to the body. We shall use the verb "electrize" to describe any action (surface charging, charge injection, polarization, etc.) that results in an electrization of a body.

† An exception is an article by B. Gross, "On the Experiment of the Dissectible Condenser," *Am. J. Phys.*, **12**, 324 (1944), which established a connection between Franklin's dissectible capacitor, Faraday's studies of dielectric absorption, and electrets.

As a result, several "new" discoveries and "new" theories have been reported and proposed in connection with electret research which, upon closer examination, are merely minor variations (and sometimes repetitions) of the observations and theories published a long time ago.

Therefore, a brief review of the most important early works in the area of long-lasting electrization is included here. It is hoped that this review will not only help readers to gain a better insight into the very interesting and complex phenomena involved in the electret mechanism, but will also provide stronger foundation for individual investigations.

6.2   EARLY EXPERIMENTS ON ELECTRIZATION BY MELTING AND SOLIDIFICATION

Stephen Gray, best known as the discoverer of electric conduction, was probably the first person to study long-lasting electrization. In a letter published in 1732 he wrote about his "enquiry whether there might not be a way found to make the property of electrical attraction more permanent in bodies."[1] For this purpose Gray experimented with various dielectric materials (rosin, shellac, beeswax, sulfur, and their mixtures), which he melted in iron or glass vessels. After allowing the melt to cool and solidify, he reheated the vessel with the hardened dielectric until the latter melted at the bottom and sides of the vessel, thus becoming removable, and then he inverted the vessel to remove the dielectric.

Gray reported that the dielectrics so formed exhibited no unusual properties at first. As they cooled, however, they began to show signs of having acquired an electrization, and finally they appeared to be strongly electrized.

Gray found that the electrization of his dielectric bodies could be preserved for at least several months by wrapping them in paper, flannel, or wool.

He also found that when sulfur was cast in a glass vessel, both the sulfur and the glass acquired electrization.*

Another of Gray's findings was that the external manifestations of the electrization fluctuated with the weather, the degree of electrization appearing to diminish during wet weather.

It is interesting to note that with one exception all Gray's basic observations were "rediscovered" some 200 years later in connection with electret studies. The exception is Gray's statement that his dielectrics acquired electrization merely by melting and solidification. It is not certain that electrization (other than a molding charge) can be produced by such a simple treatment; it is possible therefore that Gray left out something from

---

* This effect is now known as the appearance of "molding charges."

his report. On the other hand, it was found in 1950 by J. Costa Ribeiro[2] that electric charges always appear on the interface between the liquid and solid phases in a dielectric; he named this effect the thermodielectric effect. It remains to be seen whether the phenomenon of electrization described by Gray and the thermodielectric effect are one and the same thing.

Gray's investigations were continued in 1757 by the German physicist Johann Karl Wilcke, who gave the name "spontaneous" to the electricity produced during solidification of molten dielectrics. Another German physicist, Franz Ulrich Aepinus, also followed Gray's path. Later these investigations were continued by the Dutch scientist Martin van Marum (ca. 1785). For some reason, however, nothing new seems to have been done in this field during the next 134 years. Then, in 1919, the study of modern electrets began.

6.3 ELECTROPHORUS

An early device based on long-lasting electrization is the *electrophorus*, or, more accurately, the *electrophorus perpetuum*, as it was named in 1775 by Volta, who is usually regarded as its inventor (its principle was described by Wilcke as early as in 1762).

The essential parts of an electrophorus are a dielectric plate, known as the "cake" and an insulated conducting cover plate (Fig. 6.1). The cake is

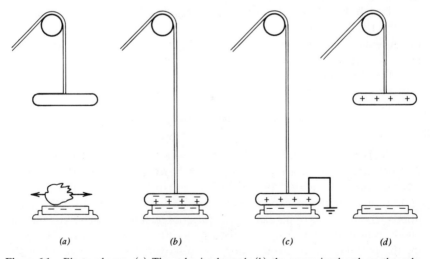

(a)   (b)   (c)   (d)

Figure 6.1. Electrophorus: (*a*) The cake is charged, (*b*) the cover is placed on the cake, (*c*) the like-induced charge is conducted to the ground, (*d*) the cover with the opposite charge is lifted from the cake. With the cake replaced by an electret and the cover plate connected to an electrometer, the apparatus is used to measure effective surface charges of the electret.

first electrized, usually by friction, and the cover plate is then placed on it. The cake (assumed to have a negative charge in Fig. 6.1) induces an opposite charge on the inner surface of the cover plate and a like charge on the outer surface. The latter charge is conducted to the ground by temporarily grounding the cover plate, usually by touching it with a finger. The plate with the remaining charge (positive in Fig. 6.1) is then lifted from the cake, and this charge is used for the desired purpose.

Since the cake, being an insulator, cannot give off its charge to the cover plate or to the base by conduction under ordinary circumstances, its electrization, once produced, can persist for a very long time (several months). Therefore the instrument can be used a great many times before the cake needs to be electrized again.*

In the past the electrophorus was used mainly as an apparatus for dispensing relatively large quantities of charge. The charge obtainable from an electrophorus 20 cm in diameter was approximately 1 μC, which is about the same charge that could be obtained from a medium-sized Leyden jar (500-pF capacitance) charged at 2000 V. An interesting early device utilizing an electrophorus was a table-top electric lighter, in which the spark from a small electrophorus contained in the base of the lighter ignited an oxygen–hydrogen mixture whenever a lever was depressed.[3] A similar but simpler hand-held device was later used as an electric gas lighter.[4] A number of very interesting early experiments with the electrophorus were described by Langenbucher in 1780.[5]

Although the electrophorus is seldom employed today, its principle constitutes the basis for the most widely used method of electret charge measurements. This method utilizes an arrangement similar to that of Fig. 6.1, but with the electret to be measured occupying the position of the cake (see Section 6.7).

6.4   DIELECTRIC ABSORPTION

In 1748 Benjamin Franklin performed several experiments to determine whether the seat of the electric "power" of a Leyden jar was in the conducting coatings or in the glass.[6] For one of these experiments he placed a large plate of glass between two smaller plates of lead, charged the lead plates oppositely by means of his electrostatic friction machine, discharged the lead plates, removed them from the glass plate, and examined the glass plate for the presence of an electrization. He discovered that as a result of this treatment the glass plate became strongly electrized, and that if it was placed

---

* It was known as early as 1737 that a dielectric disc preserves its charge when covered by a conducting plate. See, for example, Granville Wheeler, *Phil. Trans.*, **41**, 113 (1739).

between the lead plates again, a strong spark could be produced while later establishing a connection between the lead plates by means of a conductor.

On the basis of this and some other experiments, Franklin concluded that the "power" of a Leyden jar resided mainly in the glass rather than in the coatings, and that electric charges in a Leyden jar (or in the glass plate) were "imbibed" in the "pores" of the glass near its surface.

Franklin's experiment was repeated by various investigators and became known as the experiment with the "dissectible capacitor." The phenomenon of the acquisition of electrization by the dielectric in a capacitor became known as the "dielectric absorption." Dielectric absorption was found to be a relatively slow process. This was deduced from the two following effects. First, a capacitor with a solid dielectric usually required a longer time to become fully charged than an equivalent capacitor with a gaseous dielectric. Second, when the capacitor with a solid dielectric was discharged by a spark, the capacitor usually partially "recovered" after a certain time, so that a second (then a third, etc.) spark could be obtained because of the "residual" charge of the capacitor. It was also found that once the absorption had taken place, the electrization of the dielectric could persist for months. Finally, it was learned that the absorbed charge in a dielectric was positive at the surface adjacent to the positive conducting plate (or coating) during charging and was negative at the surface adjacent to the negative conducting plate (or coating) during charging.[7]

In spite of the lucidity of Franklin's explanation of the basic phenomenon,* dielectric absorption obviously required a much more detailed explanation. It is not surprising, therefore, that some of the most eminent physicists of the past century conducted both experimental and theoretical studies of it. Some of their findings are summarized below.

6.5 INVESTIGATIONS OF HIGH VOLTAGE DIELECTRIC ABSORPTION BY FARADAY AND MATTEUCCI

Faraday's investigations of the dielectric absorption were reported by him in 1837.[8] In one of the experiments he used a spherical dissectible capacitor having a hemispherical shell of shellac as the dielectric. He found that, immediately after the capacitor was discharged, the dielectric exhibited an electrization whose polarity was opposite to that of the capacitor

---

* It is interesting to note that penetration of charges into a dielectric and a subsequent "capture" (Franklin's "imbibing") of these charges in so-called traps (Franklin's "pores") are now considered to be the primary phenomena responsible for the formation and properties of electrets; see Section 6.10.

during charging; that is, the surface of the dielectric that had been in contact with the positive conductor during charging exhibited a negative charge, and the surface that had been in contact with the negative conductor during charging exhibited a positive charge.* However, this state of electrization quickly disappeared and was followed by a *reversal of polarity* of the dielectric, so that the surface in contact with the positive conductor during charging now appeared positive and the surface in contact with the negative conductor during charging now appeared negative.† After the polarity reversal, the apparent charges of the dielectric gradually increased in intensity for some time.

Faraday interpreted these results as a combination of two effects: (*a*) polarization (alignment of molecular dipoles) by induction and (*b*) penetration of electric charges from the capacitor plates into the dielectric by conduction.

The polarity reversal of the dielectric observed by Faraday can be explained in terms of the difference of the rates of these two effects. At first the polarization is predominant and the dielectric exhibits therefore a "heterocharge" state. However, the polarization rapidly decreases, while the charges that have penetrated into the dielectric, being relatively immobile, stay in the dielectric. Hence these charges begin to predominate after a certain period of time, and the dielectric exhibits a "homocharge" state (cf. Heaviside's description of electret behavior in Section 6.7). It appears that Faraday overlooked the possibility of this explanation. Instead, he explained his observations on the basis of the variation of the depth at which the charges that penetrated into the dielectric were located at various times after the charging of the capacitor. Both theories are now very seriously considered as possible explanations of certain properties of electrets.

Very interesting experiments on dielectric absorption were performed by the Italian physicist Carlo Matteucci.[9]

By placing one end of a long sulfur rod in contact with a charged conductor (Leyden jar) for 10 min and subsequently measuring the charge distribution along the rod, Matteucci found that most of the charge penetrated only a short distance into the rod. By slicing with a thin glass plate a thick cube of stearic acid (used as the dielectric in a dissectible capacitor), he confirmed the already mentioned property of the dielectric to acquire positive charges extending to a certain depth from the surface initially in contact with the positive plate of the capacitor and negative charges extending to a certain depth from the opposite surface.

Matteucci also performed a similar experiment with a compound dielectric made of a large number of thin sheets of mica. An examination of the

---

* This type of electrization is now called the heterocharge state.
† This type of electrization is now called the homocharge state.

individual sheets after the charging and discharging of the capacitor showed that they acquired charges from the plates if the charging took place for a sufficiently long time; but if the charging was of short duration, the sheets exhibited merely a residual polarization and their net charge was zero.

In another series of experiments, Matteucci charged a sulfur cylinder first positively, then negatively, and observed the resulting electrization over a period of time. At first the cylinder exhibited a negative electrization, then no electrization, and then a positive electrization. Matteucci also learned that if a dissectible capacitor having a stearic acid plate as a dielectric was charged positively and then negatively, the resulting electrization of the dielectric plate experienced a polarity reversal, just like the sulfur cylinder in the preceding experiment. He even showed that the plate could be made to undergo several polarity reversals if the capacitor containing it had been charged several times in alternation positively and negatively.* He interpreted these experiments as indicative of charge penetration into the dielectric during charging followed by a migration of the absorbed charge to the surface of the dielectric after the charging had been stopped.

Notwithstanding the experimental evidence for the penetration of electric charges into a dielectric collected by Franklin, Faraday, Matteucci, and other scientists, the phenomenon was difficult to understand. Indeed, owing to the minute irregularities of the surface of all solid bodies, the metal plates of a dissectible capacitor come in contact with only a few points of the dielectric. Therefore, as already mentioned in connection with Volta's electrophorus, only a very small charge—if any— can possibly pass to the dielectric from the plates by ordinary conduction.

New experiments provided, however, a partial solution to this apparent discrepancy. It turned out that when a dissectible capacitor was charged at a sufficiently high voltage, electric breakdown took place in the air gap between the plates and the dielectric, so that the charges from the plates were transferred to the surface of the dielectric by minute electric sparks.[10,11] The points where the sparks struck the dielectric could be made visible by subsequently sprinkling the dielectric with a fine powder of some non-conducting material (usually a mixture of minium and flowers of sulfur); the powder particles adhered to the dielectric at and near these points, forming the so-called Lichtenberg figures.[10]

On the other hand, various experiments indicated that dielectric absorption could take place also at low voltages, in which case the breakdown could not occur. Thus it became necessary to search for an explanation of the low-voltage dielectric absorption which should be a result of phenomena other than charge penetration into the dielectric.

---

* Such a mode of charging is now known as step charging.

6.6 INVESTIGATIONS OF LOW-VOLTAGE DIELECTRIC ABSORPTION BY KOHLRAUSCH AND MAXWELL

In 1854, on the basis of a series of measurements of the residual charges in various capacitors, the German physicist R. Kohlrausch[12] noticed that the phenomenon of dielectric absorption was very similar to the elastic recovery of strained bodies. The latter phenomenon manifested itself in the slow creeping back to the original state of equilibrium of an elastic body upon removal of the applied stress responsible for the initial deformation of the body. Kohlrausch concluded that, just as the phenomenon of elastic recovery was due to molecular interactions hindering the motion of neighboring molecules, the residual charge could be due to similar (or even the same) molecular interactions. He emphasized that a dielectric body polarized in an external electric field could retain a residual polarization even after the external field had been removed. Kohlrausch described four types of possible polarization associated with the internal charges of a dielectric: (a) polarization due to migration of opposite charges to opposite surfaces of the dielectric, (b) polarization due to migration of opposite charges to various layers within the dielectric, (c) polarization due to migration of opposite charges to opposite boundaries of the molecules in the dielectric, (d) polarization due to alignment, by rotation, of polar molecules (i.e., those having opposite charges on opposite ends in their natural state).

It is the latter type of polarization (together with some secondary effects) that Kohlrausch considered to be the most likely cause of the residual charge (and hence of the dielectric absorption). According to him, the phenomenon was essentially attributable to polar molecules within a dielectric, which, having been aligned by the external field, required a long time to return to their original random orientation after the removal of the field; the dielectric thus exhibited a state of long-lasting electrization due to residual orientation of these molecules.*

The idea that dielectric absorption was caused by a redistribution of internal charges in a dielectric was later developed by James Clerk Maxwell.[13] Maxwell made detailed mathematical calculations of the polarization of heterogeneous dielectrics (Kohlrausch's case b) and showed that in such dielectrics the residual charge can be a result of charge accumulations on interfaces between regions occupied by different components of the dielectric.†

---

* J. A. Fleming and A. W. Ashton, *Phil. Mag.*, **2**, 228 (1901), constructed a mechanical apparatus capable of a very accurate imitation of the dielectric effects, thus giving even more credence to Kohlrausch's theory. See also O. Heaviside, *Electrician*, **15**, 134 (1885).

† Polarization due to such charge accumulations is now usually called interface polarization. Interface polarization is presently considered to be one of the possible effects responsible for electret formation and behavior.

More recent experimental and theoretical studies have shown that dielectric absorption is usually a superposition of many effects, which may include penetration, interface polarization, and orientation polarization. In fact, it is doubtful that any of these effects can take place other than in combination with one or more of the others.

Since about 1920 dielectric absorption has been studied mostly in connection with electrets, which are the subject of the remaining sections of this chapter. Additional information on dielectric absorption studies prior to 1912 may be found in the review by Schweidler.[14]

6.7   PREDICTION OF ELECTRETS BY HEAVISIDE AND MAKING OF FIRST ELECTRETS BY EGUCHI

In 1885 Oliver Heaviside published a study of the possibility of producing permanently electrized bodies analogous to permanently magnetized bodies.[15] He suggested calling such bodies "electrets," a word that sounded like a proper electric counterpart of "magnets."

Using essentially theoretical considerations, Heaviside described in seven numbered paragraphs how such bodies could be made and what properties they would have. Here is a brief summary of the first six paragraphs of his description.

1. Place a well-absorbing dielectric into an electric field.
2. Keep the dielectric in the field for as long as is needed for an absorption to take place.
3. Remove the dielectric from the field. If the dielectric is such that the absorption results in a long-lasting (slowly subsiding) electrization, the dielectric has become an electret: it now produces its own electric field similar to the magnetic field produced by a similarly shaped magnet.
4. Cover the electret with a conducting coating. This will produce surface charges on the electret that will neutralize its intrinsic electrization.
5. Remove the coating. The neutralized electret exhibits no external field.
6. Let the intrinsic electrization subside. As it does, an electric field reappears around the electret, but of the opposite polarity to that observed in step 3; the polarity reversal occurs because the surface charges remain constant, whereas the intrinsic electrization diminishes.

Heaviside's description of the electret was basically a theoretical exercise. But between 1919 and 1925 the Japanese scientist Mototaro Eguchi actually made the first electrets (without a prior knowledge of Heaviside's prediction).[16] Eguchi's electrets were made essentially in accordance with Heaviside's procedure, and they behaved essentially as Heaviside said they would.

As the material for his electrets, Eguchi used a mixture of equal parts of carnauba wax and rosin, with or without a certain amount of beeswax.

The mixture was melted at about 130°C and poured into a flat mold with metallic bottom and cover, between which a strong electric field was created by applying a high voltage (several thousand volts) to them. The melt was allowed to solidify in the presence of this electric field. After the solidified melt had cooled sufficiently, the field was turned off, and the wax disc so formed was removed from the mold (to facilitate the separation of the solidified wax from the mold, the bottom and the cover of the latter had metal foil linings which were peeled from the wax disc after the disc had been removed from the mold). An examination of the disc showed that it now constituted an electret. Eguchi measured its electrization by means of an electrophorus arrangement (see Fig. 6.1) in which the electret was used in place of the usual cake; the charge that the electret induced on the electrophorus cover plate was measured with an electrometer.*

A remarkable property of Eguchi's electrets was the peculiar evolution of their electric characteristics. When an electret was just removed from the mold, its polarity was of the "heterocharge" type (see Section 6.5). However, the initial effective charges of the electret surfaces decayed to zero within a relatively short period of time (usually several hours). After this, each surface began to exhibit a new effective charge of the *opposite* polarity (Fig. 6.2), and the polarity of the electret now was of the "homocharge" type. Subsequent to the polarity reversal, the effective surface charges increased for some time and finally acquired an essentially constant value. This constant value was of the order of 2×10^{-5} C/m^2 (comparable to the largest possible surface charge value obtainable by friction) and lasted for at least several years.

Eguchi attributed the long-lasting electrization of his electrets to a permanent alignment of polar molecules in them. He believed that the polar molecules of the molten wax were aligned by the applied electric field and became "frozen" in the aligned positions during solidification of the melt (cf. Kohlrausch's explanation of dielectric absorption, Section 6.6). Recent investigations showed, however, that Eguchi's explanation of the electret was in error. A more up-to-date explanation of the electret effect is presented in the concluding section of this chapter.

---

* Today this method is the standard method of electret measurement. It is usually called the lifted electrode method or the dissectible capacitor method. The method yields charge values representative only for the overall value of the electret's electrization, without revealing the nature of the electrization (i.e., it does not tell whether the electrization constitutes a surface charge, an injected charge, a polarization, etc.). The charge values obtained from this method are called the "effective surface charge" since, in the absence of other information on the charge distribution in the electret, they may be attributed to an electric charge residing only on the surface of the electret.

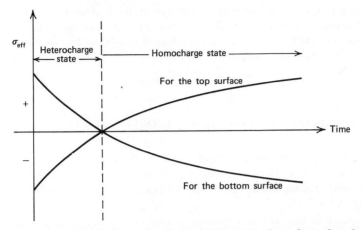

Figure 6.2. Evolution of the effective surface charge density σ_{eff} on the surfaces of an electret (the top surface was in contact with the positive plate of the mold during charging, the bottom surface was in contact with the negative plate). The time is measured from the instant when the electret was removed from the mold.

6.8 ELECTRETS AND DIELECTRIC ABSORPTION

When Eguchi's articles on electrets were first published, many scientists thought that electrets indicated the existence of some new, previously unknown, physical effects in dielectrics. It was especially the "mystery" of the electret's polarity reversal that appeared to be irreconcilable with the accepted concepts of dielectric behavior. Even today it is not generally realized that in electrets we have merely a more spectacular example of the same effects that were described more than 100 years ago under the collective term "dielectric absorption" (e.g., the phenomenon of polarity reversal in absorptive dielectrics was described by Faraday in 1837. Electrets are new only to the extent that their electrization lasts for an exceptionally long time. But then, Gray's electrized dielectric bodies and Volta's electrophorus preserved their electrization for many months, too.

The brief historical review presented in the preceding sections shows clearly that electrets must be regarded not as something basically new, but merely as another development (a very important one, of course) in the exploration of dielectric properties associated with dielectric absorption, which began two centuries ago. It may be noted that Heaviside himself introduced the word "electret" in connection with the phenomenon of dielectric absorption. Here is what he said in his article:[15]

There may also be residual electrization, namely, when absorption occurs. This may tend naturally to wholly subside, or a part of it may remain. The body is then permanently electrized . . . . A word is evidently wanted to describe a body which is naturally permanently electrized by internal causes . . . [the] word that suggests itself is *electret*. . . .

6.9 MATERIALS AND TECHNIQUES FOR MAKING ELECTRETS

Considerable progress has been made in electret research since Eguchi's pioneering work. Various substances have been tested as potential electret materials. Tables listing these substances have been published by several authors.[17-20,41]

Carnauba wax is still regarded as one of the best electret materials. Its obvious disadvantage lies in very poor mechanical characteristics. Therefore carnauba wax electrets are widely used for laboratory experiments but are not generally considered for industrial applications.

Various plastics (Lucite, Mylar, Teflon, etc.) possess desirable mechanical properties and can be used as electret materials. They are particularly well suited for thin-film electrets.[21,22,23]

Also certain ceramic materials make good electrets.[24] These materials are mechanically very strong, but their machinability is not as good as that of plastics.

A number of different methods have been developed for making of electrets. The usual technique for carnauba wax electrets is essentially the same as that employed by Eguchi. A variation of this technique is to heat wax blanks to a temperature just below the temperature at which they begin to lose their shape and then to apply the electric field to these blanks; the field is switched off when the blanks are sufficiently cool. The latter method also may be used for making electrets from plastic materials, including thin films and foils. Ceramic electrets are likewise made by heating and subsequent cooling in the presence of an electric field.

Electrets from thin materials can be made simply by spraying the materials with electric charges. This can be done by inserting the material between two electrodes and establishing either a corona discharge or a spark breakdown from the electrodes onto the material.[25] Another way involves exposing the material to an electron beam, which is one of the methods that has been given considerable attention in recent years.[26] Effective surface charge densities as high as 10^{-2} C/m² can be obtained with these methods.

A very interesting type of electret is the so-called photoelectret discovered in 1937 by the Bulgarian scientist G. Nadjakoff.[17,27] These electrets are made by exposing a suitable substance (such as sulfur) to an electric field with a simultaneous strong illumination of the material. Photoelectrets retain their electrization only when stored in the dark. If such an electret (or a part of it) is exposed to light, the electret (or the exposed part) is neutralized. Thus photoelectrets can be used as photosensitive materials for electrophotography; the picture is developed by sprinkling the electret with a powder, just as when developing Lichtenberg figures (see Section 6.5).

Several other methods for electret making have been reported (e.g., simultaneous application of electric and magnetic fields[28]), but it is not yet quite certain how reliable those methods are.

6.10 APPLICATION OF ELECTRETS AND ELECTRET THEORIES

The first practical utilization of electrets is attributed to the Japanese engineers who built condenser-type microphones in which a carnauba wax electret was used in place of the conventional auxiliary high-voltage power supply.[29] Electret microphones are now commercially available and are noted for their exceptionally excellent electroacoustical characteristics.

A number of other electret devices have been suggested and many have been patented.[17-20,29,30] Among them are radiation detectors, dosimeters, memory storage units, humidity meters, air cleaners, vibration detectors, pressure gauges, electrostatic relays, dc and ac motors, current generators, voltage generators, and electron beam deflectors. The most recently described electret devices include the electret voltmeters (or coulombmeters) and motors based on the so-called slot-effect (see Chapter 7).[31] An electret motor of this type has been successfully operated from an antenna extracting electric charges from the earth's atmosphere.[32] Another recently suggested application of electrets is to use them as probes for exploring and plotting of electric fields.[33] Still another recently suggested application of electrets is to use them (preferably in the shape of hollow spherical or cylindrical shells) as active elements in electrostatic current generators.[34] Such generators (as well as motors) are expected to function particularly well when filled with a gas of high electric breakdown strength.

A serious hindrance to the practical application of electrets is the lack of rigorous theory of the fundamental phenomena responsible for the electret effect. Therefore, we still do not know what methods, materials, and techniques may yield electrets with prescribed characteristics. Neither do we know the optimal operation conditions or the limits of the electrization attainable with electrets. For these reasons there is a certain reluctance on the part of both physicists and engineers to accept or to incorporate electrets as bona fide components for instruments and devices.

Thus far the electret effect has been interpreted in terms of the various effects discussed earlier in connection with the dielectric absorption. It now appears certain that the primary mechanism responsible for the long-lasting electrization of carnauba wax electrets is the capture of electric charges in so-called traps within the interior of the electrets. Recent experimental investigations revealed that within carnauba wax electrets there is a characteristic distribution of electric charges (Fig. 6.3) which never changes

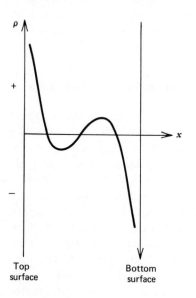

Figure 6.3. Space charge density ρ in the interior of carnauba wax electrets (corresponding to the electret whose effective charge evolution appears in Fig. 6.2). Just below the top surface there is a strong positive space charge; somewhat deeper there is a layer of negative charge extending to the midplane of the electret. Next there is a layer of positive charge, and finally a strong negative charge near the bottom surface of the electret. This charge distribution remains unchanged over the entire lifetime of the electret.

throughout the entire lifetime of the electret,[35] except, perhaps, when the electret is exposed to some extreme conditions or environment. The only certain way to destroy this internal distribution of electric charges is to melt the electret. Indirect measurements indicate that charge trapping is responsible also for the electret effect in plastic electrets[25,36] and in electrets made from other materials.[37]

Research on electrets is a highly promising and rapidly advancing branch of electrostatics. Readers who wish to acquaint themselves in detail with the various phases of this research are referred to monographs and reviews[17–20,29,38–40] on this subject and also to the 1962 publication by V. A. Johnson[41] containing a "state of the art" survey together with abstracts of more than 300 technical papers devoted to electrets.

REFERENCES

1. S. Gray, *Phil. Trans.*, **37**, 285 (1732).
2. J. Costa Ribeiro, *Acad. Bras. Sci. An.*, **22**, 325 (1950).
3. J. Gavarret, *Traité d'Electricité*, Vol. I, Victor Masson, Paris, 1857, p. 76.

4. —— *Elect. World*, **9**, 167 (1887).

5. J. Langenbucher, *Beschreibung einer beträchtlich verbesserten Elektrisiermaschine, etc.*, Mathäus Riegers, Augsburg, 1780.

6. J. Sparks, Ed., *The Works of Benjamin Franklin*, Vol. 5, Whittemore, Niles, and Hall, Boston, 1856, pp. 201–202, 245, 349.

7. T. Cavallo, *Elements of Natural or Experimental Philosophy*, Towar and Hogan, Philadelphia, 1829, p. 505.

8. M. Faraday, *Experimental Researches in Electricity*, Vol. I, Taylor and Francis, London, 1839, §§1234–1250, 1269.

9. C. Matteucci, *Ann. Chim. Phys.*, **27**, 133 (1849).

10. M. J. Jamin, *Cours de Physique*, Vol. I, Mallet-Bachelier, Paris, 1863, pp. 399–404.

11. Müller-Pouillets, *Lehrbuch der Physik*, Vol. IV, Part 1, Friedrich Vieweg, Braunschweig, 1932, p. 152.

12. R. Kohlrausch, *Ann. Phys.*, **91**, 56–82, 179–214 (1854).

13. J. C. Maxwell, *A Treatise on Electricity and Magnetism*, Vol. I, Clarendon Press, Oxford, 1881, §§ 325–334.

14. E. Schweidler, "Die Anomalien der dielektrischen Erscheinungen," in L. Graetz, Ed., *Handbuch der Elektrizität und des Magnetismus*, Vol. I, J. A. Barth, Leipzig, 1918, pp. 232–261.

15. O. Heaviside, *Electrician*, **15**, 230 (1885).

16. M. Eguchi, *Phil. Mag.*, **40**, 178 (1925).

17. V. M. Fridkin and I. S. Zheludev, *Photoelectrets and the Electrophotographic Process*, Van Nostrand, New York, 1966.

18. J. Euler, *Neue Wege zur Stromerzeugung*, Akademische Verlagsgesellschaft, Frankfurt am Main, 1963, pp. 159–191.

19. A. N. Gubkin, *Electrets*, Academy of Sciences, Moscow, 1961, in Russian.

20. O. A. Myazdrikov and V. E. Manoilov, *Electrets*, G. E. I., Moscow, 1962, in Russian.

21. G. M. Sessler and J. E. West, *J. Acoust. Soc. Am.*, **34**, 1787 (1962).

22. M. M. Perlman, "Production and Charge Decay of Film Electrets," in *Electrets and Related Electrostatic Charge Storage Phenomena*, Electrochemical Society, New York, 1968, pp. 86–89.

23. T. Takamatsu and E. Fukada, *Rep. Prog. Polym. Phys. Japan*, **15**, 393 (1972).

24. A. N. Gubkin and G. I. Skanavi, *Zh. Eks. Teor. Fiz.*, **32**, 140 (1957); transl. *Soviet Physics–JETP*, **5**, 140 (1957).

25. G. M. Sessler and J. E. West, *J. Appl. Phys.*, **43**, 922 (1972).

26. G. M. Sessler and J. E. West, *Appl. Phys. Lett.*, **17**, 507 (1970).

27. G. Nadjakoff, *Compt. Rend.*, **204**, 1865 (1937).

28. C. S. Bhatnagar, *Indian J. Pure Appl. Phys.*, **2**, 331 (1964).

29. F. Gutmann, *Rev. Mod. Phys.*, **20**, 457 (1948).

30. V. A. Johnson Carbauh, "Electrets—Applications Unlimited?" in *Electrets and Related Electrostatic Charge Storage Phenomena*, Electrochemical Society, New York, 1968, pp. 100–103.

31. O. Jefimenko and D. K. Walker, IEE Conference on Dielectric Materials, Measurements, and Applications, Lancaster, England; Institution of Electrical Engineers, London, Publ. 67, 1970, pp. 146–149.

32. O. Jefimenko, *Am. J. Phys.*, **39**, 776 (1971).

33. G. M. Gershtein, *Electrostatic Induction Method of Field Modeling*, Nauka, Moscow, 1970, pp. 106–110, in Russian.

34. O. Jefimenko and C. N. Y. Sun, "Spherical Carnauba Wax Electrets," a paper presented at the Electrochemical Society Electret Symposium, Miami Beach, 1972 (Abstract No. 142).

35. O. Jefimenko and D. K. Walker, *Bull. Am. Phys. Soc.*, **14**, 634 (1969); D. K. Walker and O. Jefimenko, "A Technique for Measuring Real Volume Charges in Wax Electrets," a paper presented at the Electrochemical Society Electret Symposium, Miami Beach, 1972 (Abstract No. 140).

36. M. M. Perlman and R. A. Creswell, "Thermal Current Study of Charge Traps in Insulators," in *1970 Annual Report, Conference on Electrical Insulation and Dielectric Phenomena*, National Academy of Sciences, Washington, D.C. 1971, pp. 30–37.

37. J. D. Brodribb, D. M. Hughes, and T. J. Lewis, "The Energy Spectrum of Traps in Insulators by Photon-Induced Current Spectroscopy," a paper presented at the Electrochemical Society Electret Symposium, Miami Beach, 1972 (Abstract No. 104).

38. C. L. Stong, *Sci. Am.*, pp. 120–124 (November 1960).

39. B. Gross, *Charge Storage in Solid Dielectrics*, Elsevier, Amsterdam (1964).

40. M. M. Perlman, "Review of Phenomenological Theories of Electrets," in *Electrets and Related Electrostatic Charge Storage Phenomena*, Electrochemical Society, New York, 1968, pp. 3–5.

41. V. A. Johnson, *Electrets*, Parts I and II, Office of Technical Services, U.S. Department of Commerce, Washington, D.C., 1962.

CHAPTER 7

Electrostatic Motors

OLEG D. JEFIMENKO
Department of Physics
West Virginia University
Morgantown, West Virginia

7.1 OVERVIEW

Electrostatic motors are electric motors in which mechanical motion is created as a result of electrostatic forces acting between electric charges.

Electrostatic motors can be subdivided into several types or classes, in accordance with the mode of operation or other characteristic features of the corresponding motors. Thus, depending on the method employed for delivering electric charges to the active part of a motor, we distinguish between *contact motors, spark motors, corona motors, induction motors,* and *electret motors.* Depending on the medium in which the active part of a motor is located, we say that a motor is *liquid-* or *gas-immersed.* Depending on the material and design (operation mode) of the active part of a motor, we distinguish between *dielectric motors* and *conducting-plates,* or *capacitor, motors.* Finally, in the case of ac-operated motors, we have *synchronous* and *asynchronous motors,* depending on the rate of rotation of a motor relative to the period of the applied voltage.

This chapter presents a brief description of the basic types and principles of operation of electrostatic motors. The essentially chronological sequence of presentation allows us to progress from simpler to more complex types of motors in the most natural way.

This chapter is essentially a condensation of author's publication *Electrostatic Motors; Their History, Types, and Principles of Operation,* Electret Scientific, Inc., 1972, to which the readers are referred for a more detailed presentation of the subject of electrostatic motors.

131

7.2   PENDULUM–TYPE MOTORS: CONTACT MOTORS

The two earliest devices that can be considered to constitute electrostatic motors (i.e., electrostatic devices converting electric energy into mechanical motion) were invented in about 1742 by Andrew Gordon, a professor of philosophy at Erfurt.[1] Gordon's first motor, a device known as the "electric bells," functioned as follows. A metallic clapper was suspended by a silk thread between two oppositely charged bells. When pushed toward one of the bells, the clapper struck the bell and from the contact with the bell acquired a charge of the same polarity as that of the bell. By the ensuing electrostatic forces the clapper was then repelled from this bell and attracted to the second bell. Striking the second bell, the clapper gave off its initial charge and acquired a charge of the same polarity as that of the second bell. Now the clapper was repelled from the latter and attracted to the first one, which it struck again, and so on.

Many variations of Gordon's bells have been described by later authors. In 1752 Benjamin Franklin used similar bells connected to an insulated lightning rod as a warning device that would "give notice when the rod should be electrified."[2]

A very interesting small motor derivable from Gordon's bells was an "electric pendulum" powered by an early high-voltage chemical battery or "dry pile," also known as the Zamboni pile. The dry pile was invented in 1806 by Georg Behrens, and in 1810 it was adapted by Giuseppe Zamboni[3] to drive a light pendulum that functioned on the same principle as the clapper in Gordon's bells. Inasmuch as a dry pile could remain active for many years, the pendulum could maintain its motion also for a very long time; thus it became known as an "electric perpetuum mobile" (see Section 7.4 for another electric perpetuum mobile by Zamboni). One such pendulum operated without interruption for at least 86 years.[4]

A fascinating electrostatic motor also derivable from Gordon's bells was the reciprocating motor (Fig. 7.1) built in 1880 by Howard B. Dailey and Elijah M. Dailey. The motor was described by one of its builders as follows:

This machine operates by the direct action of static electric attractions and repulsions. It is constructed entirely of fine wood, glass and hard rubber, there being no magnetic materials used. The flywheel is of laminated, soft wood and runs in journal bearings of very small diameter. The moving balls, mounted on the walking beam of vulcanite, are made of wood, hollowed out so that the walls are about 2 millimeters thick. They are covered with aluminum foil for static conductivity. The stationary balls are of solid wood.

To operate the engine the stationary balls are charged with electricity from a static electric generator, such as a Holtz machine, the upper balls being connected through

Figure 7.1. Dailey's electrostatic motor. Designed and built in 1880 by Howard B. Dailey and Elijah M. Dailey, this motor is now at the Museum of History and Technology of the Smithsonian Institution.

the brass ball to one pole of the machine while the lower stationary balls are connected through the binding post on the bed frame to the opposite pole of the machine. Under proper conditions, when charged, the engine will make about 375 revolutions per minute.

The walnut base upon which the engine is mounted is 14 in. long, 4 in. wide and 1.75 in. thick. The movable balls are about 1.5 in. in diameter, the upper stationary balls are 1.75 in. in diameter; and the lower stationary balls 1.5 in. The four glass rods, mounted vertically, are about 6 in. high and spaced 6 in. apart along the bed. The diameter of the flywheel is 5.75 in. It is gilded and has small wire spokes. The connecting rod is 7 in. in length.*

Closely related to Gordon's bells was a motorlike device known as the electric racing ball. Its stationary part consisted of an insulated horizontal conducting ring located above a conducting disc of larger diameter, with a vertical rim concentric with the ring. The rim and the ring were approximately of the same height and constituted the walls of a track for a light dielectric sphere placed on the disc. If the ring and the disc were connected

* The author is grateful to Elliot N. Sivowitch, Smithsonian Institution Museum Specialist, for kindly providing this description and the accompanying photograph of Dailey's motor. A less detailed description of the motor was given by H. B. Dailey in the article "Static Electric Motor," *Modern Electrics*, **5**, 916–917 (1912).

to an electrostatic machine, the sphere ran along this track due to attraction and repulsion exerted by the ring and the disc on the points of the sphere that acquired charges from contact with the disc and the ring.[5]

Since the moving elements in the foregoing motors obtained their charges from the stationary elements by direct contact, the motors may be classified as "contact motors." Charging by contact, however, is neither the most efficient nor the most expedient method for charging moving elements of electrostatic motors, and it is not likely that contact motors will ever occupy a particularly prominent position relative to electrostatic motors of other types.

7.3 ELECTRIC WIND MOTORS

As already mentioned, Andrew Gordon invented two devices that can be considered to constitute electrostatic motors. His second motor was a device known as the "electric fly,"[1,6] also called the "electric whirl" or the "electric reaction wheel." The fly consisted of one or more light horizontal metal arms pivoted on an insulated needle and having sharp-point ends bent in the same circumferential direction. When the needle was connected to an electric machine, a corona discharge occurred from the sharp ends. Thus the air near these ends acquired charges of the same polarity as that of the fly and was therefore repelled from the points by electrostatic forces (this effect is known as the "electric wind"). Similarly, the sharp ends of the fly were repelled from the charges in the air; thus the fly rotated in the opposite sense to that in which the points were directed.

In an early application of the electric fly, the device was employed for turning an orrery representing the orbital motions of the sun, the earth, and the moon.[7]

An interesting electrically operated toy utilizing the principle of the electric fly was constructed in 1761 by Ebenezer Kinnersley, a friend of Benjamin Franklin. The toy was known as the "electrical horse-race" (Ref. 2, p. 371). It consisted of a light wooden cross pivoted horizontally on a central pin. Each end of the cross carried a figurine of a rider on a horse. The motion was produced by corona discharge from the spurs of the riders.

To demonstrate that the electric fly converted electric energy into mechanical energy, electric flies were arranged to lift small weights or to climb inclined rails.[8]

Forces acting on an electric fly and on similar devices were studied by Kämpfer[9] and by Bichat,[10] who constructed an electrometer based on the principle of the electric fly (a similar electrometer was suggested much earlier by Hamilton[11]).

In recent years the possibility has been suggested of harnessing the force experienced by sharp points due to corona discharge from them for lifting a flying machine.[12] However, the achievable lift is far too small to be of practical significance.[13]

Related to the electric fly were small electrostatic motors known as the "electric-wind wheels" or the "electric-wind turbines" driven by the electric wind. Benjamin Franklin described them in 1747 (Ref. 2, p. 184) as "light windmill-wheels made of stiff paper vanes; also . . . little wheels, of the same matter, but formed like water-wheels." Franklin discovered that the function of these motors was mainly attributable to repulsion and attraction of electric charges on their moving and stationary parts, rather than to mechanical action of the electric wind. It is likely that this discovery led him to the invention of his big electrostatic motors.

7.4 FRANKLIN'S SPARK MOTORS AND MOTORS DERIVABLE FROM THEM

Benjamin Franklin was the first person who designed and built electrostatic motors of appreciable power (ca. 0.1 W). In 1748 he constructed two big motors, which he called electrical wheels (Ref. 2, pp. 204–207).

His first motor was an impressive machine, about 40 in. in diameter. The main part of this motor was a wooden disc mounted horizontally on a vertical axle and carrying 30 glass spokes with brass thimbles on their ends (Fig. 7.2). Two oppositely charged Leyden jars were placed in close proximity to the thimbles on the spokes.

As the thimbles moved past the Leyden jars, a spark jumped between a jar knob and the passing thimble, charging the latter with a charge of the same polarity as that of the knob. Therefore each knob attracted the oncoming thimbles and repelled the departing thimbles, causing the motor to turn. The motor, being perfectly symmetric, could turn in either direction; it usually required a starting push by hand, to initiate the charging of the thimbles.

Not entirely pleased with this motor because it required Leyden jars for its operation, Franklin built another motor, which operated from the electric energy stored in the motor itself.

The second motor (Fig. 7.3) was constructed as follows. A thin disc of glass was gilded on each side, except near the edge. Several lead spheres were fixed to the edge of the disc and were connected in alternation to the two gold layers. The disc was mounted on a vertical axle consisting of two pieces of strong wire, the upper and lower pieces being in contact with the top and the bottom layers of gold, respectively. Several glass pillars topped

Figure 7.2.   Replica of Franklin's first motor, built by the author.

with thimbles were placed around the disc close to its edge. The electricity for operating this motor was stored in the disc, which (because of the presence of the two layers of gold) constituted a capacitor capable of holding an appreciable quantity of charge. The motor functioned on the same principle as Franklin's first motor, expect that now the thimbles were stationary and received their charges by sparks from the spheres moving past them.*

Franklin's motors depended for their operation on charging by means of sparks, and therefore they may be classified as "spark motors."

Several authors have reported various modifications of Franklin's motors. Early simplified versions of such motors were described by Guyot (Ref. 5, pp. 290–293, pp. 307–309, Plates 29, 30). An interesting improved version of Franklin's first motor had a cylindrical rotor with electrodes in the

---

* For a more complete description of Franklin's motors see O. Jefimenko, *Am. J. Phys.*, **39**, 1139–1140 (1971).

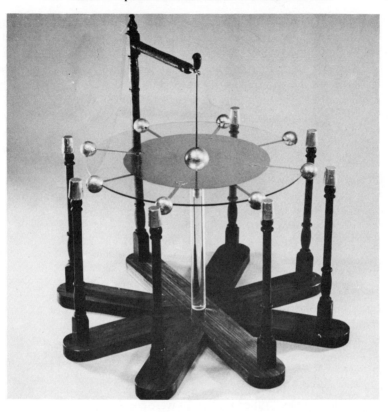

Figure 7.3.   Replica of Franklin's second motor, built by the author.

shape of long strips.[14] Such electrodes accepted more charge than thimbles or spheres, and the power of the motor was therefore considerably greater than that of the original motor.

Miniature Franklin motors have been constructed and described by A. D. Moore.[15] One such motor consists of a Plexiglas disc about 4 in. in diameter with small conducting balls embedded in the rim. The disc turns on a horizontal axis supported by a base behind the disc. The motor runs when placed between the poles of an electrostatic machine. To make the motor self-starting and one-way running, the poles of the machine may be provided with short auxiliary rods tangential to the disc. Since the rods spray charges predominantly in the direction of their ends closest to the disc, the disc also turns in this direction.

Small motors related to both Franklin's first motor and to pendulum motors were used as rotary-type " perpetum mobile " (see Section 7.3) devices operated from a Zamboni pile.[16] An analogous "millifleapower motor" has been recently built for operation from a nuclear battery.[17] In this motor a light aluminum vane bent in the form of an inverted V and having pointed ends is pivoted on a sharp needle protruding from the positive terminal of the battery. As the vane turns, its ends become attracted by two spheres connected with the ground through a high resistance. A discharge from the ends of the vane charges the two spheres, which now repel the two ends. During the time it takes the vane to complete one-half of its revolution, the charges of the spheres leak off to the ground; thus it becomes possible for the spheres to again attract the vane, accept new charges from its ends, and again repel them. The motor is expected to run for many years.

A motor derivable from Franklin's first motor and from pendulum motors has been suggested for operation from an electrophorus.[18]

7.5 CORONA MOTORS

In 1867 W. Holtz, the inventor of the electrostatic "influence" machine carrying his name, discovered that his machine could operate as a motor when powered by another similar machine.[19] This effect (known as Holtz's rotation phenomenon) attracted the attention of J. C. Poggendorff, who studied it in 1867 with the help of a special device, analogous to Holtz's machine but designed to operate only as a motor.[20,21]

The rotor of this motor was a vertical glass disc placed between metallic sharp-needle "combs." When the combs were connected in alternation to an electrostatic machine, the corona discharge from each comb deposited on the rotor a charge of the same polarity as that of the corresponding comb. Therefore each comb repelled the segment of the rotor carrying charges which it sprayed on the rotor and attracted the segment carrying charges sprayed on the rotor by the preceding comb.

Since almost the entire surface of the rotor was sprayed with electric charges, the torque exerted by the combs on the rotor was much stronger than the torque acting on the rotor of Franklin's motors, where only the electrodes were charged. According to Poggendorff, his motor used from 1200 to 1800 times as much current as Franklin's motors. In this case, assuming that the operating voltage of his motor was the same as that of Franklin's motors, the power of his motor must have been of the order of 100 W.

Holtz's rotation phenomenon was also studied in 1869 by the Danish physicist C. Christiansen, who described two motors based on Holtz's discovery.[22] Cylindrical and spherical motors operating on the same principle were described in 1871 by W. Gruel, who called them "electric tourbillons."[23] A small disc motor of the same type (attributed to Ruhmkorff) was described in 1876 by M. E. Mascart.[24] A disc motor basically the same as Poggendorff's original motor was patented in the United States in 1891 by J. W. Davis and J. B. Farrington.[25] Experiments with Poggendorff-type motors were reported in 1921 by V. R. Johnson,[26] who stated that his largest motor could run at about 2000 rpm and had a power of about 90 W.

Just as Poggendorff's motor was derived from Holtz's electrostatic machine, a series of electrostatic motors were similarly derived from Wimshurst's electrostatic machine.

In 1891 five such motors were constructed by William McVay.[27] One of McVay's motors consisted essentially of two horizontal glass discs—one stationary, the other rotating about a vertical axis just above the first. The lower disc had two quadrants of tinfoil, and the upper disc had 16 tinfoil sectors. The power (from a Wimshurst machine) was supplied to the motor by means of two insulated arms; each arm terminated in two brushes, one brush being in permanent contact with one of the lower quadrants, the other charging (usually by corona, but sometime by contact) a sector on the upper disc just clear of the edge of the quadrant. Charges of the same polarity were thus deposited on the quadrant and on the sector, causing them to repel each other. The simultaneous charging of the stationary quadrants and of the moving sectors resulted in a relatively strong starting torque and in a unidirectional operation. McVay constructed also cylindrical motors. Instructions for building a simple McVay motor may be found in Ref. 28.*

Since in all motors described in this section the charging of the rotor occurred by means of a corona discharge, these motors may be classified as "corona motors." Corona motors have been further developed in recent years, and we shall speak of them again in the concluding section of this chapter.

7.6 CAPACITOR MOTORS

In 1889 Karl Zipernowsky constructed a new type of electrostatic motor[29] derived from Thomson's quadrant electrometer.[30] The rotor of this motor consisted of two pairs of insulated aluminum sectors. The stator consisted of four double sectors inclosing the rotor. The rotor was fitted with a commutator by means of which the sectors of the rotor were charged oppositely to those sectors of the stator into which they entered and identically to those sectors of the stator which they were leaving. This motor could operate from dc as well as from ac.

* The author is grateful to Thorn L. Mayes of Mayes Engineering for this information and for other communications on electrostatic motors.

Since Zipernowsky's motor utilized electric forces exerted by one charged conducting plate on a second charged conducting plate (like the forces acting on the two plates of a capacitor), it constituted what is now called an electrostatic "capacitor motor."

A simple capacitor motor, described in 1904 by van Huffel,[31] was based on the so-called Thomson's replenisher.[32]

Capacitor motors possess a number of important advantages as compared with motors described in the preceding sections: supplied with slip rings, they can operate from relatively low voltages; they can operate from dc and ac sources alike; when powered by ac sources, they can operate both as synchronous and as asynchronous motors (Zipernowsky's original motor operated asynchronously from ac). Because of these and other useful properties, capacitor motors have been studied in considerable detail in recent years. A great deal of attention has been devoted to synchronous capacitor-type motors.

A synchronous capacitor-type electrostatic motor is essentially a multi-electrode capacitor motor without a commutator, the proper charging of the rotor being accomplished by continuously supplying an ac voltage of proper frequency between the stator and the rotor. If the rotor moves by one electrode in one period of the supply voltage, then the ac voltage accomplishes the same effect as that accomplished by a dc voltage with a commutator. The synchronous angular speed is therefore $2\pi f/N$, where f is the frequency of the supply voltage and N is the number of electrodes. Precision-made synchronous motors were described in 1969 by B. Bollée.[33]

According to Bollée, the maximum average torque on the rotor in a capacitor-type electrostatic motor is $T_{max} = kNV_0^2$, where k is a geometrical constant and V_0 is the peak voltage applied. A motor with many electrodes is therefore slow but produces greater torque. The power of the motor is proportional to the square of the applied voltage and to the frequency but does not depend on the number of electrodes.

An important parameter in capacitor-type electrostatic motors is $C_{max} - C_{min}$, where C_{max} is the greatest capacitance between the rotor and the stator and C_{min} is the smallest capacitance, as they occur when the rotor turns. The larger this parameter is, the greater are the torque and the power. The main objective in the design of such motors is therefore to obtain the greatest possible variation of the capacitance.

An interesting apparatus which can be classified either as a modified Franklin's second motor or as a modified capacitor-type motor was described by A. D. Moore (Ref. 15, p. 109), who called it the "interdigital motor." The stator is a glass bowl with strips of aluminum foil glued to it. The rotor is a conducting ball inside the bowl; the ball runs along the sloping side of the bowl when the strips are connected to an electrostatic generator (positive and negative strips are connected in alternation).

7.7 INDUCTION MOTORS

Between 1892 and 1893, Riccardo Arno[34,35] and W. Weiler[36] reported electrostatic motors operating on the principle that the polarization of a dielectric in a variable electric field lags behind the field inducing the polarization or, more specifically, that a dielectric placed into a rotating electric field experiences a force acting on the induced polarization charges causing the dielectric to follow the rotation of the field.

Arno's motor had a cylindrical stator consisting of four insulated copper segments surrounding a hollow cylindrical rotor of ebonite (the rotor weighed 40.33 g, was 18 cm long, and had a diameter of 8 cm). The stator produced a rotating electric field by means of a 3800-V transformer, an RC-circuit, and a mercury commutator. The rotor attained a speed of about 250 rpm and developed a torque of 176 cm^2-g/sec^2.

In Weiler's motors the rotating field was produced by a hand-operated commutator delivering high voltage from an electrostatic generator in sequence to the four segments of the stator. He described four different motors utilizing rotating fields and experimented also with motors having noncylindrical rotors.*

Since Arno's and Weiler's motors operated as a result of induced dielectric polarization, they may be classified as "induction motors."

Induction motors have two very attractive features: they can operate from low-voltage dc or ac and, since they require no brushes or slip rings, the friction losses in them are very small.

Several modern induction motors (including a linear motor) have been described by Bollée.[33] One of his motors operated from a three-phase power supply at 220 V and 50 Hz.

A mathematical study of induction motors was presented in 1970 by Soon Dal Choi and D. A. Dunn,[37] who also tested their theoretical conclusions on a large motor that turned at about 1000 rpm when powered by a 10-kV power supply.

7.8 LIQUID–IMMERSED MOTORS

In 1893 Weiler observed that a dielectric placed in a poorly conducting liquid between two electrodes rotated when the electrodes were connected to an electrostatic machine.[38] He later used this effect for operating a small motor. His motor was similar to a corona motor, except that the rotor obtained its charges not from a corona discharge but by the conduction current in the liquid.

---

* Rotating field electrostatic motors with disc-shaped rotors were built and studied between 1894 and 1901 in Japan by Hiderato Ho; for a résumé of his work see the article "A Rotary Field Electrostatic Induction Motor," *Elect. World*, **37**, 1012–1014 (1901).

In 1896 G. Quincke reported observations[39] similar to those of Weiler. It is interesting to note that Weiler's work remains practically unknown and that Quincke is universally considered as the discoverer of the rotation of dielectrics in poorly conducting liquids.

In recent years liquid-immersed dielectric motors have been studied by P. E. Secker and co-workers.[40,41] Their motors operated in liquids such as hexane, hexane doped with amyl or ethyl alcohol, and isoamyl alcohol. The maximum speed of the motors was about 2500 rpm. One of their motors had a stator with six electrodes surrounding a Perspex rotor 7/8 in. in diameter and 2 in. long. The surface of the rotor was covered with a layer of high-permittivity material. At 30 kV the motor turned at 1700 rpm. The total power input for this motor was 5.4 W, of which 2.7 W appeared at the rotor shaft.

A discussion of electrohydrodynamic effects taking place in liquid-immersed motors has been presented by J. R. Melcher and G. I. Taylor.[42]

7.9   ELECTRET MOTORS

In a narrow sense of the word, an electret is a permanently polarized dielectric body (see Chapter 6) and may be considered to constitute the electrical counterpart of a permanent magnet.

In 1961 the Russian physicist A. N. Gubkin described an electrostatic motor[43] using electrets for its operation. The motor consisted of a stator formed by two horizontal parallel-plate capacitors and a rotor with two flat, horizontal, oppositely polarized electrets mounted on a vertical axle and capable of passing between the plates of the capacitors. When a voltage was applied to the two capacitors, the capacitors attracted the electrets as soon as the latter came close to their plates, and the electrets were pulled into the capacitors. A commutator changed the polarity of the capacitors just as the electrets were coming out from the space between the plates, and the electrets were then repelled from capacitors, and so on.

Electrets gradually lose their polarization in the absence of adequate shielding. Electret motors with almost perfect shielding were described by the author and D. K. Walker in 1970.[44] These motors were based on the so-called electret slot effect,[45] which ensured both a nearly perfect shielding and a relatively large driving force.

A simple electret motor* utilizing the slot effect is shown in Ref. 21 (Fig. 14). It uses a carnauba wax electret rotor consisting of two oppositely polarized half-discs. The thickness of the rotor is 0.5 in., the diameter is 5 in. The motor operates from an 8-kV power supply and rotates at 1500 rpm.

---

* The motor was designed by the author and built by one of his students, Charles Lynn Walls, in 1966.

A later slot-effect motor[44] uses a stationary electret (Ref. 21, Fig. 15) in the shape of a hollow cylinder with four sections of opposite radial polarization. The rotor is made of four internal and four external aluminum electrodes forming two cylinders with four slots in each. The inner electrodes are cross-connected with the outer ones, and all electrodes are supported by a Plexiglas disc mounted on a steel axle. This motor uses no commutator; the power is delivered to two adjacent external electrodes by means of two sharp points which charge the rotor through a corona discharge. The overall diameter of this motor is 3 in., the operating voltage is 6 kV, the speed is up to 5000 rpm, and the power is about 20 mW.

Synchronous electret motors have been recently developed by the General Time Corporation. The motors are about 1.25 in. in diameter and 0.25 in. thick. In these motors a thin plastic electret disc with 15 active sectors and 15 equally large cutouts is the rotor. The rotor is placed between two stator plates, each plate having 30 electrodes connected in alternation to the two input terminals. The operation of this motor is similar to that of the first slot-effect motor described above, except that the reversal of polarity of the electrodes is accomplished directly by the applied ac voltage rather than by a commutator.

7.10 WHAT MAY BE EXPECTED OF ELECTROSTATIC MOTORS?

Bollée[33] has noted that, since electromagnetic motors are quite inefficient in scaled-down versions, very small capacitor-type and induction-type electrostatic motors may be a better choice for miniaturized systems. We can expect, therefore, that miniature electrostatic motors of these types will find applications in various sensor and control devices where only very small torques and powers are needed.

Experiments conducted in author's laboratory[44] indicate that electret motors can be useful in systems where powers of up to 1 W are needed.

However, the most promising electrostatic motors appear to be corona motors. These motors are extremely simple in design and require no expensive materials. Having only few metal parts, they possess a very good power-to-weight ratio; they are fully capable of developing appreciable amounts of power; and they can attain very high speeds.

Experiments with relatively advanced corona motors were reported in 1958 and 1960 by the Russian engineers Yu. Karpov, V. Krasnoperov, and Yu. Okunev.[46,47] One of their motors was a 6-W cylindrical motor operating from a 7-kV power supply and turning at a rate of 6000 rpm. This motor had a hollow Plexiglas rotor 10.5 cm in diameter and 17 cm long with a conducting lining on its inner surface. The stator consisted of 16 knifelike

Figure 7.4. A corona motor developed under the author's direction. The Plexiglas cylinder, lined with conducting foil, is driven by slanted corona blades having alternate polarities. The motor is about 5 in. long, and can produce 0.10 hp. It operates from a 6-kV power supply or from an earth–field antenna.

electrodes inclined relative to the surface of the rotor in the direction of the desired rotation. The lining of the rotor increased the electric field in the gap between the electrodes and the surface of the rotor, thus enhancing the corona discharge from the electrodes.

A high-speed corona motor with a disc rotor and circumferential electrodes was described by J. D. N. Van Wyck and G. J. Kühn of South Africa in 1961.[48] The motor had a rotor 1.5 in. in diameter turning in jewel bearings. The stator had six electrodes. Operating from 8 to 13 kV, the motor attained speeds of up to 12,000 rpm.

A similar motor was studied by the Polish scientists B. Sujak and W. Heffner in 1963.[49]

A number of advanced corona motors have been studied in author's laboratory,[50-52] and one is shown in Fig. 7.4.* On the basis of this study it appears that corona motors with input power of 100 to 1000 W and efficiency of at least 60% can be constructed without much difficulty. There appears to be no reason why even more powerful corona motors

---

* A diagram of a linear corona motor designed by the author is reproduced in Ref. 21 (Fig. 11).

could not be built, especially with rotors immersed in a gas other than the air at atmospheric pressure.

An important property of electrostatic motors is that they can operate from a much greater variety of sources than the electromagnetic motors. One very interesting source of electricity for electrostatic motors is the ordinary capacitor. Another possible source of power for electrostatic motors is a high-impedance high-voltage battery of the type of Zamboni's pile.

Electrostatic generators constitute a potentially very important source of electricity for electrostatic motors. Considerable advances in development of such generators have been made in recent years,[53,54] and it is conceivable that electrostatic motor-generators will be used to convert the high-voltage dc produced by such generators into the conventional low-voltage dc or ac.

Finally, a very interesting source of power for electrostatic motors is the atmospheric electric field. Experiments on operating electrostatic motors from this source have been conducted by the author.[55] In the initial experiments an electret motor and a corona motor were used; the electricity was extracted from the atmosphere by means of earth–field antennas. The corona motor was the one shown in Fig. 7.4. These experiments indicate that it is entirely possible to operate small electrostatic motors directly from atmospheric electricity. Whether it will be possible to operate large motors in this manner depends on how successful we are in designing and building earth–field antennas capable of extracting appreciable power from the atmospheric electric field.

Electrostatic motor research is still at a very rudimentary stage, and only a few scientists are engaged in it. Therefore it is one of the potentially most rewarding research fields in electrostatics and is well accessible to both professional scientists and to amateurs. In view of this, it may be appropriate to conclude this chapter with a quotation from one of the letters written to the author by A. D. Moore:

Youngsters are especially intrigued by things that *move*. Various electrostatic motors are so easy to construct that adult readers should encourage high school students to build them, as their science projects. The builders should keep in mind that one imperative is to have bearings with very low friction, such as needles put through holes in Plexiglas. Some of the youngsters thus stimulated will eventually find their way into electrostatics as a career.

REFERENCES

1. P. Benjamin, *A History of Electricity*, Wiley, New York, 1898, pp. 506, 507.
2. J. Sparks, Ed., *The Works of Benjamin Franklin*, Vol. 5, Whittemore, Niles, and Hall, Boston, 1856, p. 301.

3. Gilbert, *Ann. Phys.*, Ser. 1, **49**, 35–46 (1815).

4. Müller-Pouillet, *Lehrbuch der Physik und Meteorologie*, 10th ed., Vol. 4, Friedrich Vieweg, Braunschweig, 1909, p.338.

5. M. Guyot, *Nouvelles Recreations Physiques et Mathematiques*, Vol. I, Guffier, Paris 1786, pp. 272–274, and Plate 27.

6. V. K. Chew, *Physics for Princes*, Her Majesty's Stationery Office, London, 1968, p. 13.

7. G. Adams, *An Essay on Electricity*, 5th ed., London, 1799, pp. 580–581 and Plate 4, Fig. 79.

8. H. J. Oosting, *Z. Phys. Chem. Unterricht*, **9**, 84–85 (1896).

9. D. Kämpfer, *Ann. Phys.*, Ser. 3, **20**, 601–614 (1883).

10. E. Bichat, *Ann. Chim. Phys.*, **12**, 64–79 (1887).

11. Benjamin Wilson, *Phil. Trans.*, **51**, 896 (1760).

12. A. Christenson and S. Moller, *AIAA J.*, **5**, 1768–1773, (1967).

13. L. B. Loeb, *Electrical Coronas*, University of California Press, Berkeley, 1965, pp. 402–406.

14. J. W. Draper, *Text-Book on Chemistry*, Harper, New York, 1847, p. 105.

15. A. D. Moore, *Electrostatics*, Doubleday, Garden City, New York, 1968, pp. 104–108.

16. M. J. Jamin, *Cours de Physique*, 2nd ed., Vol. 3, Gauthier-Villars, Paris, 1869, p. 39.

17. R. H. Dressel, *High Sch. Sci. Bull. N. M. State Univ.* **6**, 3–4 (April 1, 1964).

18. H. Wommelsdorf, *Ann. Phys.*, Ser. 4, **70**, 135–138 (1923).

19. W. Holtz, *Ann. Phys.*, Ser. 2, **130**, 168–171 (1867).

20. J. C. Poggendorff, *Ann. Phys.*, Ser. 2, **139**, 513–546 (1870).

21. O. Jefimenko and D. K. Walker, *Phys. Teacher*, **9**, 121–129 (1971).

22. C. Christiansen, *Ann. Phys.*, Ser. 2, **137**, 490 (1869).

23. "Elektrischer Tourbillon," *Ann. Phys.*, Ser. 2, **144**, 644 (1871).

24. M. E. Mascart, *Traité d'Électricité Statique*, Vol. I, G. Masson, Paris, 1876, p. 179.

25. J. W. Davis and J. B. Farrington, U. S. Patent No. 459678, Sept. 15, 1891.

26. V. E. Johnson, *Modern High Speed Influence Machines*, E. and F. N. Spon, London, 1921, pp. 175–205.

27. "A group of static motors," *Elect. World*, **18**, 418 (1891).

28. "A static electric motor," *Elect. Exp.*, p. 137 (January 1914).

29. "Zipernowsky electrostatic motor," *Elect. World*, **14**, 260 (1889).

30. W. Thomson, British Association Report, 1855 (2), p. 22.

31. N. G. van Huffel, *Z. Phys. Chem. Unterricht*, **17**, 316–317 (1904).

32. W. Thomson, British Association Report, 1867, pp. 489–513.

33. B. Bollée, *Phillips Tech. Rev.*, **30**, 178–194 (1969).

34. R. Arno, *Rend. Atti Reale Accad. Lincei*, **1** (2), 284 (1892).

35. R. Arno, *Electrician*, **29**, 516–518 (1892).

36. W. Weiler, *Z. Phys. Chem. Unterricht*, **7**, 1–4 (1893).

37. S. D. Choi and D. A. Dunn, *Proc. IEEE*, **59**, 737–748 (1971).

38. W. Weiler, *Z. Phys. Chem. Unterricht*, **6**, 194–195 (1893).

39. G. Quincke, *Ann. Phys.*, Ser. 3, **59**, 417–486 (1896).

40. P. E. Secker and I. N. Scialom, *J. Appl. Phys.*, **39**, 2957–2961 (1968).

41. P. E. Secker and M. R. Belmont, *J. Phys. D: Appl. Phys.*, **3**, 216–220 (1970).

42. J. R. Melcher and G. I. Taylor, *Ann. Rev. Fluid Mech.*, **1**, 111–146 (1969).

43. A. N. Gubkin, *Electrets*, Academy of Sciences, Moscow, 1961, pp. 130–133, in Russian.

44. O. Jefimenko and D. K. Walker, *Conference on Dielectric Materials, Measurements and Applications*, The Institution of Electrical Engineers, London, 1970, pp. 146–149.

45. O. Jefimenko, *Proc. W. Va. Acad. Sci.*, **40**, 345–348 (1968).

46. Yu. Karpov, V. Krasnoperov, and Yu. Okunev, *Tekh. Molod.*, **26**, 36–37 (September 1958), in Russian.

47. Yu. Karpov, V. Krasnoperov, Yu. T. Okunev, and V. V. Pasynkov, "On the Motion of Dielectrics in an Electric Field," in *Dielectric Physics*, Academy of Sciences, Moscow, 1960, pp. 124–131, in Russian.

48. J. D. N. Van Wyck and G. Kühn, *Nature*, **192**, 649–650 (1961).

49. B. Sujak and W. Heffner, *Acta Phys. Polon.*, **23**, 715–726 (1963).

50. O. Jefimenko and H. Fischbach-Nazario, *Proc. W. Va. Acad. Sci.*, **42**, 216–221 (1970).

51. W. Aston, *W. Va. Univ. Mag.*, **3**, No. 4, Spring 1971, pp. 6–11.

52. C. P. Gilmore and W. J. Hawkins, *Pop. Sci. Month.*, **198**, 95–97, 114 (May 1971).

53. N. J. Felici, *Elektrostatische Hochspannungs Generatoren*, G. Braun, Karlsruhe, 1957.

54. J. Hughes and P. Secker, *New Sci. Sci. J.*, **49**, 468–470 (1971).

55. O. Jefimenko, *Am. J. Phys.*, **39**, 776–778 (1971).

CHAPTER 8

Electrostatic Generators

M. J. MULCAHY, *Director*
High Voltage Consulting and Testing
Ion Physics Corporation
Burlington, Massachusetts

W. R. BELL
Ion Physics Corporation
Burlington, Massachusetts

8.1 OVERVIEW

Electrostatic machines do not yet offer a serious threat to electromagnetic machines, but in certain specialized areas they are becoming the best choice. These are the areas where high-voltage supplies are needed (100 kV to 10 MV) with currents up to 100 mA. Examples of applications are power supplies for paint sprayers, electron microscopes, ion implantation, separators, test equipment, and irradiation equipment. Most industrial electrostatic power supplies are of the "charge transporter" type where mechanical work is done in carrying positive or negative charge against an electric field, so raising its potential. This process is analogous to carrying buckets of water to an elevated storage tank. Mechanical work is done against gravity (i.e., a gravitational field) and is converted to potential energy at the tank. This potential energy can be converted into work when the water is drawn from the tank. For a practical water tank, adding water raises the water level and thus the potential, until the tank eventually overflows. The electrical equivalent has a similar limitation when flashover discharges the stored electrostatic energy. Common examples are the Van de Graaff belt-type and the cylindrical-type generators; the former is best suited for 0.5 to 10 MV and up to about 5 mA and the latter for 50 to 800 kV and up to 50 mA.

8.2 INTRODUCTION

Electrical generators are energy converters and as such, simply change the form of the energy in which it is available to electrical energy. Thus, heat may be converted directly, as in the thermionic anode generator, and chemical energy as in the fuel cell.[1] However, by far the most common type of generator is the electromechanical generator in which mechanical energy is converted to electrical energy; examples are the alternator and the dc generator, and the electrostatic generator, which is generally unknown.

Until about 1950, electrostatic (ES) generators appeared to have died a natural death, despite the determined efforts of notable workers such as Felici at Grenoble and Trump and Van de Graaff at M.I.T. Application was limited to high-voltage, low-current devices such as accelerators, and x-ray devices in which the simplicity and reliability offset the low power density. Interest was revived, however, with man's entry into space, where efficient high-voltage generators are required which can operate reliably in this environment. Desirable features include high power density, high-temperature operation, and the ability to operate under high-vacuum space conditions. The vacuum-insulated ES generator appears to be an obvious choice, converting efficiently the mechanical energy from a prime mover into the high-voltage power supply for the propulsion unit (ion and colloidal engines, e.g.). It is hoped that, ultimately, devices will be available for direct conversion of heat to electrical energy.

For terrestrial applications ES generators do not yet offer serious competition to transformer–rectifier assemblies, except perhaps when the high-voltage load necessitates a low-current, highly stabilized supply with low ripple voltage, when currents up to 100 mA are required at high voltage, or when the supply must be inherently safe. The ES generator is therefore useful for high-voltage physics and for special industrial applications such as paint spraying, electron microscopes, ion implantation, separators, various high-voltage test and irradiation equipment, and for x-radiation and radiotherapy in the medical field.

8.3 BASIC PRINCIPLES OF OPERATION AND EFFICIENCY OF ELECTROSTATIC GENERATORS

8.3.1. Basic Principles of Operation

Electrostatic generators are based on Coulomb's law that electric charges attract or repel (depending on their relative sign) by an amount that can be written as

$$F = \frac{Q_1 Q_2}{4\pi\varepsilon d^2} \text{ N} \qquad (1)$$

where, Q_1, Q_2 are the point charges (C) d the separation (m), and $\varepsilon = \varepsilon_0\,\varepsilon_r$. Coulomb's law applies to point charges but may be adapted to include free charges in an electric field $\overline{E}$ V/m. Work is done moving the charge q a distance d against the electric field, raising the potential of the charge by V V (Fig. 8.1). Motor action involves the reverse process, electrical energy being converted into mechanical work.

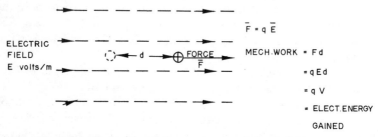

Figure 8.1. Mechanical to electrical power by way of electric field.

The dual nature of both electromagnetic and electrostatic generators has been discussed in some detail by various authors,[2-4] and we know that they are voltage and current generators, respectively, with voltage and current proportional to rate of change of magnetic flux linkages and electric flux, respectively.

8.3.2 Efficiency

The ES machine can be considered as a force field E_T interacting with an electric charge σ or a force field E_T interacting with an induction field D_N (Figs. 8.2a and 8.2b, respectively). The machine power can be increased by raising σ or E_T. Both σ and E_T depend on the dielectric strength of the medium, and this sets an upper limit analogous to saturation. Thus, for example, if the maximum electric field is 10^5 V/cm, the gap energy density is 0.44×10^{-3} J/cm^3. This is much lower than electromagnetic (EM) machine energy density (e.g., 0.4 J/cm^3) and explains why the latter is more commonly used for power generation. An increase in the dielectric strength by a factor of 30 would make the ES machine competitive. The surface charge density σ is applied by an exciting voltage V_{ex} and can be increased by raising the relative permittivity of the gap (i.e., raising the capacitance).

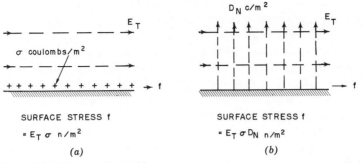

Figure 8.2.   Electrostatic machine.

Neglecting losses, the differential equations for the electrostatic generator are

$$\text{current } i = \frac{d}{dt} (\text{electric flux}) = \frac{d}{dt}(q)$$

$$= \frac{d}{dt}(cv) = c\frac{\partial v}{\partial t} + v\frac{\partial c}{\partial t} \tag{2}$$

$$\text{power} = vi = cv\frac{\partial v}{\partial t} + v^2\frac{\partial c}{\partial t} \tag{3}$$

$$\underset{\substack{\text{stored in}\\ \text{electric}\\ \text{field}}}{\qquad\qquad} \underset{\substack{\text{mechanical}\\ \text{power}\\ \text{conversion}}}{\qquad}$$

where capacitance c is a function of position (and hence time).

Thus the ES generator relies on the capacitance being a function of position and, hence, by motion to vary with time. The three-element or charge-transporter type of ES generator described in the next section is often referred to as a fixed-capacitance machine in that the capacitance of the output terminal remains constant. However, as far as the charge on the transporting element is concerned, it is situated on a variable capacitor. It is worthy of note that ES and EM machines have a common principle in that the conversion of mechanical energy into electrical energy, or conversely, is possible only if the system capacitance or inductance is changed.

8.4   TYPES OF ELECTROSTATIC GENERATORS

8.4.1   Introduction

Electrostatic generators can be conveniently classed into two main groups, namely, three-element and two-element machines. The three-element machine (sometimes referred to as the fixed-capacitance or charge-transporter type) consists of an inductor element which causes charge to be deposited on the transporting element; motion of the transporting element away from the inductor causes the potential of the charge to be raised, and the charge is removed by the collecting element. The transporting element may have a variety of forms including metallic elements, insulating belts, discs and cylinders, gas, oil, water droplets, and dust. The best known example is probably the Van de Graaff belt-type generator. The two-element machine is simply one in which the capacitance varies cyclically between maximum and minimum values. To achieve a net output with a dc machine, some form of commutation is essential because the machine, by its very nature, is of alternating type. This machine is less well developed than the former but holds more promise when high efficiency and power density are sought.

Electrostatic machines have been classified in many ways, and each method has its own advantages and disadvantages. A recent classification, aimed at reducing the shortcomings of previous systems, has been proposed by Polotovskiy.[5] It is based on the main properties and features of ES machines, such as the nature of movement, excitation, and rotor design. Table 8.1 gives this classification in tabular form. Trump[6] in 1933

Table 8.1.   Classification of ES Machines Proposed by Polotovskiy[5]

No.	Feature	Type of Machine	Example/Comment
1	Method of communicating the charge	Conduction Induction	Variable capacitance Van de Graaff
2	Nature of the field in which rotor moves	Unipolar Bipolar	Unipolar does useful work only each half cycle
3	Excitation system	Independent Self Parallel Series Mixed	—
4	Nature of commutation	With apparent commutation; conducting rotor	
		With latent commutation; dielectric rotor	Van de Graaff
5	Rotor design	Solid rotor (disc, cylindrical)	Cylindrical, Wimshurst Electrohydrodynamic,
		Flexible rotor	belt
6	Nature of motion	Rotary Reciprocating	

analyzed the basic variable-capacitance machine for various circuit configurations. He suggested classifying machines by a type number according to Table 8.2. The Type III generator has received by far the most attention. Felici[7] gave a historical review of ES machines in his book published in 1957, as well as descriptions and principles of several machines and their industrial applications.

Table 8.2. Classification of ES Generators According to Trump[6]

No.	ac/dc	Motor–Generator	Description
I	ac	Synchronous motor-generator	Line excitation; may be brushless
II	ac	Motor–generator	dc excitation; capacitor bridge arrangement
III	dc	Generator	Separate excitation, rectifiers, may be brushless; most reliable for dc
IV	dc	Generator	dc line excitation; charge transfer; commutators and rectifiers
V	dc	Generator	Separate excitation; charge transfer; commutators and rectifiers
VI	dc	Generator	dc line excitation; charge transfer; commutators and rectifiers
VII–X	dc	Motors	Commutators and rectifiers

It was shown in Section 8.2 that the power output of an ES machine depends on $D_n E_t$ as well as on velocity. Now $D_n = \varepsilon_0 \varepsilon_r E_n$ (where $\varepsilon_0 = 8.85 \times 10^{-12}$ F/meter, ε_r = relative permittivity of the medium, and E = electric field strength, V/m). Thus to increase the output of an ES machine, the designer seeks materials that will withstand a high electric field (since power is proportional to E^2); the materials must also have a high relative permittivity. The velocity limitation is usually mechanical—set by bearings, seals, and the strength of the structural materials.

8.4.2 Brief Survey of Dielectrics

As discussed earlier, the power density of an electrostatic machine is limited by the dielectric strength E_B and the relative permittivity ε_r of the insulating medium. A figure of merit for this medium is thus $\varepsilon_r E_B{}^2$. The resistivity is also important and should be high enough to prevent significant loss of charge by conduction. It is thus appropriate to examine, very briefly, the possible insulating media.[8]

8.4.2.1   *Solids*

Solids are ideal electrically, since modest values of ε_r (~ 5) and high electrical breakdown strength (1 MV/cm) can be achieved. Unfortunately, the necessity of relative motion means that a liquid or gaseous medium must also be present.

8.4.2.2   *Liquids*

High breakdown strengths (0.25 MV/cm) are possible, but most insulating liquids have low values of ε_r (3–5). Liquids with high values of ε_r (e.g., water, ethyl alcohol, nitrobenzene, acetone) exhibit strong ionic conductivities and relaxation times much shorter than 10^{-2} sec. However, progress[9] with ion-exchange membranes indicates that high permittivity dielectrics with resistivities up to 10^{12} Ω-cm are possible. For example, nitrobenzene ($\varepsilon_r = 37$) has operated at 250 kV/cm in a 0.1-in. gap. There remain two disadvantages of liquids:

1. Under continuous electric stress, impurities align to form bridges which seriously reduce the electric strength.

2. The viscosity introduces large mechanical losses at any appreciable velocity, which offsets the gain due to high ε_r. The efficiency is thus low.

8.4.2.3   *Gases*

Compressed gases, particularly the electronegative gases such as sulfur hexafluoride, can be operated at high electric fields but have ε_r essentially unity. Where processes such as ionic commutation are present, it has been found advantageous to accept the reduced breakdown strength of compressed hydrogen because of its better ionic conduction and low windage loss properties. Gases introduce large windage losses particularly at high pressures and small gaps, in addition to requiring expensive and bulky pressure vessels. A comparison of gases based on energy density indicates that hydrogen is the best, with a value of 6 (taking air as 1).

8.4.2.4   *Vacuum*

The natural choice for space vehicles is the vacuum, but this may be costly for terrestrial applications, where vacuum pumps, seals, and special bearings are needed. We might expect the electrical breakdown strength of high vacuum to be infinite, but in practice it is relatively low, particularly at long gaps or with large-area electrodes. For example, by using special alloys and polishing techniques, electric field strengths approaching 10^6 V/cm can be maintained across a 1-mm gap if the electrodes are small, but this value is reduced by a factor of 3 for an electrode area of 1000 cm^2. With small electrodes and 1-cm gap, the electric strength is reduced also by approximately 3. Electrode material is important; high-melting-point materials usually superior. The relative permittivity is unity, but there are no windage losses. There are problems with degassing of materials and bearings, however, particularly at elevated temperatures, as well as dust problems caused by brushes.

8.4.3   Three-Element Generators (Fixed-Capacitance or Charge-Transporter Type)

8.4.3.1   *General*

When certain substances are rubbed together, charge transfer occurs between materials which can result in a buildup of potential difference if this unipolar charge is allowed to accumulate. This phenomenon, a curse to many industries, was the basis of the early "friction" machines. Frictional charging tends to be unreliable as well as inefficient and is thus rarely used today. Corona charging is perhaps the most common method of unipolar charge generation for use in the three-element class of transporter-type generators.

The principle behind this class of ES generators is illustrated in Fig. 8.3.

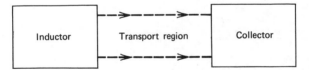

Figure 8.3.   Basic principle of three-element ES generator.

Unipolar charge is produced at the inductor, and it is carried from there to the collector by the transporter. As described earlier, we can either imagine that the capacitance of the individual charges falls as it moves away from the inductor (hence raising its potential) or that work is done by the transporter in carrying the charge against a retarding electric field. Both approaches lead to the same result—that the work done by the transporter is converted to electrical energy, which is deposited at the collector.

We can conveniently further divide this class of generators into two: (1) those with fluid transporters (electrofluiddynamic, or EFD, generators) and (2) those with solid dielectric transporters (e.g., belt-type machines). Early forms of these machines such as the Wimshurst machine and the Kelvin water dropper are not included in the present discussion of these two subgroups, since they have been largely replaced by improved devices.

Similarly, the EFD dust generators of Pauthenier,[10, 11] Moreau-Hanot,[12] and Morand et al.,[13] although quite successful, have been omitted. The reader will, however, find it interesting and valuable to study these machines.

8.4.3.2   *Van de Graaff Generator* (*Belt Machine*)

Review articles on the belt type of generator have been written by Van de Graaff et al,[14] Shire,[15] and Fortescue.[16] The main elements of the generator appear in Fig. 8.4. Charge is sprayed on to an insulating moving belt from corona points mounted below the ground plane and is removed from the belt by discharging points inside the high-voltage terminal. The terminal has capacitance C and thus charges up at a rate given by

$$\frac{dV}{dt} = \frac{I}{C} \tag{4}$$

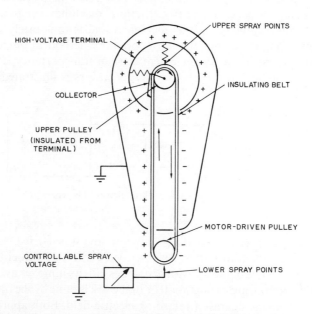

UPPER SPRAY POINTS

HIGH-VOLTAGE TERMINAL

INSULATING BELT

COLLECTOR

UPPER PULLEY
(INSULATED FROM
TERMINAL)

MOTOR-DRIVEN PULLEY

CONTROLLABLE SPRAY
VOLTAGE

LOWER SPRAY POINTS

Figure 8.4.   Schematic diagram of Van de Graaff electrostatic belt generator.

where I is the net charging current. Thus if constant current is supplied and no current is lost, the terminal potential will increase linearly with time. Ultimately flashover will limit the potential. The steady terminal potential is normally achieved when the total load current is equal to the charging current. The shape of the terminal must be such that the surface gradient of

Figure 8.5.   Final assembly of the transuranic accelerator, with an MP system in the background. (Courtesy of High Voltage Engineering Corporation.)

electric field is always less than the corona onset field at the design voltage; otherwise, discharges occur. The best shape is a sphere surrounded by a concentric ground sphere with the intervening space filled with compressed gas. (At present, electronegative gas mixtures up to pressures of 400 psi are a popular choice.) The field between the terminal and ground along the belt is kept nearly uniform by grading rings held close to the belt to reduce potential variations caused by the belt charge. With this technique, tangential belt gradients up to 1 MV/ft are possible.

The maximum belt charge density is governed by electrical breakdown, and the maximum charge permissible is approximately half the theoretical maximum value $\varepsilon_0 E_B$, where E_B is the breakdown stress normal to the belt. The belt current is given by $I = bv\sigma$, where b is the belt width, v is the belt velocity, and σ, the charge density; therefore, I is increased by raising b, v, and σ as far as possible. For example, a 20-in. belt moving at 60 ft/sec delivers 1 mA in a generator pressurized to 400 psi. Extra current is achieved by double charging (as shown) and by using multiple belts.

Van de Graaff type generators have been designed with current capabilities up to several milliamperes, a limit which is set by belt and charge density considerations. Also, these machines lead all other generators in voltage performance—indeed, in 1970 at High Voltage Engineering Corporation (HVE), a Van de Graaff generator for the transuranic accelerator was operating at more than 20 Million V (Fig. 8.5). It is of interest to note that a dramatic

Table 8.3. Van de Graaff High-Voltage Generators and Accelerators

Characteristics	Single-Stage Accelerators for Positive Ions, Basic Models					
	AN 400	AN 700	An 2500	KN 3000	KN 4000	CN
Max voltage (MV)	0.5	0.7	2.5	3.8	4.8	7.5
Max energy (MeV)	0.5	0.7	2.5	3.8	4.8	7.5
Max current (μA)	150	150	150	400	400	70
Neutrons/sec	$>2 \times 10^{10}$	$>3 \times 10^{10}$	4×10^{10}	$>5 \times 10^{11}$	$>10^{12}$	$>5 \times 10^{11}$
Stability ($\pm$ kV)	5	5	2	2	2	2

Characteristics	Single-Stage Accelerators for Electrons, Basic Models				
	AS 400	GS	AS 2500	KS 3000	KS 4000
Max voltage (MV)	0.4	1.5	2.5	3.0	4.0
Max energy (MeV)	0.4	1.5	2.5	3.0	4.0
Max current (μA)	100	1700	250	1000	1000
x-ray output (rpm at 1 M)	1	210	150	1350	3500
Stability ($\pm$ kV)	10	10	10	10	10

Characteristics	Two-Stage Tandem Accelerators, Basic Models				
	TTT-1	TTT-2	EN	FN	MP
Max voltage (MV)	1	2	7	10	14.4
Max proton energy (MeV)	2	4	14	20	28.0
Current at Max MeV (μA)	1	2	4	4	10.0
Stability ($\pm$ kV)	1	1	1	1	1

Characteristics	Three-Stage Tandem Accelerators, Basic Models			
	EN	EN-CN	FN	MP
Max voltage (MV)	6	6	7.5	10
Max proton energy (MeV)	18	17.5	21.5	30
Current at Max MeV (μA)	0.5	0.5	0.5	0.5
Stability ($\pm$kV)	5	5	5	5

upgrading of the energy performance of Van de Graaff accelerators was achieved by using a voltage doubling or charge exchange technique suggested by Alvarez[17] and others. This is described elsewhere,[18] but essentially it provides for additional stages of acceleration for the particle beams, and accelerators using this principle are termed tandem accelerators.

Table 8.3 summarizes the performance characteristics of some of the principal generators and accelerators produced by HVE for applications in many fields, including high-energy, solid-state, and thermonuclear physics; chemistry; radiobiology; medicine; nondestructive testing; space research; and radiation physics. To date about 500 of these accelerators have been commisioned around the world and are being used in a variety of ways, both to benefit mankind and further his understanding of the world around him—a fitting tribute and memorial to the genius and pioneering spirit of Robert J. Van de Graaff.

Anton[19] applied the foregoing techniques to a disc-type machine (constant oblique field generator) and predicted 10 kW of power from a 40-cm diameter disc rotating at 24,000 rpm. To secure better field control, a semiconducting material is used between high- and low-voltage terminals instead of grading rings.

8.4.3.3 Cylindrical Machines

Based on Felici's work at Grenoble, a French company,[20] SAMES, has developed a complete line of ES generators with cylinders rotating in hydrogen under pressure.[21, 22] The cylindrical form has no special advantages for very high voltages but is superior for larger powers and currents because it has higher peripheral speeds, good commutation, and better field control. The general form of the dielectric cylinder generators is shown in Fig. 8.6. The rotating dielectric cylinder (several millimeters of resin) acts like the belt in the Van de Graaf machine. The stator is inside the rotor and is made of slightly conductive glass (10^{12} Ω-cm) to give good field distribution. Thin steel strips charge and discharge the rotor by ionization as a result of inductors behind the glass cylinder. Using hydrogen at 15 to 20 atm, rotor power densities of 2 W/cm^2 are possible with an efficiency of better than 90%.

Although many other gases have superior electrical strength, hydrogen is superior for ionization charging owing to the high mobility of the ions; corona damage of the rotor is also considerably reduced. Typical charge densities are 1 to 2 μC/cm^2 with tangential fields about 15 kV/cm. The output current is proportional to the active rotor surface, the speed, the number of poles, and the surface charge density, whereas the maximum voltage is proportional to the pole spacing. Because of the latter limitation, present single-stage machines do not exceed 1 MV, since mechanical

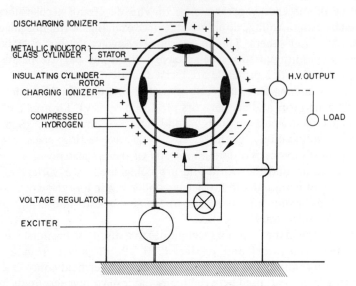

DISCHARGING IONIZER

METALLIC INDUCTOR
GLASS CYLINDER   STATOR

INSULATING CYLINDER
          ROTOR
CHARGING IONIZER

COMPRESSED
HYDROGEN

H.V. OUTPUT

LOAD

VOLTAGE REGULATOR

EXCITER

Figure 8.6.   Schematic of cylindrical machine (after Felici).

problems limit the maximum rotor diameter. Similarly, mechanical re-
sonances restrict the axial length of the rotor to 1.5 rotor diameters for
an open-bell rotor design and 3 to 4 rotor diameters for a drum type. These
restrictions result in approximate design limits of 800 kV, 14 mA for a bell
rotor and 500 kV, 50 mA for a drum type.

The present range of powers is from 20 to 3000 W, but it is likely that this
will be extended to 10,000 W. Mechanical tolerances are small and limit
the maximum rotor diameter to 45 cm at 3000 rpm. Hence for a 100-kV
machine (using a rotor 3 ft long), 20 kW seems to be the optimum rating.

Both the Van de Graaff and the cylindrical-type machines lend themselves
quite readily to automatic voltage control. Unlike an EM machine, which is
a voltage generator, an uncontrolled ES generator attempts to produce
constant current. Since most practical applications operate at approximately
constant voltage, some form of electronic control is necessary. The output
voltage is compared with a reference voltage and the error used to modify
the current from the charge spray device. Stabilities of better than 1% are
readily achievable, and values of 0.01% are claimed.

8.4.3.4   *Liquid-Immersed Conducting-Segment ES Generator*

The liquid-immersed conducting-segment ES generator has been included
as an example of recent developments in the field and changes in thinking
to overcome some of the disadvantages and limitations of the more

established belt and cylindrical machines. In principle, a conducting-segment machine is similar to the Felici cylindrical machine; it differs in that metallic bars are set in an insulating drum rotor and conduction is charged and discharged by rollers. In the charge position, the grounded rotor segment has capacitance C_{in} to the exciting stator electrode (at potential V_{in}) and capacitance C_L to the output terminal (V_{out}). The charge applied to the rotor segment is thus

$$Q_{in} = V_{in} C_{in} - V_{out} C_L \tag{5}$$

At the output, most of the charge is removed. However, because of C_L an amount of Q_R remains on the rotor segment

$$Q_R = V_{out} C_L \tag{6}$$

The net output current is therefore

$$I_{out} = (V_{in} C_{in} - 2V_{out} C_L) \, pn \tag{7}$$

where n = revolutions/sec and p = number of rotor segments.

In the steady state, with the machine having a load resistance R, the output voltage V_{out} is $I_{out} R$. substituting in Eq. 7 gives

$$I_{out} = \frac{pn V_{in} C_{in}}{1 + 2RC_L pn} \tag{8}$$

Figure 8.7 shows the essentials of a machine developed by Secker et al.[23, 24] at the University of North Wales, producing 1 mA at 200 kV. This is achieved by incorporating reverse charging at the output terminal which doubles the current output. A number of series-connected neon tubes make up the necessary bias impedance, and a slightly polar liquid with $\varepsilon_r = 6$ (Arachlor) is used as the main insulation.

The advantages of this type of machine are its compactness, low cost, and rugged construction. However, the large viscous losses in the liquid result in low efficiency (13%) and the necessity to cool the liquid. Commutation also produces gradual degradation of the oil due to carbonization. The generator is thus poorly suited for powers in excess of 1 kW.

8.4.3.5 *Electrofluid Dynamic Generation*

a. GENERAL. Electrofluiddynamic (EFD) high-voltage generation is based on the direct conversion of kinetic energy to electrical energy by making use of the retarding force exerted by an electric field on charge carriers, whether these be electrons, ions, or larger particles. This is an old concept[25,26] which has recently been revived because of the many potential applications for this type of high-voltage generator,—for example, in areas where overall conversion efficiency is not a prime factor, where there is an abundance of

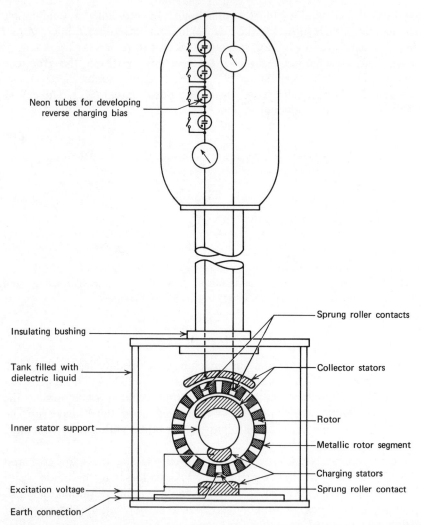

Figure 8.7. Liquid-immersed conducting-segment ES generator. (Courtesy of Industrial Development, Bangor, N. Wales.)

kinetic energy available, or where the generation process itself makes EFD generation particularly suitable for the application (e.g., paint spraying and particle precipitation).

The basic principle of operation may be understood by referring to Fig. 8.3. There is an injector, a conversion or transport region, and a collector, with fluid flow in the direction from the injector to the collector. The charge carriers are injected into the insulating fluid (at low potential) at the injector, transported by the mechanical forces of the fluid across the conversion region, and removed by the collector, which is well insulated from ground and which then assumes a high potential.

For successful operation it is necessary that there be negligible slip velocity of the particles as compared with the velocity of the carrier fluid. Thus if ions or electrons are used they may be suspended in an insulating liquid, resulting in very small mobilities. This is termed electrohydrodynamic (EHD) high-voltage generation. Another method that also results in low slip velocity involves increasing the size of the particles carrying the charge (e.g., aerosols or solid particulates) and suspending them in a high-velocity gas stream. This is termed electrogasdynamic (EGD) high-voltage generation.

b. ELECTROHYDRODYNAMIC GENERATORS. The prime development work in the field of EHD generation was carried out by Steutzer[27, 28] and later by Secker and his colleagues.[29,30] The basic elements as well as the general layout of an EHD generator are presented in Fig. 8.8. Initially the charge was injected into the working fluid using corona from needle points at high potential relative to the surrounding metal surfaces. However, because the corona action ultimately led to degradation of the liquid, this injection system was replaced[29] with a field-emission razor blade system which, for multiblade arrays, provided currents up to about 500 μA without damage to the liquid (typically hexane). For best performance the razor blades are mounted such that their edges are parallel to the liquid flow and the emission is normal to same. The charge is removed from the liquid by bringing the charge carriers into intimate contact with the collector surface, which, for this reason, is usually a honeycomb or multiplate etched structure of large surface area through which the liquid is pumped.

The following simplified analysis, due to Secker,[29] show that quite high voltages can be generated over relatively short conversion regions. Assume that unipolar charge is released into the moving liquid at the injector and is swept toward the collector as a result of the momentum transfer during fluid-molecule–charge-carrier collisions. When charge accumulates on the collector, its potential rises and the charge carriers moving toward it are slowed by the opposing field. The resultant carrier velocity v_c can be written in the form

$$v_c = v_f - \mu E \qquad (9)$$

where v_f = fluid velocity
μ = charge carrier mobility
E = opposing field

The maximum potential is obtained when the charges can no longer reach the collector and remain effectively stationary in the conversion space (i.e., the electrostatic forces due to E balance the viscous drag of the moving liquid). If some simplifying assumptions are made, this voltage is given by

$$V = \frac{v_f d}{\mu} - \left(\frac{8}{9}\frac{Jd^3}{\varepsilon\mu}\right)^{1/2} \tag{10}$$

where J = charge carrier current density
$\quad\quad \varepsilon$ = permittivity of the liquid
$\quad\quad d$ = length of the conversion space

The first term on the right-hand side of Eq. 10 represents the basic generating action and the second term the space charge distortion.

For a given output current density, there is an optimum value of the conversion space, corresponding to zero field at the collector and a maximum generated voltage. This critical d value is given by

$$d_c = \frac{1}{2}\frac{V_f^2\varepsilon}{\mu J} \tag{11}$$

and the associated maximum output voltage is

$$V_{max} = \frac{1}{6}\frac{\varepsilon}{\mu^2 J}v_f^3 \tag{12}$$

Thus low carrier mobility is desirable. For example, if electrons or ions are the charge carriers, the value of μ for hexane is 1.5×10^{-4} cm^2/(V)(sec). This may be used to find values of d_c and V_{max}, respectively, for a typical injector design for J of 1.5 μA/cm^2. With a flow rate of 1 m/sec, d_c is about 4 cm and V_{max} is approximately 800 kV. The EHD generator in Fig. 8.8 was designed to produce voltages approaching 1 MV. For this reason, the collector was surrounded with a large stress-distributing sphere and mounted on top of two coaxial glass tubes, the latter to provide adequate clearance from ground in the open laboratory. The maximum voltage achieved by this unit was 500 kV, with maximum short-circuit current of 30 μA. A later design of generator, also due to Secker et al., yielded 800 kV before being limited by external flashover at a total machine height of 8 ft.

c. ELECTROGASDYNAMIC GENERATORS. The development of EGD generators to the practical system stage has taken place only relatively recently. Significant contributions to this development have been made by many workers, including Gourdine,[31-33] Marks,[34] Barreto,[35] Lawson,[36] Whitby,[37] and Cox.[38] Just as for EHD, the basic elements (Fig. 8.9) consist of an ionizing or injector section, a conversion section, and a collector section. However, the most convenient method for producing the charged particles is a corona discharge between a sharp-point electrode and a ring electrode, called the attractor.

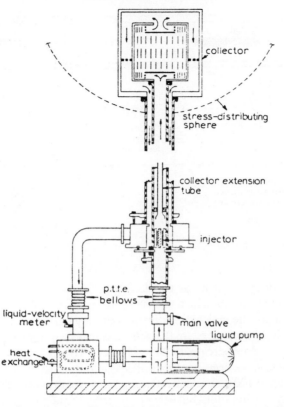

Figure 8.8. General arrangement of liquid-filled ES generator. (Courtesy of the Institution of Electrical Engineers, London.)

The needle and the attractor are usually located at the throat of a supersonic nozzle or in another high-velocity duct. Gas is then blown through this nozzle with the objective that the mechanical body forces (μE) act on the charged particles and blow them downstream, through the conversion region to the collector. It was found that in practice, when gases only were used, most of the charge drifted to the attractor ring; this effect is attributable to the high mobility of the ions and electrons. Fortunately, this problem can be overcome by injecting condensible vapors upstream under conditions that permit them to become supersaturated in the nozzle and to condense on the ions that serve as condensation nuclei. The mobility of the latter then drops (by orders of magnitude) to a level at which there is negligible slip between the charged particles and the gas stream. In this way, clouds with a relatively high charge density (0.03 C/m^3) can be produced, and very little is lost to the attractor ring, as evidenced by the approximate

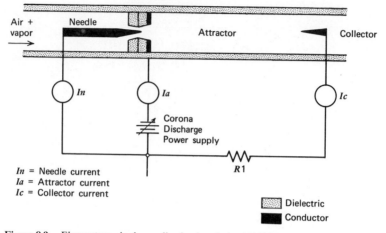

Figure 8.9.   Elementary single-needle slender-channel EGD generator.

equivalence of the ionizing or needle current and the collector current and by the attractor current, which drops almost to zero.[39] As an alternate, solid particles can also be injected upstream, and they perform the same function as the condensing vapors in that the electrons or ions attach to them and are swept downstream to the collector. To date, point, sphere, and ring electrodes have been used for collectors and all have yielded collection efficiencies up to 100%. The conversion section usually consists of a dielectric cylinder of constant radius, but other geometrical configurations are possible; a truncated cone, for example, serves to continue the expansion throughout the region.

Basically, EGD generators may be termed broad or slender channel depending on whether $L/R \gg 1$ or $L/R \ll 1$, where L is the conversion channel length and R is the channel radius. The maximum voltage that can be generated depends on the breakdown characteristics of the attractor–collector gap, and hence on the breakdown field strength of the gas–vapor mixture used. Furthermore, analysis of the energy conversion process reveals that the electrical body force term F_Z in the axial direction, and the pressure drop Δp across the conversion region, which determine the conversion of energy from gas flow to electrical, also depend on the axial E_R and radial E_Z fields as follows:

$$\Delta p = \frac{F_Z}{\pi R^2} = \frac{\varepsilon_0 E_Z^2}{2} \qquad \text{for} \quad \frac{L}{R} \ll 1 : \text{broad channel} \qquad (13)$$

$$\Delta p = \frac{F_Z}{\pi R^2} = \varepsilon_0 E_Z E_R \frac{2L}{R} \qquad \text{for} \quad \frac{L}{R} \gg 1 : \text{slender channel} \qquad (14)$$

where ε_0 is the permittivity of free space.

Clearly, the electric field strength factor is critical to the performance of EGD generators, and thus the latter can be upgraded by increasing the pressure or using high-dielectric-strength gases. Also, to yield reasonable output for some applications, multistaging may be used, both series and parallel. It is beyond the scope of this brief review to discuss in detail all the aspects of EGD generation—thermodynamic cycles, overall efficiency, power densities, limitations, advantages, and others. For this as well as for the development of practical systems, the reader is referred to the bibliography (Refs. 31–44).

It is relevant to note, however, that, although most of the applications for EGD generators appear to be for relatively low-voltage (<200 kV) systems, future needs and developments may well extend the operating range to the megavolt region. The present limit seems to be about 500 kV with currents up to 300 μA.

Finally in this section, mention should be made of the controlled growth colloidal ion generator developed by Cox.[38] This is similar in operation to the conventional EGD high-voltage generator, in that vapor condenses on charged condensation nuclei under supersaturation conditions after passing through a supersonic nozzle, thus forming the charged colloids. It differs in that expansion, condensation on the ions, and growth of the colloids take place under near-vacuum conditions.

The main applications to date of EGD high-voltage generators have been in the following areas: paint spraying, crop spraying and other coating applications, power supplies for propulsion or instrumentation in space, and electrostatic precipitators. The EGD principle has also been considered for high-voltage power generation, but the feasibility of this application has not yet been demonstrated.

8.4.3.6 *Table-Top Generators*

It is worthwhile, also, in this discussion of ES generators to mention the table-top or instructional generators. This is because students are invariably introduced to ES generators by way of these rather than by the larger machines. Two table-top generators are well known—the Van de Graaff, which has already been described, and the Wimshurst wheel induction machine. Another more recently developed induction machine is that devised by A. D. Moore, which he named the Dirod; to date, he has produced six models. Figure 8.10 shows the first, Dirod I. An insulation disc carries conducting rods, which are mounted parallel or radially, depending on the design. There is a neutral, whose conducting rubber brushes connect opposite rods when they come under the inductors, which then induce opposite charges in the two rods. The rods carry the charges to collector plates which, in turn, are connected to the inductors. Dirods typically have

Figure 8.10. Dirod I, the original Dirod. First of a series developed by A. D. Moore.

maximum voltages to 70 to 95 kV, they are both rugged and reliable, and the first five are fully described in Prof. Moore's book.[45] The latest, Dirod VI, is double, with two complete units on one shaft. The rear unit serves as an exciter, to keep the front unit's inductors fully charged, even when the front unit is shorted. The generator can be reversed instantly, merely by touching the front inductors.

8.4.4 The Variable-Capacitance Generator

The principles of the variable-capacitance machine have been presented in detail by several authors,[6,46–53] the most notable being Trump at M.I.T. The force and simplified energy equations are derived in the next section, where it is shown that the output energy per cycle of capacitance variation is

$$W = \frac{V^2}{4}\frac{(C_m - C_0)^2}{(C_m + C_s)}\ \mathrm{J} \tag{15}$$

Clearly, to raise the power from such a machine it is necessary to:

1. Raise the frequency by increasing the number of poles and the cycles per second. Thus a rotary machine is better than a reciprocating.
2. Increase the capacitance variation $(C_m - C_0)$.
3. Operate with maximum voltage across the capacitor.

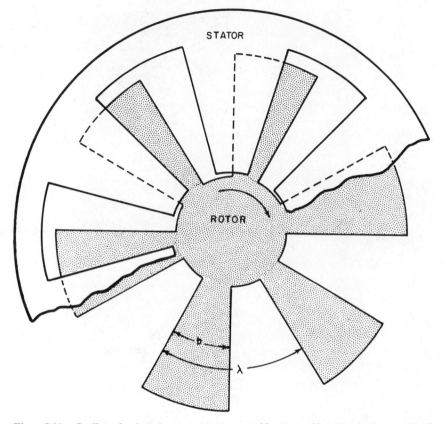

Figure 8.11. Outline of a six-pole rotor stator system. (Courtesy of Ion Physics Corporation.)

The optimum geometry to satisfy items 1 and 2 is to make the electrodes in the form of blades on circular discs. The blades are then stacked in cylindrical form with alternate blades connected to rotor and stator (Figs. 8.11 and 8.12). Increasing the number of poles is limited because a point is reached when the capacitance variation per cycle falls rapidly to a small value. It has been found by electrolytic tank techniques that the best value of b/λ is approximately 0.3 to 0.4. Felici[54] has shown that the blade thickness should be twice the rotor–stator gap; an optimized blade profile is given in Fig. 8.13.

The choice of gap spacing between rotor and stator is complicated by two conflicting results.

1. The breakdown voltage usually increases with gap spacing, but less than proportionately.

Figure 8.12.   Experimental vacuum-insulated generator. (Courtesy of Ion Physics Corporation.)

2. Capacitance is inversely proportional to gap spacing. The maximum speed is limited by mechanical design—for example, by blade strength and bearings. The latter is particularly important if the generator is operated under high vacuum when lubrication presents a serious problem.

The choice of insulating medium is probably compressed gas for terrestrial machines operating at 50 kV and beyond; vacuum is a better choice for space applications and outputs of less than 50 kV. To illustrate machine size and performance, some data are given for Trump's machine and a similar machine built by Ion Physics Corporation (IPC). Both were vacuum insulated.

Trump Machine
   3 rotors, 4 stators of 16-in. O.D.
   16 poles per disc with 0.062-in. gap
   at 3600 rpm the machine delivered 60 W at 25 kV

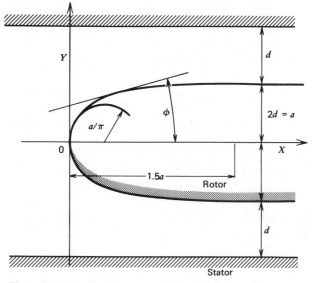

Figure 8.13.   Profile of rotor edge (after Felici).

IPC Machine
   1 rotor, 2 stators of 20-in. O.D.
   40 poles with 1-mm gap
   at 10,000 rpm, the machine delivered 430 W at 24 kV

Two significant points emerge from the IPC work. First, a vacuum-tight rotary seal was developed which operated at 10,000 rpm. Second was the first reported appearance of an electrode area effect, that is, a reduction in dielectric strength with increasing electrode area. This is obviously of great practical significance in large-power ES generators. High-pressure gases have a similar disadvantage.

8.5   FORCES AND ENERGY IN VARIABLE CAPACITANCE MACHINES

8.5.1   Dependence of Force on Capacitance Variation

In general, when the capacitance varies with time, both the charge and the potential difference of the system will vary simultaneously. The complete cycle of an ideal variable-capacitance generator can be approximated to two periods of constant charge and two periods of constant voltage. Since the electrostatic forces we are considering have their origin in the electric charge separation, the force equation should be independent of the external circuit and should depend only on the capacitance, the voltage, and the charge on the system. Consider two extreme cases—one with constant charge on the capacitor and another with the capacitor connected to a constant voltage supply.

8.5.1.1 *Constant Charge*

Consider an isolated capacitor which has impressed on its plates constant charges $+Q$ and $-Q$. The capacitance C will be a function of the geometry and the medium. Let us assume a general position coordinate ζ and calculate the electrostatic forces in this direction by the principle of virtual work.

If the capacitor increases in the ζ direction by a small amount $\Delta\zeta$, there will be an increase in stored energy ΔW_e where, neglecting higher order terms, we have

$$\Delta W_e = \Delta\left(\frac{1}{2}\frac{Q^2}{C}\right) = \frac{\partial\left(\frac{1}{2}\frac{Q^2}{C}\right)}{\partial C}\frac{\partial C}{\partial \zeta}\Delta\zeta$$

$$= -\frac{1}{2}\frac{Q^2}{C^2}\frac{\partial C}{\partial \zeta}\Delta\zeta = -\frac{1}{2}v^2\frac{\partial C}{\partial \zeta}\Delta\zeta \qquad (16)$$

where v is the instantaneous voltage at ζ.

Since there is no external electrical energy source or sink connected to the capacitor, this change of energy can be caused only by mechanical work. Neglecting losses, the work done is, therefore, written as

$$F(\zeta)\,\Delta\zeta = -\frac{1}{2}v^2\frac{\partial C}{\partial \zeta}\Delta\zeta \qquad (17)$$

where $F(\zeta)$ is the total force in the ζ direction:

$$F = -\frac{1}{2}v^2\frac{\partial C}{\partial \zeta}\text{ N} \qquad (18)$$

This implies that the attractive forces in a capacitor depend on the rate of change of capacitance with position and act to increase the capacitance. The work done in changing the capacitance of a system from C_1 to C_2, with fixed Q C on the plates, must, therefore be the change in stored energy; that is,

$$\frac{Q^2}{2}\left(\frac{1}{C_2} - \frac{1}{C_1}\right)\text{J} = \frac{Q}{2}(V_2 - V_1) \qquad (19)$$

when the voltage changes from $V_1 = Q/C_1$ to $V_2 = Q/C_2$. Note that this only depends on the initial and final conditions. Knowledge of this change is useful when calculating the power output from a machine.

8.5.1.2   *Constant Voltage*

Consider the capacitor just described, now connected to an ideal constant voltage supply V V, which can accept or supply charge to the capacitor.

Let the capacitor move by an elemental amount $\Delta\zeta$. The change in stored energy is now

$$\Delta W_e = \Delta\left(\frac{1}{2}\,CV^2\right) = \frac{\partial}{\partial C}\left(\frac{1}{2}\,CV^2\right)\frac{\partial C}{\partial\zeta}\,\Delta\zeta$$

$$= \frac{1}{2}\,V^2\,\frac{\partial C}{\partial\zeta}\,\Delta\zeta\text{ J} \tag{20}$$

The positional change $\Delta\zeta$ will produce a change in capacitance ΔC and in charge Δq where

$$\Delta q = V\,\Delta C = \frac{V\,\partial C}{\partial\zeta}\,\Delta\zeta\text{ C} \tag{21}$$

The voltage supply provides energy as follows:

$$V\,\Delta q = V^2\,\frac{\partial C}{\partial\zeta}\,\Delta\zeta\text{ J} \tag{22}$$

and mechanical work done is $F(\zeta)\,\Delta\zeta$ J.

Neglecting losses, the change in stored energy is equal to the electrical input plus the work done on the system, that is,

$$F(\zeta)\,\Delta\zeta + V^2\,\frac{\partial C}{\partial\zeta}\,\Delta\zeta = \frac{1}{2}\,V^2\,\frac{\partial C}{\partial\zeta}\,\Delta\zeta \tag{23}$$

and from this we have, as before,

$$F(\zeta) = -\frac{1}{2}\,V^2\,\frac{\partial C}{\partial\zeta} \tag{24}$$

It can be seen that for a mechanical input of W J, the stored energy *falls* by W J and $2W$ J of energy is returned to the electrical supply, where we can write

$$W = \tfrac{1}{2}V^2(C_1 - C_2) \tag{25}$$

using the previous notation.

8.5.2   Analysis of a Simplified Typical Generator

Consider the Trump type IIIa generator (Fig. 8.14b) operating around the cycle shown in Fig. 8.15. Neglecting C_A, we can demonstrate that the electrical input from V_1 is $V_1(Q_1 - Q_2)$ and the electrical energy delivered to V_2 is $V_2(Q_1 - Q_2)$; that is, a net electrical output per cycle can be written

$$(V_2 - V_1)(Q_1 - Q_2) = (C_m + C_0)V_1 V_2 - C_m V_1{}^2 - C_0 V_2{}^2 \qquad (26)$$

after substituting from C_m and C_0. This is equal to the net mechanical work done on the system. A more detailed analysis introducing C_A gives the output per cycle as

$$\frac{(C_m - C_0)^2}{C_m + C_A} \frac{V_2{}^2}{4} \, \text{J} \qquad (27)$$

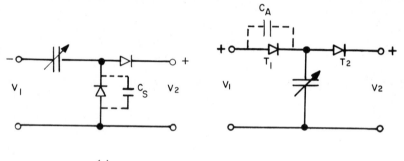

(a)   (b)

Figure 8.14.   Electrostatic generators: (a) type III, (b) type IIIa.

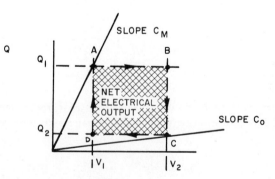

Figure 8.15.   Simplified energy cycle.

If C_m, C_0, and V_2 are fixed, it can be shown that the optimum value of V_1 is given by

$$\frac{V_1}{V_2} = \frac{C_m + C_0 + 2C_A}{2(C_m + C_A)} \rightarrow \frac{1}{2} \qquad \text{if} \quad C_m \gg C_0 C_A \qquad (28)$$

A similar analysis of the Trump type III generator (Fig. 8.14a) reveals that there is *no* net electrical energy per cycle, abstracted from V_1; the electrical output (equal to work done) is expressed as

$$V_2 V_1(C_m - C_0) - V_2{}^2(C_0 + C_S) \qquad (29)$$

and optimum conditions are

$$\frac{V_1}{V_2} = \frac{C_m + C_0 + 2C_S}{C_m - C_0} \rightarrow 1 \qquad \text{if} \quad C_m \gg C_0 C_S \qquad (30)$$

This assumes that the maximum voltage which can be sustained across the variable capacitor (i.e., $V_1 + V_2$) is V_d, the safe maximum operating voltage. The energy per cycle is then given by

$$W = \frac{V_0{}^2}{4} \frac{(C_m - C_0)^2}{(C_m + C_S)} \text{ J} \qquad (31)$$

8.5.3   EFFECT OF DIELECTRIC STRENGTH AND RELATIVE PERMITTIVITY ON THE POWER DENSITY

A variable-capacitance machine can be analyzed from the pressure $\overline{P}$ exerted on a metal surface subjected to an electric field $\overline{E}$, V/m. The pressure $\overline{P}$ is normal to the surface A and is given by

$$\overline{P} = \tfrac{1}{2}\varepsilon_0 \, \varepsilon_r \, E^2 \text{ N/m}^2 \qquad (32)$$

Thus if $\overline{v}$ is the velocity of an elemental area dA on which the pressure is $\overline{P}$, the total power associated with the motion is given by summating over all the active elemental areas; that is,

$$\text{power} = \int \overline{P} \cdot \overline{v} \, dA \qquad (33)$$

This integral is quite difficult to carry out for most geometries. The capacitance variation approach gives the power much more simply and is, therefore, the one normally used. In practice, both approaches are necessary because the breakdown voltage of an electrode system depends on the geometry as well as the medium. Good design attempts to minimize the maximum electric field occurring in the system (see Fig. 8.13).

The effect of dielectric strength and relative permittivity on the power density of a variable-capacitance machine can now be shown by taking the capacitance variation approach. Taking the energy per cycle result derived in Section 8.5.2., we can write

$$W = \frac{V_0{}^2}{4} \frac{(C_m - C_0)^2}{C_m + C_S} \text{ J} \tag{34}$$

Let us, for simplicity, take $C_0 = C_S = 0$, and the effective blade area A m². If the rotor–stator gap is d m and the insulation dielectric strength is E V/m, the maximum gap voltage V is approximately Ed V. Substituting and making the foregoing assumptions, the parallel-plate capacitor formula gives

$$W = \frac{V^2}{4} C_m$$

$$= \frac{E^2 d^2}{4} \frac{\varepsilon_0 \varepsilon_r A}{d} \tag{35}$$

Rearranging, we have

$$W = \tfrac{1}{2}(\tfrac{1}{2}\varepsilon_0 \varepsilon_r E^2) \, Ad$$

$$= \text{constant(max. ES energy density } W_e)\text{volume} \tag{36}$$

The deliverable energy per second per unit volume (i.e. the power density) is proportional to $W_e \cdot np$, where n is revolutions per second and p is poles per disc. We may call $\varepsilon_r E^2$ a figure of merit for the medium.

The following values are given as an indication of the magnitude of the energy density in various insulating media. These should be compared with that in the air gap of an electromagnetic machine of 4×10^5 J/m³ at a flux density of 1 Wb/m². The equivalent electric field is 3 MV/cm for ε_r equal to unity.

Insulating Medium	W_e (J/m³)
Vacuum at 100 kV/cm	500
Kerosene at 150 kV/cm	2×10^3
Compressed gas at 620 kV/cm	1.7×10^4
Ferroelectric at 100 kV/cm ($\varepsilon_r = 1000$)	4×10^5

8.6 THE FUTURE OF ELECTROSTATIC GENERATORS

Electrostatic dc power supplies are now becoming attractive where voltages in excess of 100 kV are required with currents up to tens of milliamperes (i.e., high-voltage dc power supplies up to ca. 20 kW). Outside this range, transformer–rectifier assemblies are superior in availability, performance, and cost. The ES generator has its greatest appeal in the range mentioned, when the supply must be safe, well stabilized, and easily controlled. Unlike the ES generator, EM assemblies require expensive and bulky inductors and capacitors when the dc output must be smooth.

We have seen that the power density of working ES machines is low in comparison with EM machines (a factor of 100 or so) because of the limitation of dielectric strength and ε_r, the relative permittivity. In Section 8.5, it was indicated that for $\varepsilon_r = 1$, a dielectric strength of 3 MV/cm is necessary to achieve comparable gap energy densities. Present insulation research suggests that this stress level is not achievable at voltages higher than 10 kV or so, and it is certainly not possible to reliably operate large electrode areas at anything like this stress level; 200 kV/cm is nearer the truth. However, the energy density can be increased by raising the relative permittivity ε_r. Considerable progress has been made in raising the resistivity and hence the suitability of high ε_r liquids. This approach will probably increase the competitiveness of ES machines for low to medium power applications. However, it will not be an efficient device, owing to the inherent viscous drag forces.

The variable-capacitance multidisc holds long-term promise where higher power and increased efficiency are required. For space applications, the vacuum-insulated version seems to be the best system because of its low weight, its high efficiency, and its ability to operate at high temperature and in a radiation environment. Dry, high-speed bearings, operating under vacuum, have been considerably improved in the 1960s, and high-speed, pressure-to-vacuum seals are now available for ground-based devices. The main problem with this type of machine is the relatively low dielectric strength at voltages above 100 kV. A breakthrough here will have an immediate effect on the machine's competitiveness. Where efficiency is not of prime importance, the use of compressed gas in a variable-capacitance machine results in useful power densities at the expense of windage losses.

Probably for the first time in the history of electrical engineering, industry is extremely interested in ES devices as medium-power high-voltage dc power supplies. This is mainly because of the rapid growth of new industrial processes such as electron beam irradiators, ion implantation, separators, and paint spraying, and new instruments such as the electron and ion microscopes. Space vehicles have a small but ever-increasing high-voltage power requirement. Thus ES machines, which have received limited and inadequate attention in the past, owing to the ready availability of efficient EM power supplies, are now being actively developed. However, if ES machines are to be rapidly developed to the same degree of perfection as EM machines, then a considerable increase in effort and resources will have to be made.

It is appropriate to conclude this chapter by quoting from Professor Felici.[22]

No quick progress can be achieved in a field which does not rouse collective enthusiasm, and, moreover, the actual results always seem, to the scientifically minded critic, to fall far below reasonable expectancies, . . . This sober appraisal should be kept in mind when comparing present-day electrostatic machines, which are still waiting for their Watt or their Siemens, to more fully developed kinds of equipment. It also means, however, that most of the potentialities of electrostatic generation are still untapped, . . .

REFERENCES

1. G. W. Sutton, *Direct Energy Conversion*, McGraw-Hill, New York, 1967.

2. N. J. Felici, *Direct Current*, **4**, No. 7 (December 1959).

3. —— in *La Physique des Forces Electrostatiques*, CNRS Colloquium, Grenoble, Section 75, 1960.

4. F. Ollendorf, *Arch. Electrotech.*, **12**, 297 (1923).

5. L. S. Polotovskiy, "High Voltage Direct Current Capacitance Machines," Moscow, 1960. [English transl. from U.S. Department of Defense, Armed Services Technical Information Agency. AD264 505 (1961)].

6. J. G. Trump, "Vacuum Electrostatic Engineering," Ph.D. thesis, Massachusetts Institute of Technology, Cambridge, 1933.

7. N. J. Felici, *Elektrostatische Hochspannungs-Generatoren*, G. Braun, Karlsruhe, 1957.

8. M. J. Mulcahy, Ed., *Proc. High Voltage Technology Seminar*, Ion Physics Corporation, Boston, 1969.

9. N. J. Felici, *Electron. Power (IEE)*, 169 (May 1965).

10. M. Pauthenier, et al., *Comp. Rend.* **201**, 1332 (1935); **202**, 1915, (1936); *J. Phys. Radium.*, **8**, 193 (1937); **3**, 590 (1932); **6**, 257 (1935); *Rev. Gen. Elect.*, **45**, 583 (1939).

11. A. L. Cox, and S. Harrison, in *La Physique des Forces Electrostatiques*, CNRS Colloquium, Grenoble, Section 9, 1960.

12. M. Moreau-Hanot, *Compt. Rend.*, **206**, 1168 (1938).

13. M. Morand, et al., *Compt. Rend.*, **234**, 2450 (1952).

14. R. J. Van de Graaff, et al., *Rep. Progr. Phys.*, **11**, 1 (1946).

15. E. S. Shire, Institute of Physics Symposium, London, 30 (1950).

16. R. L. Fortescue, *Progr. Nucl. Phys.*, **21** (1950).

17. L. W. Alvarez, *Rev. Sci. Instr.*, **22**, 705 (1951).

18. R. J. Van de Graaff, *Nucl. Instr. Meth.* **8**, 195 (1960).

19. H. F. Anton, *Proc. Symposium of Electrostatic Energy Conversion*, University of Pennsylvania, Section 6 (May 1963).

20. Technical Report, Société Anonyme des Machines Electrostatiques, Grenoble, France.

21. R. Morel, in *La Physique des Forces Electrostatiques*, CNRS Colloquium, Grenoble, Section 137, 1960.

22. N. J. Felici, "Cylindrical Electrostatic Generators," in *Radiation Sources*, Pergamon Press, New York, 1964.

23. P. E. Secker, K. W. Starr, and G. D. Read, "A Liquid Immersed High Voltage Electrostatic Generator," 3rd National Electrostatic Conference, May 1971.

24. Technical Report, Industrial Development Co., Bangor, North Wales.

25. A. P. Chattock, *Phil. Mag.*, **48**, 401 (1899).

26. R. E. Vollrath, *Phys. Rev.*, **42**, 298 (1932).

27. O. M. Steutzer, *J. Appl. Phys.*, **30**, 984 (1959); **31**, 136 (1960).

28. ———, *Rev. Sci. Instr.* **32**, 16 (1961).

29. P. E. Secker, and J. F. Hughes, *Proc. IEE*, **116**, 1785 (1969).

30. J. F. Hughes, and P. E. Secker, *Electron. Lett.*, **2**, 174 (1966).

31. M. C. Gourdine, E. Barreto, and M. P. Khan, "On the Performance of EGD Generators," *Proc. 5th Symposium on Engineering Aspects of MHD*, 1964.

32. B. Khan, and M. C. Gourdine, *AIAA J.*, **2**, No. 8, 1423 (1964).

33. P. L. Cowan, M. C. Gourdine, and D. H. Malcolm, IEEE Winter Power Conference Paper 68-CP-137, 1968.

34. A. M. Marks, Heat–Electrical Energy Conversions through the Medium of a Charged Aerosol, U.S. Patent 2,638,555, 1953.

35. A. Marks, E. Barreto, and C. K. Chu, *AIAA J.*, **2**, 45 (1964).

36. M. O. Lawson, "Performance Characteristics of EFD Energy Conversion Processes Employing Viscous Coupling," VIth AGARD Combustion and Propulsion Colloquium, France, 1964.

37. K. T. Whitby, *Rev. Sci. Instr.*, **32**, 1355 (1961).

38. A. L. Cox, *AIAA J.*, **1**, 2491 (1963).

39. E. Barreto, and M. J. Mulcahy, *J. Geophys. Res.*, **70**, No. 6, 1303 (1965).

40. E. L. Collier, and M. C. Gourdine, *AIAA J.*, **6**, No. 12 (1968).

41. J. A. Decaire, and M. O. Lawson, "EFD Power Generation, Trends and Expectations," *Proc. 8th Symposium on Engineering Aspects of MHD, Stanford*, 1967.

42. *Proc. International Symposium on EHD*, M.I.T., 1969.

43. H. H. Woodson, and J. R. Melcher, *Electromechanical Dynamics*, Vol. III, Wiley, New York, 1968, Chapter 12.

44. E. L. Collier, M. C. Gourdine, and D. H. Malcolm, *Ind. Eng. Chem.*, **58**, No. 12, 26 (1966).

45. A. D. Moore, *Electrostatics*, Doubleday, Garden City, N.Y., 1968; a Doubleday Anchor Book S57, 240 pages.

46. J. G. Trump, "Electrostatic Sources of Electric Power," *Elect. Eng.*, **66**, 525 (June 1947).

47. A. F. Joffe, et al., "The Electrostatic Generator," *J. Tech. Phys. (USSR)*, **2**, 243 (1940).

48. F. Ollendorf, "Über Kapazitätsmaschinem," *Arch. Electro-tech.*, **12**, 297 (April 1923).

49. A. W. Simon, "Theory of Electrostatic Generator," *J. Franklin Inst.*, **237**, (March 1944).

50. J. G. Trump, in *La Physique des Forces Electrostatiques*, CNRS Colloquium, Grenoble, Section 23 1960.

51. A. S. Denholm, J. G. Trump, and A. J. Gale, *Progress in Astronautics and Rocketry*, Vol. 3, Academic Press, 1961, p. 745–766.

52. F. J. McCoy, C. N. Coenraads, and A. S. Denholm, "Design of Electrostatic Generators for Operation in Space," ARS Reprint 2052, 1961.

53. C. N. Coenraads, A. S. Denholm, J. Lavell, and F. J. McCoy, "Electrostatic Generators for Space," *Progr. Astronaut. Aeronaut.*, **11**, 917 (1963).

54. R. Morel, and N. J. Felici, "Contribution a l'Étude Rationnelle des Machines Electrostatiques," *Ann. Univ. Grenoble*, **23**, 1971 (1947) [transl. from SLA Translation Center].

CHAPTER 9

Electrostatic Precipitation

MYRON ROBINSON
Health and Safety Laboratory
U.S. Atomic Energy Commission
New York, New York

9.1 OVERVIEW

9.1.1 Precipitator Fundamentals

The electrostatic precipitator is a device for removing fine suspended particles from a gas by charging the particles in a corona discharge and separating them from the gas by means of an electric field. There are an estimated 4000 industrial precipitators in the United States today, treating a greater volume of contaminated gas than any other type of particulate collector. Major precipitator installations are individually large enough to treat upward of 10^6 ft³/min of flue gas, and the quantity of coal fly ash alone that is precipitated annually in this country—primarily in electric-power generation—is of the order of 2×10^7 tons.

Fly-ash collection in electric-power generating stations accounts for about 60% of total precipitator capacity. The balance is principally divided among steel and cement production ($\sim 10\%$ each) and the processing of paper and nonferrous metals ($\sim 7\%$ each). Precipitation in the chemical industry, including, for example, sulfuric acid manufacture, phosphate processing, petroleum refining, fuel-gas detarring, and carbon-black collection, is responsible for most of the remainder ($\sim 6\%$).

Precipitators are available in a wide range of sizes, depending on application. A new concept of commercial importance is the modular or package unit, a development that has made the electrostatic precipitator, traditionally a high capital-cost item, available to some of the smaller producers of particulate polutants. Modular units, typically handling 10,000 to 25,000 ft^3/min, can be installed in segments and expanded to meet increased gas flows or efficiency demands. Still smaller designs serve as indoor air cleaners, and miniature units small enough to hold in the hand find application as aerosol samplers.

Electrostatic precipitators are commonly built in one of two basic forms (Fig. 9.1). In the simpler device (Fig. 9.1a), the precipitator comprises a grounded cylinder designated, in accordance with its function, the collecting electrode; coaxial with it, there is a high-potential wire, the corona-

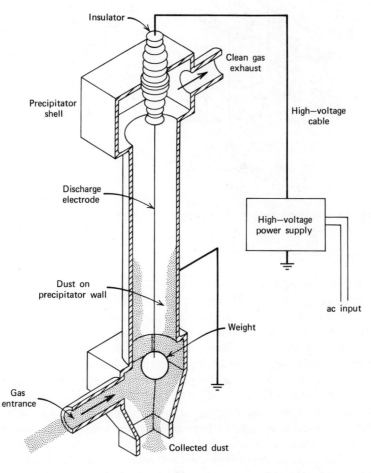

(a)

Figure 9.1. Single-stage electrostatic precipitators: (a) Tubular. (Used with permission of Addison-Wesley Publishing Company.)

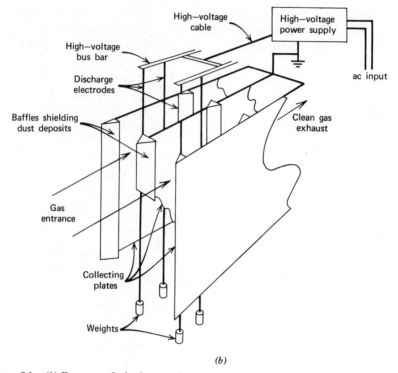

Figure 9.1. (*b*) Duct type. In both cases the corona discharge and precipitating field extend over the full length of the apparatus.

discharge electrode. An alternative basic design (Fig. 9.1*b*) consists of two grounded parallel plates (the collecting electrodes) together with an array of parallel discharge wires mounted in a plane midway between the plates. Gas with suspended solid or liquid particles is passed either through the tube or between the plates. If a sufficient difference of potential exists between the discharge and collecting electrodes, a corona discharge will form about the wire(s). The corona serves as a copious source of unipolar ions of the same sign as that of the discharge electrode(s). The ions, in migrating across the interelectrode space under the action of the impressed electric field, in part attach themselves to aerosol particles moving with the gas through the system. The charged particles, in turn, are attracted to the collecting surface; they adhere, and so are extracted from the gas stream. Solid particles build up a layer on the collecting surface, from which the accumulated deposit is dislodged periodically (by rapping or flushing) and allowed to fall into a hopper or sump for subsequent removal. Liquid particles (droplets) form a

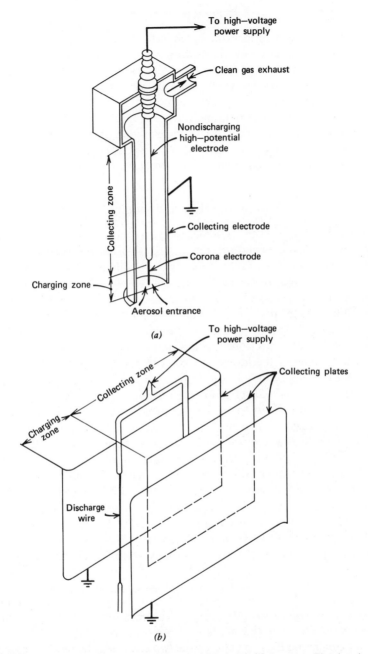

Figure 9.2. Two-stage electrostatic precipitators: (*a*) Tubular, (*b*) duct type. The charging zone in each model is confined to the region about the corona wire. The downstream collecting zone provides an electrostatic field in which the previously charged particles are precipitated on the collecting surface.

film on the collecting surface, the precipitated liquid then dripping off into a sump. The cleaned gas is discharged at the outlet of the precipitator.

Various modifications of the "classic" designs are possible. The two-stage precipitator, for example, which finds its principal use in domestic and commercial air-cleaning and aerosol-sampling applications, divides the particle-charging and precipitating functions into adjacent segments of the apparatus (Fig. 9.2). However, single-stage models of the varieties illustrated in Fig. 9.1 have so far found almost universal application in the cleaning of contaminated industrial gases by electrostatic precipitation.

9.1.2   Historical Origins

Although we know that the existence of electrostatic forces of attraction had been noted in classical antiquity, the earliest reported observations in aerosol electrostatics date from about 1600, when William Gilbert observed that frictionally charged dielectrics "entice smoke sent out by an extinguished light." Closely associated with studies of smoke and other particulate suspensions was the discovery of the man-made corona discharge after 1670 by von Guericke. Among the early investigators (and frequent rediscoverers) of electroseparation, electrodeposition, and the high-voltage discharge are to be counted some of the foremost names in the history of science: Boyle, Newton, Franklin, Coulomb, and Faraday. By their work and by the efforts of others, the essential phenomena of fine-particle electrostatics were demonstrated in principle long before an advancing technology provided the accessory equipment required for successful commercial application.

Modern electrostatic precipitation is closely connected with the name of Frederick G. Cottrell. In 1907, by adapting the newly developed mechanical rectifier and high-voltage transformer to his needs, Cottrell became the first to build a workable electrostatic precipitator. With the subsequent development of the fine-wire discharge electrode, the demonstration of the general superiority of negative corona over positive, and continuing improvements in rectifying equipment to give high voltages at reasonable currents, the three basic ideas to dominate the art, even today, were soon established.

9.1.3   Mechanisms of Corona Formation

Consider a system of two juxtaposed electrodes immersed in a gas such as atmospheric air, one electrode having a much smaller radius of curvature than the other, and the gap between the two electrodes being large compared with the radius of the smaller. Typical examples of such a configuration are a point opposite a plane, a coaxial wire and pipe, and a wire parallel to a plane. As the potential difference between the electrodes is raised, the gas near the more sharply curved electrode breaks down at a voltage less than the spark-breakdown value for the gap length in question. This incomplete breakdown, called corona, appears in air as a highly active region of glow, extending into the gas a short distance beyond the discharge electrode surface. The corona on a positive wire has the appearance of a more or less uniform sheath covering the wire surface facing the opposite electrode. On a negative wire, the corona is concentrated in tufts of glowing gas spaced at intervals along the wire.

The initiation of the corona discharge requires the availability of free electrons in the gas in the region of the intense electric field surrounding the discharge electrode. In the case of a negative discharge wire, these free electrons gain energy from the field to produce positive ions and other electrons by collision. The new electrons are in turn accelerated and produce further ionization, thus giving rise to the cumulative process termed an electron avalanche. The positive ions formed in this process are accelerated toward the wire. By bombarding the negative wire and giving up relatively high energies in the process, the positive ions cause the ejection from the wire surface of secondary electrons necessary for maintaining the discharge. In addition, high-frequency radiation originating in the excited gas molecules within the corona envelope may photoionize surrounding gas molecules, likewise contributing to the supply of secondary electrons. Electrons of whatever provenance are attracted toward the anode and, as they move into the weaker field away from the wire, tend to form negative ions by attachment to neutral oxygen molecules. These ions, which form a dense unipolar cloud filling most of the interelectrode volume, constitute the only current in the entire space outside the region of corona glow. The effect of this space charge is to retard the further emission of negative charge from the corona and in so doing, limit the ionizing field near the wire and stabilize the discharge. However, as the voltage is progressively raised, complete breakdown of the gas dielectric, that is, sparkover, eventually occurs.

Should the discharge electrode be positive, the physical mechanisms are quite different. In this case, electrons formed by chance ionizing events in the high-field space in the neighborhood of the wire establish electron avalanches that move in toward the wire. As with the negative discharge, these avalanches are responsible for the visible, highly ionized state of the gas close to the wire. Positive ions formed in the avalanches are left behind by the electrons and drift from the region of the wire into a field of progressively decreasing intensity. The ions are thus incapable of acquiring sufficient energy between molecular collisons to produce either significant ionization in the gas or electron ejection at the cathode surface. Photons originating in the corona glow presumably give rise to ionization in the gas and perhaps cause photoemission at the cathode. In this manner, the secondary electrons

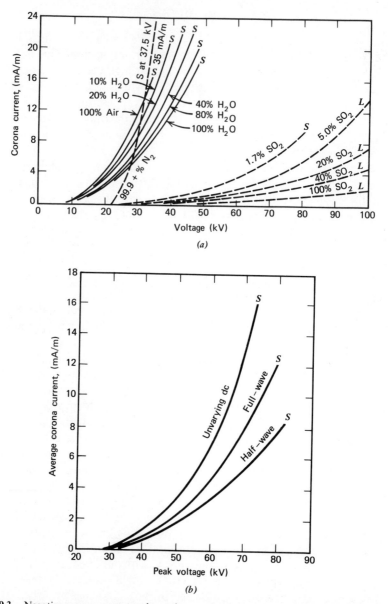

Figure 9.3. Negative corona current–voltage characteristics of wire–tube systems: L designates the limiting voltage of the power supply and S sparkover. (a) The effect of adding to the gas a constituent of relatively low ion mobility: solid curves, water vapor in dry air (tube diameter, 7.6 cm; wire diameter, 0.25 mm; temperature, 200°C; pressure, 1 atm); broken curves: sulfur dioxide in nitrogen (tube diameter, 15.2 cm; wire diameter, 2.8 mm; room temperature and pressure). In fly-ash precipitation, the sulfur dioxide content of the gas is of the order of 0.1% and, in the presence of relatively large proportions of electronegative oxygen and water, causes no noticeable electrical effect in the gas. Conversely, in acid–mist precipitation, the

on which a self-sustaining discharge depends are generated. Nevertheless, virtually all the current in the space external to the corona envelope is carried by the positive ions.

When the corona-starting voltage is reached in either polarity, the current increases slowly at first and then more rapidly with increasing voltage. As spark-over is approached, relatively small increments of voltage give sizable increases in current (see, e.g., Fig. 9.3).

A negative discharge under conditions prevailing in electrostatic precipitators generally (although not invariably) yields a higher current at a given voltage than positive, and the sparkover voltage, which sets the upper limit to the operating potential of the precipitator, is also usually greater. The only common exceptions to the use of negative corona occurs in domestic air-cleaning applications. For even though increased current and voltage alike make for superior efficiency of precipitation, negative corona usually produces physiologically objectionable ozone in greater quantities.

The foregoing description of the corona has been given with specific reference to the discharge in air at approximately room conditions. The same pattern remains essentially valid for a much broader range of temperatures and pressures, and for a variety of gases other than air.

When an electron collides with a neutral gas molecule it may attach itself and form a negative ion. The likelihood of doing so depends on the so-called electronegative nature of the gas and on the energy of the electron, slower moving electrons remaining for a longer time within the range of the atomic field of the gas molecule. In addition, the probability of electron attachment is markedly affected by the presence of certain gases and vapors as impurities. The inert gases and hydrogen, if *very* pure, do not form negative ions at all by electron attachment. Negative corona is impossible in such cases, sparkover alone occurring as the voltage is raised. Electrostatic precipitation is, however, feasible in nonattaching gases. We circumvent the difficulty merely by employing positive corona. In practice, nominally nonattaching gases of commercial grade are likely to contain sufficient electronegative impurities to obviate the problem.

No references to the literature are cited in the text of this chapter, but further details or sources can be found in the References, which constitute a bibliography of the principal works on the subject in English.

---

sulfur dioxide content may be about 6%; in this case, the effect of sulfur dioxide cannot be ignored, even in the presence of appreciable oxygen. (*b*) Current–voltage relations in room air as a function of voltage waveform. The electrodes are a concentric tube and wire, diameters 15.2 cm and 2.8 mm, respectively. In agreement with field observations, sparking voltage generally increases with waveform in the order unvarying dc, full-wave, and half-wave. This phenomenon is presumably due to the presence of dust on the positive collecting surface, for under exceptionally clean conditions all waveforms result in the same sparkover voltage.

9.2   CURRENT—VOLTAGE—FIELD RELATIONS

9.2.1   Wire-Pipe Geometry

9.2.1.1   *Particle-Free Condition*

The mathematical theory of the corona discharge relevant to the design and performance of electrostatic precipitators is developed here. The simplest precipitator geometry for purposes of analysis is the coaxial wire-cylinder combination. For this arrangement, the electric field E (V/m) is given in terms of the potential V (V) by

$$E = -\frac{\partial V}{\partial r} \tag{1}$$

where r (m) is the radial distance from the tube axis. In the case at hand, Poisson's equation, the fundamental electrostatic relation between field and charge, can be written

$$\frac{1}{r}\frac{d}{dr}(rE) = \frac{\rho_i}{\varepsilon} \tag{2}$$

where ρ_i (C/m^3) is the ion space-charge density and ε (F/m) is the permittivity of the gas. The permittivity may be expressed as

$$\varepsilon = \kappa\varepsilon_0 \tag{3}$$

where ε_0 is the permittivity of free space (8.85×10^{-12} F/m) and κ (dimensionless) is the relative dielectric constant. Since this quantity is essentially unity for gases for all realistic conditions of precipitation, ε_0 is hereafter used as the gas permittivity. Prior to the onset of corona ρ_i is zero. Taking· V_0 as the potential of the wire surface r_0 and grounding the tube ($V = 0$ at $r = r_1$), integration yields the electrostatic field

$$E = \frac{V_0}{r \ln(r_1/r_0)} \tag{4}$$

It is assumed throughout these calculations that the passive electrode is at ground potential. This procedure entails no loss of generality; it corresponds to usual practice and, moreover, permits the potential of the discharge electrode to be equated to the voltage across the system.

The motion of an ion acted on by an electric field is impeded by repeated collisions with gas molecules in its path. The average drift velocity of the ion v_i (m/sec) is related to the field by the equation

$$v_i = bE \tag{5}$$

where the constant of proportionality b $(m^2/sec\text{-}V)$ is called the mobility. Although experimental mobilities are tabulated in the literature, the exact quantity to use in an actual situation is subject to uncertainty, for impurities have a pronounced effect in reducing the value. This is particularly true of pure nonattaching gases, whose negative mobility is the electron mobility.

In accordance with the theoretical requirement that mobility be proportional to the mean free path, experiments indicate that mobility is almost inversely proportional to the gas density over a wide range of temperature and pressure. Thus we have

$$b = \frac{b_0}{\delta} \tag{6}$$

where b_0 is the mobility at standard conditions and δ is the gas density relative to standard conditions (i.e., $\delta = 1$ at $0°C$ and 1 atm). The relative gas density is found from

$$\delta = \frac{273 p_a}{T + 273} \tag{7}$$

where T $(°C)$ is the temperature and p_a (atm) is the pressure.

Above the corona threshold, the static expression for the field (Eq. 4) remains a valid approximation only as long as the interelectrode space charge, whether due to ions or charged dust particles, is insignificant compared with the surface charge on the electrodes. When this requirement is not satisfied—the usual case in an electrostatic precipitator—we must introduce a nonzero space charge density into Eq. 2. This may be done by expressing the current per unit length of conductor j_l (A/m), as follows:

$$j_l = 2\pi r \rho_i b E \tag{8}$$

Thus, eliminating ρ_i from Eq. 2 and integrating gives

$$E = \left[\frac{j_l}{2\pi\varepsilon_0 b} + \left(\frac{r_0}{r} \right)^2 \left(E_c^{\,2} - \frac{j_l}{2\pi\varepsilon_0 b} \right) \right]^{1/2} \tag{9}$$

where E_c represents the minimal field intensity required to effect local breakdown of the gas close to the wire surface $(r \simeq r_0)$. This corona-starting field E_c is characteristic of the gas and wire radius but is independent of the size or shape of the outer electrode (Section 9.2.3). For sufficiently large r and j_l, Eq. 9 simplifies to the convenient approximation

$$E = \left(\frac{j_l}{2\pi\varepsilon_0 b} \right)^{1/2} \tag{10}$$

which is independent of r. Although this formula is widely quoted in the literature, it should be noted that it may seriously underestimate the field at the relatively low discharge currents commonly encountered in industrial usage (see Fig. 9.5). At the higher current levels which are generally the rule in laboratory work, Eq. 10 may give tolerable accuracy.

The corona current–voltage relation is found by integrating Eq. 9. Neglecting second-order terms, we find

$$\frac{V_0 - V_c}{V_c} \ln\left(\frac{r_1}{r_0}\right) = (1 + \phi)^{1/2} - 1 - \ln\frac{1 + (1 + \phi)^{1/2}}{2} \tag{11}$$

where V_c (V) is the corona-starting potential related to the starting field E_c by the following form of Eq. 4:

$$E_c = \frac{V_c}{r_0 \ln(r_1/r_0)} \tag{12}$$

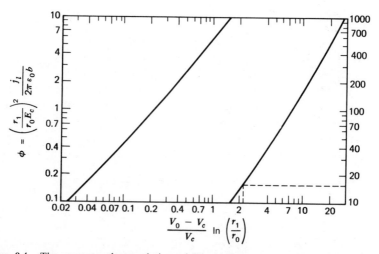

Figure 9.4. The current–voltage relation of Eq. 11 in terms of dimensionless variables. *Example*: Estimate the negative corona current to be expected in clean dry air at 25°C and 1 atm at a voltage of 50 kV using a 12-ft long wire-tube precipitator of respective diameters 109 mils and 9 in. The mobility of singly charged negative ions in dry air at standard conditions is 2.1×10^{-4} m²/V-sec.

From the known quantities $r_0 = 1.38 \times 10^{-3}$ m, $r_1 = 0.114$ m, $V_0 = 5.0 \times 10^4$ V, and $V_c = 3.36 \times 10^4$ V (example of Section 9.2.3), calculate the abscissa $[(V_0 - V_c)/V_c]\ln(r_1/r_0) = 2.16$. As shown by the dashed lines, this value corresponds to the ordinate $\phi = 15.6$. Setting $\varepsilon_0 = 8.85 \times 10^{-12}$ F/m, $b = 2.3 \times 10^{-4}$ m²/(V-sec) (Eq. 6), and $E_c = 54.9 \times 10^5$ V/m (example of Section 9.2.3), we then have $j_l = 2\pi\varepsilon_0 b\theta(r_0 E_c/r_1)^2 = 0.88 \times 10^{-3}$ A/m. The current in a 12-ft (3.67 m) system is, therefore, 3.2 mA.

The quantity ϕ is defined as

$$\phi = \left(\frac{r_1}{E_c r_0}\right)^2 \frac{j_l}{2\pi\varepsilon_0 b} \tag{13}$$

The most direct way of solving Eq. 11 is to plot the dimensionless variables $(V_0 - V_c)/V_c \ln(r_1/r_0)$ and ϕ against each other, as in Fig. 9.4, and read off the appropriate quantities.

For low currents near the corona threshold (i.e., $\phi \ll 1$), Eq. 11 may be represented by the simple approximation

$$j_l = \frac{8\pi\varepsilon_0 b}{r_1^2 \ln(r_1/r_0)} V_0(V_0 - V_c) \tag{14}$$

9.2.1.2   Particle Space Charge

The cylindrical corona field described by Eq. 9 makes no allowance for the presence of suspended charged particles in the interelectrode space. Such particles contribute to the total space charge, and owing to their low drift velocity relative to that of the gas ions, the space-charge effect due to the particles is often much greater than that of the ions, at least near the precipitator inlet.

The influence of particle space charge on the field may be approximated by assuming that the particle concentration N_p m^{-3} is constant over a given cross section of the precipitator, although decreasing with downstream distance. The aerosol's specific surface, that is, the particle surface per unit volume of gas S (m^2/m^3), is

$$S = 4\pi a^2 N_p \tag{15}$$

where a (m) is the radius of assumed spherical particles. Section 9.3.1 indicates that a particle charged by the ion-bombardment process (the dominant particle-charging mechanism in most applications) acquires a charge of $4\pi\varepsilon_0 pEa^2$ (C), where p (dimensionless) is given by Eq. 33. The charged aerosol thus increases the initially present ion space-charge density ρ_i by an amount ρ_p (C/m^3)

$$\rho_p = \varepsilon_0 pES \tag{16}$$

Including ρ_p in Eq. 2 and solving gives

$$E = \left\{ \left[\left(\frac{r_0}{r}\right)^2 \left(E_c^2 + \frac{j_l}{2\pi\varepsilon_0 b}\right) + \frac{j_l}{4\pi\varepsilon_0 b(pSr)^2} \right] e^{2pSr} - \frac{j_l}{4\pi\varepsilon_0 b} \left[\frac{2}{pSr} + \frac{1}{(pSr)^2} \right] \right\}^{1/2} \tag{17}$$

If consideration is restricted to relatively large currents, to the region $r \gg r_0$, and to common situations in which the dimensionless term pSr is of the order of a few tenths or less, we have

$$E = \left[\frac{j_l}{2\pi\varepsilon_0 b} \left(1 + \frac{2pSr}{3}\right) \right]^{1/2} \simeq \left(\frac{j_l}{2\pi\varepsilon_0 b} \right)^{1/2} \left(1 + \frac{pSr}{3}\right) \tag{18}$$

The electric field as a function of radial distance, with and without space-charge effects, is shown in Fig. 9.5. It is seen that space charge works

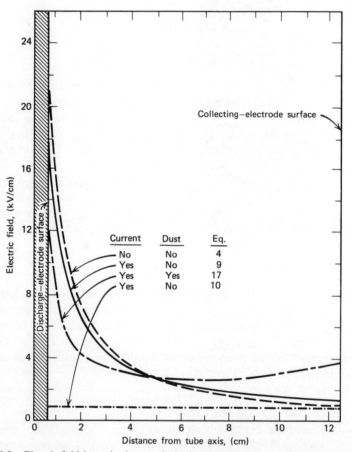

Current	Dust	Eq.
No	No	4
Yes	No	9
Yes	Yes	17
Yes	No	10

Figure 9.5. Electric field intensity in a typical wire-pipe precipitator. The current density is $j_l = 82\ \mu A/m$ (corresponding to a potential difference of 40 kV) and the specific surface is $S = 8.3\ m^2/m^3$ (corresponding to a dust concentration of 18 g/m³ for particles of diameter 6.6 μm and specific gravity 2). (Used with permission of H. J. Lowe, Institute of Physics and Physical Society.)

to decrease the field near the wire and raise it near the tube. Indeed, with adequate space charge, the minimum field no longer occurs at the collecting electrode surface. The reduction in field at the wire comes about because the space charge partly shields the discharge wire from the oppositely charged tube. Initiation of corona on a wire entails a critical corona-starting field at the wire surface (Section 9.2.3). In order to attain this field in the presence of space charge, it is necessary to impress a higher voltage across the electrodes than would otherwise be required.

Space-charge effects can be considered from an alternative point of view which also yields useful results. To a first approximation—and unlike the case of Eq. 8—both particle and ion space-charge densities, ρ_p and ρ_i, respectively, are assumed to be independent of position in the tube cross section. Poisson's equation, upon integration, yields

$$E = \frac{V_c}{r \ln(r_1/r_0)} + \frac{(\rho_i + \rho_p)}{2\varepsilon_0} \frac{(r^2 - r_0^2)}{r} \tag{19}$$

The field at any point r in the tube is the sum of two components: the electrostatic field prevailing in the presence of space charge, and the supplementary space-charge field.

Integration of Eq. 19 gives the following potential of the wire:

$$V_0 = V_c + \left(\frac{\rho_i + \rho_p}{4\varepsilon_0}\right) r_1^2 \tag{20}$$

where the potential of the tube is zero and r_0 is neglected relative to r_1. Expressing ρ_i in terms of j_l (Eq. 8), the preceding equation can be written as

$$V_0 = V_c + \frac{\rho_p}{4\varepsilon_0} r_1^2 + \frac{j_l \ln(r_1/r)}{8\pi\varepsilon_0 b V_0} \tag{21}$$

Except for the particle space-charge term, Eq. 21 is identical to the low-current approximation of Eq. 14. Evidently, the presence of particle space charge raises the apparent corona-onset potential from the particle-free level V_c to a higher effective value

$$V'_c = V_c + \frac{\rho_p}{4\varepsilon_0} r_1^2 \tag{22}$$

The particle space-charge density may be calculated using the relations of Section 9.3.

In most applications, the contribution of ρ_p to current is inconsequential.

9.2.2   Wire–Plate Geometry

9.2.2.1   *Particle-Free Condition*

A solution of Poisson's equation for a wire–plate electrode system (Figs. 9.1*b* and 9.6) along the lines described earlier for the particle-free wire–tube configuration is beset by formidable mathematical difficulties. Considerable simplification results, however, if we assume that the current is low and that the resultant alteration of the potential by the space charge can be represented by an additive correction analogous to that obtaining in Eq. 21. It can then be shown that

$$j_l = 4cj_s = \frac{4\pi\varepsilon_0 b}{s^2 \ln(d/r_0)} V_0(V_0 - V_c) \tag{23}$$

where j_s (A/m^2) is the average current density at the plate, $2c$ (m) is the wire-to-wire spacing, and s (m) is the wire-to-plate spacing (wires measured from centers; see Fig. 9.6). The parameter d (m) is represented closely by

$$d = \frac{4s}{\pi} \quad \text{for} \quad \frac{s}{c} \le 0.6 \tag{24}$$

$$d = \frac{c}{\pi} e^{\pi s/2c} \quad \text{for} \quad \frac{s}{c} \ge 2.0 \tag{25}$$

and Fig 9.7 for $0.6 < s/c < 2.0$.

The corona-starting voltage V_c in Eq. 23 is

$$V_c = r_0 E_c \ln \frac{d}{r_0} \tag{26}$$

where E_c is given by Eq. 31.

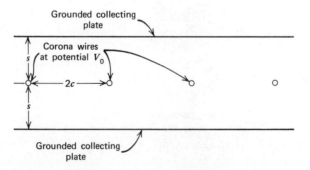

Figure 9.6.   Wire-plate (duct-type) electrode arrangement.

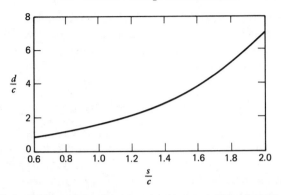

Figure 9.7. Curve for determining the parameter d in terms of the dimensions of a wire-plate precipitator (Eqs. 23ff). (Used with permission of the Institute of Electrical and Electronic Engineers.)

Consideration of the foregoing relations leads to the following conclusions:

1. As the wire-to-wire spacing increases, j_l tends to become constant.

2. There is a wire-to-wire spacing, depending on the other dimensions, for which j_s is a maximum. This comes about because reduced spacing raises the starting voltage and so tends to lower the current, but at the same time more wires are introduced between the plates, tending to raise j_s. The current maximum is broad and the associated electrode dimensions are not critical.

3. The value of j_l depends on wire-to-plate spacing inversely as a power lying between 2 and 3.

9.2.2.2 Particle Space Charge

Drawing an analogy between Eqs. 14 and 23 governing wire–tube and wire–plate systems, respectively, in the latter case the ion space charge is given by the expression

$$\rho_i = \frac{j_l \ln(d/r_0)}{2\pi b V_0} \tag{27}$$

Assuming uniformly distributed particle space charge, the following relation, corresponding to Eq. 21, can be written:

$$V_0 = V_c + \frac{\rho_p}{2\varepsilon_0} s^2 + \frac{j_l \ln(d/r_0)}{4\pi\varepsilon_0 b V_0} \tag{28}$$

where the apparent corona starting voltage has been raised to

$$V_c' = V_c + \frac{\rho_p}{2\varepsilon_0} s^2 \tag{29}$$

Comparison of Eqs. 22 and 29 reveals that for equal tube diameter and plate-to-plate spacing, particle space charge elevates the effective duct starting voltage twice as much as for the tube, and this increase is independent of wire-to-wire spacing.

9.2.3   Corona Onset and Sparkover

In the absence of an adequate theory of breakdown in nonuniform fields, it is impossible to calculate from atomic data either the corona-starting field and voltage or the sparkover field and voltage. Furthermore, accuracy of measurement in the laboratory, and even more so in the field, is marred by surface asperities and dust deposits on the electrodes. Surface irregularities in combination with misaligned electrodes inevitably lower both corona-onset and sparkover levels in practical installations. Maximum sparkover voltages to be anticipated are conveniently determined in the laboratory; practical values, on the other hand, are best established through observation of similar installations. One indication of the uncertainties encountered is given by the empirical relation

$$V_{sn} = V_{s1} - C_1 \ln n \tag{30}$$

where V_{s1} and V_{sn} are the respective sparkover voltages for 1 and n wires energized by a single power supply, and C_1 is a constant. The physical basis of Eq. 30 becomes clear when it is remembered that (1) the instantaneous potential of all parallel-connected wires is set by the reduced potential of whatever wire in the system is then experiencing sparkover, and (2) the greater the total length of the precipitator, the higher is the probability of occurrence, somewhere in the system, of those conditions tending to lower the sparkover potential.

Attempts to compensate for poor performance by adding wires—the new wires also being fed by the original electrical set—are apt to be self-defeating. For effective precipitation the length of corona wire energized by a single power supply must be limited. It is primarily this consideration that leads to the general practice of dividing large precipitators into sections, each section energized by an independent supply.

Table 9.1 gives rule-of-thumb values of sparkover voltage and associated variables useful for orientation purposes. The sparking potential coincides with the cyclic peak of the imperfectly filtered unidirectional waveform.

Peak voltage gradients in the tabulated examples are typically of the order 4 to 6 kV/cm averaged across the interelectrode gap. Both corona-starting and sparkover voltages are temperature and pressure dependent. Electrical breakdown, whether partial (corona) or complete (sparkover), depends on the likelihood of electrons accelerating to ionizing energies in the space of

Table 9.1. Electrical Characteristics for Commmon Precipitator Applications

Application	Tube Diameter or Duct Width (cm)	Sparkover Voltage (kV)	Average Corona Current (mA/100 m^2)	Average Corona Energy Density (W-sec/m^3)
Fly ash	20–25 (duct)	40–60	10–50	100–250
Cement	20–25 (duct)	40–70	7–30	150–550
Paper mill	25 (duct)	70–80	7–30	100–550
Blast furnace	20 (pipe)	35–45	10–60	150–400
	30 (pipe)	65–75	10–30	70–400
Sulfuric acid	25 (pipe)	70–100	10–40	150–900
Copper and zinc smelters, converter gas	20 (duct)	40–65	10–30	~350
Roaster and reverberatory gas	30 (duct)	~75	~10	~300

a mean free path, this distance being inversely proportional to relative gas density δ.

It has been demonstrated empirically in numerous single gases and gas mixtures that combinations of round wires and outer electrodes of arbitrary shape exhibit a corona-starting field E_c given by

$$\frac{E_c}{\delta'} = A_g + \frac{B_g}{(r_0\,\delta')^{1/2}} \tag{31}$$

where A_g (V/m) and B_g (V/m$^{1/2}$) are constants. The relative gas density δ' is conventionally taken with respect to 1 atm and 25°C. For air, the values $A_g = 32.2 \times 10^5$ V/m and $B_g = 8.46 \times 10^4$ V/m$^{1/2}$ are recommended. Values for other gases are reported in the literature. Note that Eq. 31 is not applicable to negative corona in pure nonattaching gases, and it may grossly overestimate the negative corona-starting field at elevated pressures (Section 9.5.1).

Example

Determine the corona-starting voltage in room air for a duct precipitator (Fig. 9.6) of 9-in. plate-to-plate spacing ($2s = 0.229$ m), 4-in. wire-to-wire spacing ($2c = 0.102$ m), and 109-mil wire diameter ($2r_0 = 2.77 \times 10^{-3}$ m). Compare with a 109-mil diameter wire in a 9-in. diameter pipe.

The duct corona-starting voltage V_c is given by Eq. 26 for which E_c, the corona-starting field at the wire, and the parameter d are required. Therefore, the following quantities are calculated: $\delta' = 1$ (p. 197); $E_c = 54.9 \times 10^5$ V/m (Eq. 31); $s/c = 2.24$; $d = 0.548$ (Eq. 25), and $V_c = 45.7$ kV (Eq. 26). The pipe starting voltage is given by Eq. 12: $V_c = r_0 E_c \ln(r_1/r_0) = 33.6$ kV. For equal duct width and pipe diameter and identical wire sizes, duct starting voltage will always exceed wire-pipe starting voltage. Starting voltages measured in industrial precipitators are invariably lower than the calculated estimates. This effect is due to irregular electrode spacing and extraneous discharges from electrode asperities; in ducts, it also is attributable to the lower starting voltage of the end wires.

9.2.4 Power Supplies

Optimum precipitator performance requires, as a rule, the highest level of electrical energization attainable in a given set of circumstances. Long experience confirms that pulsating half- or full-wave voltages obtained from imperfectly filtered rectifier sets yield collection efficiencies superior to those resulting from unvarying direct voltage. The relatively long decay periods for pulsating waveforms allow time for sparks to extinguish between current pulses, thus permitting operation at voltage and power levels not attainable in the absence of sparking. But although sparking is generally desirable in industrial practice, too frequent sparking will detrimentally lower the useful power input. It is mainly this consideration—the need to operate at some optimum spark rate (commonly of the order of 100 sparks per minute per electrical set)—that determines the nature of the transformer–rectifier and the associated automatic control equipment used to energize the precipitator. However, when spark rate provides the only feedback signal to the control system, transformers and rectifiers are likely to be vulnerable to damage from excess current. Consequently, a double-feedback system (monitoring current in addition to spark rate) is usually recommended to assure the most favorable average spark rate despite erratic variations in line and load conditions.

Air-cleaning and sampling precipitators are run below sparkover and present no control problems of consequence.

Recent developments in precipitator power supplies and control apparatus emphasize solid-state devices (silicon and selenium rectifiers to replace vacuum tubes, thyristor controls to replace magnetic amplifiers) having advantages of improved response, lower power losses, and reduced equipment size. Pulse energization has been occasionally investigated, but this method has so far not shown sufficient merit to warrant commerical application.

9.3   PARTICLE CHARGING

Two distinct particle-charging mechanisms are generally considered to be active in electrostatic precipitation: (1) bombardment of the particles by ions moving under the influence of the applied electric field (sometimes called field charging), and (2) attachment of ionic charges to the particles by ion diffusion in accordance with the laws of kinetic theory.

9.3.1   Ion Bombardment

If a spherical particle bearing a uniformly distributed free surface charge Q (C) is placed in a uniform electric field E_0 in a gas, induced and free charge on the particle distort the original field E_0 and impart to it a radial component. This component, found from a solution of Poisson's equation subject to appropriate boundary conditions, is

$$E_g = -\frac{\partial V}{\partial r} = E_0 \cos \theta \left[2\left(\frac{\kappa_p - 1}{\kappa_p + 2}\right)\frac{a^3}{r^3} + 1 \right] + \frac{Q}{4\pi\varepsilon_0 r^2} \qquad r \geq a \quad (32)$$

where r (m) is the radius vector from the center of the particle, θ is the polar angle between r and the undistorted field E_0, κ_p is the relative dielectric constant of the particle, and a (m) is its radius. An ion of charge q (C) is attracted to the particle and imparts its charge by attachment if the ion approaches from an angle θ for which the radial force $F = qE_g$ (N) is negative. Particle charging ceases at $F = 0$.

The lines of force for an initially uncharged sphere are given in Fig. 9.8a. As the charging proceeds, the charge already present on the particle creates a repulsive force that modifies the configuration of the electric field and thereby reduces the rate of charging. Figure 9.8b shows the field configuration when the sphere is partly charged. The lines of force entering the sphere on the side facing the on-coming ions have been reduced. Eventually the lines of force will bypass the sphere completely and charging will halt. Setting $\theta = \pi$, $r = a$, and

$$p = 2\left(\frac{\kappa_p - 1}{\kappa_p + 2}\right) + 1 \tag{33}$$

we see that the maximum free charge on the surface of the spherical particle is, from Eq. 32

$$Q_{max} = 4\pi\varepsilon_0 \, pa^2 E_0 \tag{34}$$

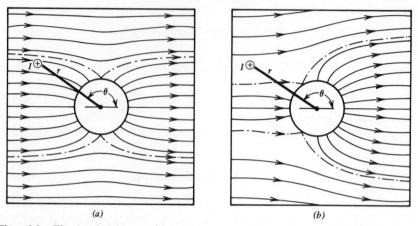

Figure 9.8. The electric fields near (a) an uncharged and (b) a partly charged dielectric sphere. The dashed lines represent the limits of the fields passing through the spheres. The fields inside the spheres are not shown. In (a) the ion I shown at (r, θ) within the dashed lines will be captured, whereas in (b) the ion is outside the region of attraction and will escape the particle.

Consideration of the gas–ion current to the particle indicates that the particle charge as a function of time is

$$Q = Q_{max}\frac{t}{t + \tau} \tag{35}$$

where τ (sec), the particle-charging time constant, is given by

$$\tau = \frac{4\varepsilon_0}{N_0 qb} \tag{36}$$

and N_0 (m^{-3}) is the ion density. The time constant represents the time in which half the limiting charge is attained.

The preceding considerations remain valid for a conducting particle. In this instance, allowing the dielectric constant κ_p to increase without limit, $p = 3$.

Industrial precipitators commonly operate at space-charge densities as low as 5×10^{12} ions/m³, one or two orders of magnitude less than the figures frequently quoted for laboratory precipitators.

9.3.2 Diffusion Charging

In addition to ion-bombardment charge, a suspended aerosol particle in an ionized gas acquires a charge by virtue of the random thermal motion of the ions and their consequent collision with and attachment to the particle. This process is called ion diffusion, and its effect can be calculated in the following approximate manner.

We know from kinetic theory that the density of a gas (in this case, an ion cloud) in a potential field varies in accordance with the expression

$$N = N_0 e^{-U(r)/kT_A} \tag{37}$$

where N (m^{-3}) is the number of ions per unit volume in the presence of a particle field, N_0 (m^{-3}) is the undisturbed uniform ion density, $U(r)$ (J) is the potential energy of an ion due to its position r in the field, k (J/$^{\circ}$K) is Boltzmann's constant, and T_A ($^{\circ}$K) is the absolute temperature. Neglecting the effect of the applied field, the potential energy of an ion of charge q (C) in the vicinity of a uniformly charged spherical particle is

$$U = \frac{qQ}{4\pi\varepsilon_0 r} \tag{38}$$

where Q (C) is the charge on the particle and r (m) is the distance from the center of the particle to the ion. The time interval t (sec) associated with the ion–particle collisions is given by

$$t = \frac{1}{\pi a^2 N \bar{v}_i} \tag{39}$$

where $\bar{v}_i$ (m/sec) is the kinetic-theory rms velocity of the ions. If it is assumed that all ions impinging on a particle are attached by image forces, then charging will proceed at the rate

$$\frac{dQ}{dt} = \pi a^2 N q \bar{v}_i \tag{40}$$

From Eq. 40, for an initially uncharged particle, we have

$$Q = \frac{4\pi\varepsilon_0 \, akT_A}{q} \ln\left(\frac{aN_0 q^2 \bar{v}_i}{4\varepsilon_0 \, kT_A} t + 1\right). \tag{41}$$

Both ion-bombardment and diffusion charging are simultaneously operative. For the dominant particle sizes met in most industrial precipitator applications, charging by ion bombardment is prevalent. A convenient, but not inviolate, rule is that ion bombardment predominates for particles of radius greater than a few tenths to 1 μm and ion diffusion for less than 0.1 μm. Table 9.2 compares the two mechanisms under typical discharge conditions in room air.

Table 9.2.   Number of Electronic Charges Acquired by a Conducting Spherical Particle

Conditions		Charges Acquired During							
		Ion Bombardment, Eq. 35				Ion Diffusion, Eq. 41			
Treatment time (sec)		0.01	0.1	1	∞	0.01	0.1	1	10
Particle radius (μm)	0.1	1	3	4	4	3	7	11	15
	1	120	340	410	420	69	110	150	190
	10	12,000	34,000	41,000	42,000	1,100	1,500	1,900	2,300

The conditions on which the charge calculations are based are $E_0 = 2 \times 10^5$ V/m, $\kappa_p = \infty$, $T_A = 300°$K, $N_0 = 5 \times 10^{13}$ ion/m^3, $b = 1.8 \times 10^{-4}$ m^2/(V-sec), $k = 1.38 \times 10^{-23}$ J/(°K-molecule), and $\bar{v}_i = 510$ m/sec. The rms ion velocity is given by $\bar{v}_i = (3p_g/\rho_g)^{1/2}$ where p_g is the gas pressure (N/m^2) and ρ_g is the gas density (kg/m^3). These values approximately represent a wire-in-tube assembly of respective inner and outer radii 3.8 mm and 12.7 cm in room air at 40 kV with a negative discharge current of 0.13 mA/m.

9.4   PARTICLE COLLECTION AND EFFICIENCY CALCULATION

9.4.1   Background of the Problem

Particle collection in an electrostatic precipitator is essentially a process of mass transfer through a moving gas, in a net direction that is normal to the collecting surface. It is necessary to distinguish at least three forms of particulate mass transfer from the main body of gas to the passive electrode: (1) electrostatic convection under the action of Coulomb forces and the electric wind, (2) turbulent diffusion of aerodynamic and electrodynamic origin, and (3) inertial drift. The Coulomb force drift mechanism has been quantitatively investigated and its main features are quite clear. In contrast, the physical nature of mass transfer by turbulent diffusion is much more complicated and, in relation to electrostatic precipitators, it has so far not proved possible to set up general and rigorous quantitative definitions. Inertial drift, another complex phenomenon, is a consequence of a particle's tendency, by virtue of its momentum, to continue moving in a straight line in the face of opposing or deflecting forces. Thus, the more massive a particle, the less likely it is to closely follow the motion of an eddy in which it is entrained.

Purely electrostatic mass transfer to the walls occurs in stationary gases and under the influence of the laminar flow regime. It is a characteristic feature of laminar flow that the direction of gas motion as a whole coincides with the direction in which any separate part of the gas moves. Therefore, there is no macroscopic motion of the gas in the transverse direction (i.e., normal to the collecting electrode) under fully laminar conditions. In such circumstances, mass transfer to the precipitator collecting surface can occur only by electrostatic drift.

The laminar-flow precipitator is, however, a laboratory novelty; all, or virtually all, commercial precipitators and all single-stage precipitators operate with various degrees of turbulence. Several attempts have been made to develop a general and comprehensive theory of turbulent precipitation, and indeed, the lack of such a theory constitutes the major obstacle in the design of precipitators today. Before considering some of these attempts, however, we must calculate a fundamental quantity for all theories of electrostatic precipitation, the particle migration velocity. This is the velocity exhibited by a charged suspended particle moving toward the collecting surface under the influence of an external electric field.

9.4.2   Theoretical Particle-Migration Velocity

The drag F (N) on a spherical particle moving at velocity w through a gas is

$$F = \tfrac{1}{2}C_D \pi a^2 \rho_g w^2 \tag{42}$$

where a (m) is the particle radius, ρ_g (kg/m^3) the gas density, and C_D the dimensionless drag coefficient. The drag coefficient is a function of the *particle* Reynolds number Re (dimensionless)

$$\mathrm{Re} = \frac{\rho_g w(2a)}{\mu} \tag{43}$$

where μ (decapoise—dP) is the gas viscosity. In laminar flow (Re less than about unity) $C_D = 24/\mathrm{Re}$ and Eq. 42 reduces to Stokes's law

$$F = 6\pi\mu a w \tag{44}$$

The electrostatic precipitating force on a charged particle is $E_p Q$, where E_p is the precipitating field. For particles charged by ion bombardment, the limiting charge $Q_{\max}$ is given by Eq. 34. Equating electrostatic and drag forces in this case, we have

$$w = \frac{2p\varepsilon_0 E_c' E_p a}{3\mu} \tag{45}$$

where E_c' designates the charging field. A corresponding relation is easily written for charging by ion diffusion (see Fig. 9.9). Appropriate combinations of large particle size, high gas density (occurring at elevated pressure), and high fields can violate the criterion Re < 1; in such an event, Eq. 45

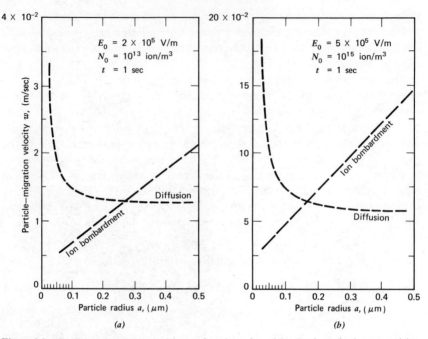

Figure 9.9. Particle-migration velocity as a function of particle size for submicron particles. The curves include the Cunningham correction and assume that the charging and precipitating fields, E'_c and E_p, respectively, equal E_0. Note the very high migration velocities possible for ultrafine particles ($a < 0.1$ μm). This explains the observation that collected material at the mouth of a precipitator is often a mixture of course and ultrafine particles. (Used with permission of Gauthier-Villas, Publishers.)

will overestimate the drift velocity. If the particle size is comparable to the mean free path λ (m) of the gas molecules ($\lambda = 6.8 \times 10^{-8}$ m in atmospheric air at 25°C), there is a greater tendency for particles to slip between the molecules, and w increases by a factor C (the Cunningham correction):

$$C = 1 + \frac{\lambda}{a}\left(1.26 + 0.400e^{-1.10a/\lambda}\right) \tag{46}$$

Various practical difficulties, which may be insurmountable if rigor is demanded, face the investigator attempting to calculate particle-migration velocities under realistic precipitating conditions. No generally satisfactory means exists for selecting appropriate "effective" values of field intensity or space charge density: both of these vary transversely and longitudinally through the precipitator, both depend on variable dust concentrations, and both vary in time unless pure dc voltage is used.

In calculating ion-bombardment w, it is sometimes argued that the maximum field near the wires be used for E_c', since turbulence may be supposed to carry most particles into the high-field region at some point before collection. Choice of E_p depends on viewpoint. According to one commonly held theory (Section 9.4.3), E_p is active only in the laminar sublayer—turbulence governing particle motion elsewhere—and therefore E_p must be assigned that value of electric intensity prevailing at the wall. In line with the diffusion theories of precipitation, however (Section 9.4.4.2), particle convection is considered active throughout the cross section; from this, it seems reasonable to set E_p equal to the average field.

On the basis of experimental work by a number of investigators, it appears that particle-charging theory, whether for submicron or larger particles, can be relied on to give at least order-of-magnitude agreement with observation.

9.4.3   The Deutsch Efficiency Equation

An exponential efficiency equation for electrostatic precipitators was discovered in 1919 and written in the empirical form

$$\eta = 1 - k_A^t \tag{47}$$

where η (dimensionless) is the fractional efficiency, t (sec) is the treatment time, and k_A is a precipitator constant. Three years afterward, Deutsch showed that the precipitator constant could be evaluated in terms of physically significant quantities. A modified derivation of Deutsch's efficiency equation follows.

The precipitator cross section is considered to consist of two zones: a laminar boundary layer very close to the collecting wall, and a turbulent core occupying virtually the entire cross-sectional area. Particle motion in the core is assumed to be completely dominated by turbulence, turbulent mixing yielding a uniform particle concentration C_p (kg/m^3) throughout a given cross section. In the boundary layer, the particle has a component toward the wall of velocity w (m/sec) which, as a reasonable first approximation, is assumed to be constant over the length of the precipitator. Particle charge is taken to reach an almost constant level in an interval that is small compared with the total treatment time. In the time dt, the particulate matter lying within the distance $w\, dt$ is precipitated on collecting area dA' where A' (m^2) is the cumulative collecting surface measured downstream of the mouth of the precipitator. This action reduces the particle concentration in the gas opposite dA' by dC_p. Equating rates of particle loss from the gas and particle accumulation on the surface in time dt yields

$$V_g\, dC_p = -wC_p\, dA' \tag{48}$$

where V_g (m³/sec) is the volume flow rate of the gas. Integrating this expression, and defining precipitator efficiency η in terms of respective inlet and outlet particle concentrations C_{in} and C_{out}, we have

$$\eta = 1 - \frac{C_{out}}{C_{in}} = 1 - e^{-Aw/V_g} \tag{49}$$

where A (m²) is the total collecting area.

Since, regardless of the charging mechanism, w is a function of particle size, the overall efficiency of a precipitator treating a distribution of particle sizes should be calculated in terms of a distribution of efficiencies. When this is done, moderate departures from the monodisperse condition are found to have decided effects in lowering efficiency, particularly for a high-performance precipitator.

It is common practice in precipitator design to employ a single effective migration velocity w_e in the Deutsch equation, even for those aerosols having broad particle-size distributions. This circumstance, coupled with various uncertainties in the derivation or use of the Deutsch equation (e.g., cross-sectional uniformity of particle concentration, appropriate value of charging field E_c', neglect of agglomeration and reentrainment), severely limits the equation's reliability. The effective migration velocity for a given aerosol is best determined empirically, and preferably at the same linear gas velocity at which it is intended to later use the data. Since sizable variations in w_e can be found in a given application, values of w_e reported in the literature without full background details must be accepted with caution and, in new situations, should be used only as very approximate guides.

Contrary to the assumption made in deriving Eq. 49, it is observed in practice that particle migration velocity calculated from efficiency measurements is dependent on the linear gas velocity. In particular, migration velocity generally rises at first with increasing gas velocity, peaks, and then decreases. The decrease is attributed to the onset of reentrainment of collected particles from the walls after the gas velocity exceeds some critical level. A number of suggestions have been put forth to explain the initial rise in migration velocity. These include the following possibilities:

1. Increased turbulence associated with higher mainstream velocities causes more particles to be carried into the region of the discharge electrodes, where they acquire a greater charge.

2. Increased turbulence improves mechanical collection efficiency by facilitating particle–wall contact.

3. Particulate mass transfer by longitudinal diffusion in the downstream direction is greater, relative to longitudinal convection, at lower mainstream gas velocities than at higher, because at the lower velocities the longitudinal

concentration gradient is higher. The residence time of a particle in the precipitator at lower mainstream velocities thus becomes less than proportional to reciprocal velocity.

9.4.4   Quality of Gas Flow and Role of Turbulence

9.4.4.1   *Introduction*

The dominant role played by conditions of gas flow in electrostatic precipitation cannot be overemphasized. Disturbed flow in the form of uneven distribution, jets, or swirls, not only increases reentrainment losses from electrodes and hoppers, but is responsible for poor collection initially. This is because if longitudinal gas velocity varies from duct to duct across a precipitator—and there may be dozens of ducts in parallel—the application of Eq. 49 to each duct individually will reveal a decay in overall efficiency. It is common experience to improve efficiency from 60 or 70% to 95% or better by corrections in gas flow.

Characteristic solutions to the problems presented by large-scale gas-flow disturbances involve the development of appropriate equipment (e.g., turning vanes, diffusion screens, transitions, and plenum chambers). But in addition, it must be recognized that poor gas flow is a system problem in total plant design.

The trend of future work in precipitation gas dynamics lies not in a further refinement of the criteria for good large-scale gas flow but rather in (1) understanding the small-scale process of particle diffusion as it affects precipitation, (2) assessing the importance of turbulence and convection due to the electric wind, and (3) attempting to harness, as far as possible, these various particle-transfer mechanisms to reenforce electrostatic collecting forces.

9.4.4.2   *Recent Theories of Precipitation*

A number of recent attempts have been made to incorporate into a modified precipitator efficiency equation particle diffusion and other mechanisms ignored in the Deutsch analysis. In each case, a condition essential to the Deutsch derivation is abandoned: uniformity of particle concentration over the precipitator cross section. The modified derivations assume that the dust is uniformly distributed over the cross section only at the inlet of the precipitator duct. As the gas proceeds downstream, the action of the wallward migration velocity promptly tends to clear the midstream section of the duct in the neighborhood of the plane of the wires. Opposing this, however, is eddy diffusion, which sweeps particles from the zone of high concentration near the walls back into the depleted midstream region. Thus turbulent eddies continuously redistribute over the cross-section particles, which the electric field tends to concentrate nearer the walls. Efficiency of collection is governed by the dominance of the latter effect over the former.

Detailed consideration of the problem is further complicated by reentrainment of collected particles from the walls. This process contributes to raising the particle concentration in the gas close to the walls and so also enhances back diffusion.

The various theoretical treatments of this problem, and the conflicting assumptions that have been made, reflect the lack of experimental data besetting the investigator who tries to develop a precipitation theory more comprehensive than that of Deutsch. Observed results are particularly scanty regarding the nature of the cross-sectional particle distribution profile and the effect of the electric wind of the corona discharge in modifying "normal" aerodynamic turbulence.

It should be noted that none of the more sophisticated theoretical approaches described previously has so far been developed to a point of practical utility.

The few experimental dust-concentration profiles that have been reported confirm the presence of a relatively dust-free zone in the region of the discharge wires, the width of the zone progressively broadening as the gas moves downstream. The extent of the central clear zone and its tendency to widen with downstream travel appear to depend on the level of turbulence, on particle size, and, in duct precipitators, on the wire-to-wire spacing. Knowledge of the particle concentration very close to the collecting wall is of prime importance in establishing the boundary conditions in any theory of precipitation, but this concentration is very difficult to establish experimentally.

9.4.4.3 *The Electric Wind*

The observation that smoke particles could be borne along in the electric wind—the movement of gas induced by the repulsion of ions from the neighborhood of a corona-discharge electrode—was made as long ago as the eighteenth century. But except for a flurry of interest in the early 1930s, the role of the electric wind has been only occasionally considered by precipitation workers. Studies using gas- and particle-tracer techniques to follow gas-flow patterns suggest that the particle-migration velocity relative to the electrodes includes an electric-wind component superposed on the velocity of the particles relative to the gas. Strong objections have been raised to this conclusion, however, on the grounds that (1) it is not necessary to propose a major electric-wind contribution to the particle-transport rate to explain observed particle-migration velocities and (2) requirements of flow continuity are not clearly satisfied. The question remains unresolved.

9.4.4.4   *Particle Adhesion and Reentrainment*

The behavior of dust particles on and near the collecting electrode is a subject of prime practical importance that has received insufficient attention. Observation of individual particle trajectories reveals that, contrary to assumptions usually made in theoretical efficiency calculations, impact phenomena at the collecting electrode or precipitate surface cannot be neglected, particularly for particles greater than about 10 μm. These larger particles may rebound on impact without losing their charge, or they may erode agglomerates of previously precipitated dust. Dust on the collecting electrode tends to acquire a like charge by induction and, if dislodged under low current conditions, can be forcefully accelerated away from that electrode. The repulsive force acting on the dust layer is opposed by an attractive electrical force due to the ion current. The net electrical force per unit area of dust surface F' (N/m^2) is

$$F' = \tfrac{1}{2}\varepsilon_0[(\kappa_p j\rho_d)^2 - E^2] \tag{50}$$

where E (V/m) is the field in the gas adjacent to the dust layer and F' is positive when attractive. In addition, various "mechanical" surface forces must be considered (e.g., van der Waals forces, adhesion due to moisture films, and reentraining forces due to wind drag).

A subject of principal concern in electrostatic precipitation is the removal of collected dust from the passive electrode and its transfer, with minimal reentrainment, to the hoppers. In modern precipitators, dust may fall as much as 12 m through a transverse gas stream before reaching the hoppers; 15-m high plates are under consideration.

American and European plate-rapping practice generally exhibits a fundamental difference in outlook. The tendency in the United States toward continuous rapping (i.e., every few minutes or less) aims at the elimination of visible rapping puffs in the stack discharge, psychologically so objectionable. Higher long-term collection efficiencies are, however, more likely when the rapping is intermittent (i.e., at intervals of as much as several hours). Ideally, the rapping interval should be adjusted to the needs of different parts of the precipitator.

A number of collecting-electrode configurations are in use, the various designs purporting to increase efficiency by (1) providing baffles to shield deposited dust from the reentraining forces of the gas stream, (2) providing catch pockets which convey precipitated dust into a quiescent gas zone behind the collecting plate, and (3) minimizing protrusions from the plate surface in order to raise sparkover voltage.

9.5 PRECIPITATION UNDER EXTREME CONDITIONS

9.5.1   High Temperature–High Pressure

In recent years, electrostatic precipitators have been used in chemical-processing, power-generation, and mass-transport applications involving temperatures and pressures well in excess of conventionally accepted limits. Successful pilot or full-scale trials have been run at pressures up to 55 atm and temperature (not simultaneous) to 800°C.

Operation at elevated temperatures or pressures must take into account unique corona phenomena not encountered in ordinary practice. For example, at high pressure, the negative corona threshold occurs unpredictably over a band of voltages; Eq. 31, giving the corona-starting field, applies only to the upper limit of the band (Fig. 9.10). With positive corona, however, Eq. 31 remains valid (Fig. 9.11). Both polarities exhibit a critical gas density at which corona-starting and sparkover curves intersect. A reduction in wire diameter or an increase in the size of the interelectrode gap raises the critical density. Conventional precipitation is possible only at gas densities below the critical level for the gas and electrode geometry in question.

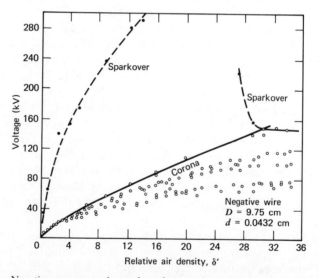

Figure 9.10. Negative corona-starting and sparkover voltages for coaxial wire-pipe electrodes in air (25°C); here D and d are the pipe and wire diameters, respectively; voltage is unvarying dc. The solid curve represents the corona-starting voltage corresponding to the field given by Eq. 31 and indicates an upper limit to the band of starting voltages observed experimentally.

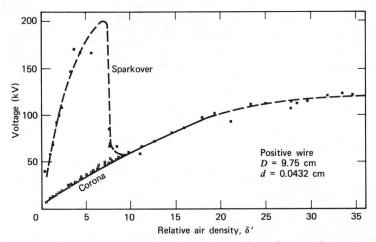

Figure 9.11.   Positive corona-starting and sparkover voltages for coaxial wire-pipe electrodes in air (25°C); D and d are the pipe and wire diameters, respectively; voltage is unvarying dc. The solid line is the corona-starting curve according to Eqs. 12 and 31 and is in good agreement with the experimental data below the critical density. (Used with permission of the American Institute of Physics.)

Upper temperature limits to the precipitation process also exist. In very hot gases the corona will be overwhelmed by thermal ionization, with resultant currents so high, and associated fields so low, that effective precipitation is impossible. Thermal ionization rates usually depend on trace quantities of the alkali metals (elements of low ionization potential) rather than on the principal gaseous constituents. To some extent, raising the pressure together with the temperature helps to maintain a stable corona.

9.5.2   High-Resistivity Precipitate

9.5.2.1   *Back-Corona Formation*

In the precipitation of high-resistivity dusts it is generally observed that, after a brief initial period of operation, collection efficiency deteriorates, current increases, and the sparkover voltage—assuming negative corona— drops. The current increase is due to secondary emission, so-called back corona, which originates in the dust deposit on the collecting electrode and assumes the form of a luminous sheath of tuftlike discharges of polarity opposite to that of the primary discharge. The porous dust dielectric appears to serve as a condenser which, first charged by primary ions, discharges when the voltage attains the breakdown value of the gas in the pores. In most practical cases, the breakdown strength E is of the order of 10^6 V/m. Industrial corona-current densities j are generally less than about 10^{-3} A/m² (Table 9.1). At first thought, we should not expect back-corona disturbances to appear until the dust resistivity exceeds

$$\rho_d = \frac{E}{j} = \frac{10^6}{10^{-3}} = 10^9 \ \Omega\text{-m} \tag{51}$$

a figure somewhat greater than the "critical" resistivity of about 10^8 Ω-m widely quoted in industrial practice. Definition of the back-corona threshold in terms of particle resistivity, however, is a complex matter that cannot be fully accounted for in terms of Eq. 51. Furthermore, dust resistivity is generally dependent on applied voltage, duration of the test, and other experimental incidentals.

9.5.2.2 *Particle Charging in a Bi-Ionized Field*

In the presence of back corona and the resultant bi-ionized inter-electrode field (the total corona current consisting of negative ions migrating in one direction and positive in the other), the equations given earlier governing particle charging are no longer applicable. Assuming that the oppositely charged ions are uniformly interspersed in the gas, it may be shown that Eq. 35 is to be replaced by

$$Q = Q_{max} \frac{1 - e^{-\alpha t}}{1 - [(1 - \xi)/(1 + \xi)]e^{-\alpha t}} \tag{52}$$

where

$$\alpha = \frac{1}{\varepsilon_0} (b_f \rho_f b_b \rho_b)^{1/2} \tag{53}$$

$$\xi = \left(\frac{b_b \rho_b}{b_f \rho_f}\right)^{1/2} = \left(\frac{j_b}{j_f}\right)^{1/2} \qquad 0 \leq \xi \leq 1 \tag{54}$$

and the limiting charge Q_{max} is given by

$$Q_{max} = 4\pi\varepsilon_0 \, pa^2 E_0 \frac{1 - \xi}{1 + \xi} \tag{55}$$

Ion mobility b, space-charge density ρ, and current density j due to "forward" and back corona are distinguished by the respective subscripts f and b. Unlike monopolar ion bombardment, which effectively ceases after a few time constants τ (Eq. 36), bipolar ion bombardment continues indefinitely. A maximum charge Q_{max} is approached because positive and negative ion currents to the particle tend to become equal and so neutralize each other.

The debilitating effect of back corona on an electrostatic precipitator is strikingly illustrated by considering a case in which the back-corona current is only one-third of the total current $j_t = j_b + j_f$. We then have $\xi = (0.5)^{1/2} = 0.71$, and Q_{max} (whence, the particle-migration velocity w) is reduced by a factor $(1 - \xi)/(1 + \xi) = 0.17$. According to the Deutsch equation, precipitator length must be increased by $1/0.17 = 5.8$ times to restore collection efficiency to the level corresponding to zero back corona. Other considerations, such as the reduction of applied voltage by sparking and the lowering of the interelectrode field by increased voltage drop across the high-resistivity precipitate, are likely to reduce efficiency still further.

Poisson's equation for a wire–pipe system containing bipolar space charge is

$$\frac{1}{r}\frac{d}{dr}(rE) = \frac{1}{\varepsilon_0}(\rho_f - \rho_b). \tag{56}$$

Experimental observations with a negative forward corona justify the assumption that the component forward- and back-corona currents per unit length, j_{fl} and j_{bl} (A/m), respectively,

$$j_{fl} = 2\pi r b_f E \rho_f \tag{57}$$

$$j_{bl} = 2\pi r b_b E \rho_b \tag{58}$$

are conserved between wire and cylinder. That is, $j_{fl} + j_{bl}$ give the total current per unit length j_{tl} as measured directly. The parameter ξ of Eq. 54 is thus independent of the radius vector r for a given set of operating conditions. Eliminating ρ_f and ρ_b from Eq. 56, we have

$$rE\frac{dE}{dr} + E^2 + \frac{j_{tl}}{2\pi\varepsilon_0}\left[\frac{b_b - \xi^2 b_f}{(1 + \xi^2)b_f b_b}\right] = 0 \tag{59}$$

whence

$$E = \left\{\frac{j_{tl}}{2\pi\varepsilon_0}\left(\frac{b_b - \xi^2 b_f}{b_f b_b(1 + \xi^2)}\right) + \left(\frac{r_0}{r}\right)\left[E_c^2 - \frac{j_{tl}}{2\pi\varepsilon_0}\left(\frac{b_b - \xi^2 b_f}{b_f b_b(1 + \xi^2)}\right)\right]\right\}^{1/2} \tag{60}$$

For j_{tl} and r not too small (cf. Eq. 10), we can write

$$E = \left[\frac{j_{tl}}{2\pi\varepsilon_0}\left(\frac{b_b - \xi^2 b_f}{b_f b_b(1 + \xi^2)}\right)\right]^{1/2} \tag{61}$$

Analysis of experimental data shows that the negative primary current in the presence of back corona is greater than it would be without back corona. This condition is partly due to the bombardment of the central electrode by back-corona ions and partly due to photoionization originating both at the wire and at the surface of the precipitate.

9.5.2.3   Conditioning; Sulfur Oxides in Flue Gas

The passage of an electric current through a layer of precipitate occurs both over the surface and through the volume of the individual particles. Surface conductivity is dependent on surface moisture and chemical films adsorbed on the particles, whereas volume conductivity is a property inherent in particle composition. For semiconducting materials usually of interest in electrostatic precipitation, volume resistivity ρ_v (Ω-m) is a decreasing function of temperature in reasonable agreement with the relation

$$\rho_v = \rho_\infty e^{E/kT_A} \tag{62}$$

where the quantity ρ_∞ (Ω-m) and the activation energy E (J) are constants of the material. Depending on the nature of the dust, the validity of Eq. 62 in a humid gas may not become apparent until the temperature exceeds 100 to 300°C or more and conducting surface films are driven off. Figure 9.12, illustrating the transition from surface to volume conduction, is qualitatively typical of curves for numerous substances.

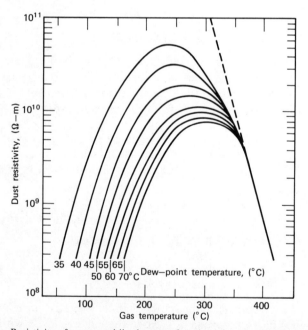

Figure 9.12.   Resistivity of a cement-kiln dust as a function of gas temperature and dew point. The broken line shows the effect of volume conduction in accordance with Eq. 61. Fly-ash resistivities commonly peak between 100 and 200°C. (Used with permission of Staubforschungsinstitut des Hauptverbandes der gewerblichen Berufsgenossenschaften e.V.)

Control of particle resistivity by conditioning the carrier gas plays an important role in many applications of electrostatic precipitation. The availability of moisture alone in some cases, or moisture with small quantities of chemical conditioners in others, often suffices to lower the particle resistivity by one or more orders of magnitude. In this fashion, by eliminating back-corona disturbances, precipitator performance can be dramatically upgraded. Inlet gas can most economically be humidified by water atomization in a spray chamber installed immediately ahead of the precipitator.

The physical adsorption of a moisture film on a chemically inert surface is characterized by relatively low binding energies. A water film alone consequently produces less effective conditioning than does a chemical binder that is strongly adsorbed on the particle surface and, in turn, strongly adsorbs moisture. This activated adsorption effect is often accomplished for weakly basic particles by strong acids (e.g., H_2SO_4) and for weakly acidic particles by strong bases (e.g., NH_3).

Sulfate deriving entirely from sulfuric acid in flue gas provides a natural conditioning substance in the combustion of coal, in which sulfur is normally present in quantities ranging from a small fraction of 1% to about 5%. Although the actual concentrations of sulfur oxides produced in combustion vary depending on the coal, the furnace design, and the operating conditions, the usual orders of magnitude are 0·1% for sulfur dioxide and 0.001% for sulfur trioxide.

The conditioning of low-sulfur flue gas by the injection of sulfur trioxide, viewed unfavorably—but progressively less so—by American public utility companies, has been successfully employed on a commercial scale in Great Britain and Australia. The small amount of sulfur trioxide supplied, generally less than 50 ppm, is completely adsorbed by the dust and does not lead to increased emission of sulfur compounds into the atmosphere.

9.5.3   Corona Quenching

The heavy particle space charge often resulting in treating very finely divided dust tends to elevate the corona threshold and thus depress the current at a given voltage. As long as it can be assumed that the current is carried essentially by free ions (unattached to aerosol particles), the treatment given in Sections 9.2.1.2 and 9.2.2.2 remains applicable. In other words, it is valid to assume that

$$j_t = j_i + j_p = E(\rho_i b + \rho_p b_p) \cong E\rho_i b \qquad (63)$$

where j_t (A/m^2) is the current density at a point, the subscripts t, i, and p designating total, free-ion, and particle, respectively. Particle mobility

b_p [(m/sec)/(V/m)] is commonly 2 or 3 orders of magnitude less than the ion mobility b; therefore, although ρ_p may be larger than ρ_i, j_p is ordinarily no greater than a few percent of j_t.

The mild corona quenching described previously gives way to a severe variety when the particle concentration is so dense that all free ions are captured. Equation 63 then takes the form

$$j_t = E\rho_p b_p \tag{64}$$

In this event, the current consists solely of ions riding on slowly migrating dust particles and j_t is characteristically reduced to 1% or less of its former value. The maximum charge Q that a particle can acquire is determined not by the relevant law of charging (Section 9.3) but by the limited availability of free ions. Specifically

$$Q = \frac{\rho_p}{N_p} = \frac{\rho_i'}{N_p} \tag{65}$$

where ρ_i (C/m^3) is the limited density of ions available for charging and N_p (particles/m^3) is the particle number density. With particularly severe quenching, thus, the charge per particle may be far below that attainable in ordinary circumstances, and the associated particle-migration velocity and collection efficiency may be correspondingly low. Over that length of the precipitator in which severe quenching prevails

$$w = w_s \frac{\rho_i'}{N_p Q_s} \qquad \rho_i' \le N_p Q_s \tag{66}$$

where Q_s is the charge that would be acquired were ρ_i' adequate, and w_s is the associated migration velocity. The Deutsch differential equation (Eq. 48) and Stokes's law then give

$$\eta = \frac{E_p \rho_i' A}{6\pi\mu a V_g (N_p)_{\text{in}}} \tag{67}$$

where $(N_p)_{\text{in}}$ is the inlet particle number concentration. Despite severe quenching, particle collection still occurs, albeit at a reduced rate.

The deposition of a coat of highly resistive dust ($\rho_p > 10^8$ Ω-m) on the discharge electrodes sometimes accounts for still another type of corona suppression. Fine particles (<10 μm), whose motion is largely controlled by turbulence, can be driven directly onto the discharge electrode or sufficiently close to acquire an opposite-ion charge in the corona region, whereupon they are precipitated onto the discharge electrode. As the resultant dust layer builds up, the potential drop across it increases. and, if the particle resistivity is high enough, the interelectrode field will drop sufficiently to materially reduce the current. Conditioning of the gas to lower the particle resistivity offers the most promising remedy in such cases.

9.6 APPLICATIONS

Coverage of the numerous practical aspects of electrostatic precipitation presents a two-fold difficulty which cannot be resolved in this chapter. First, an adequate treatment requires a compilation and critical evaluation of operational data. Much of this comes from sources of indeterminate reliability scattered throughout the literature. Second, much of the essential design data—particle-migration velocities, for instance—are regarded as confidential proprietary information by manufacturers who have accumulated it at considerable cost in time and effort. This is why published descriptions of industrial precipitators, no matter how detailed in certain respects, are commonly deficient in revealing certain critical variables of the process.

The degree to which a full-scale precipitator performs in accordance with theory is highly variable; therefore, no substitute exists in industrial precipitation practice for diversified, long-term experience. However, approaches to precipitator design that are *governed* by empiricism are likely to lead to confusion. Observations in the field, which by their nature often lack adequate controls or other experimental safeguards may, when presented injudiciously, be cited in support of almost any conclusion. Furthermore, the subjective approach that unfortunately goes hand in hand with much field work provides a very limited base for new developments and fails to permit the analysis of design and performance in terms of fundamental physical relationships.

Structural or operational details in specific applications are most readily established by a search of the literature. Particularly helpful in this regard are the comprehensive bibliographies listed in Refs. 2 (or 3) and 7.

NOMENCLATURE

a	particle radius (m)
A	total collecting surface area (m^2)
A'	cumulative collecting surface area from inlet (m^2)
A_g	constant of the gas (V/m)
b	ion mobility (m^2/V-sec)
b_b	back-corona ion mobility (m^2/V-sec)
b_f	forward-corona ion mobility (m^2/V-sec)

b_0 ion mobility at $0°C$ and 1 atm $(m^2/V\text{-sec})$

b_p particle mobility $(m^2/V\text{-sec})$

B_g constant of the gas $(V/m^{1/2})$

c one-half wire-to-wire spacing, between centers (m)

C Cunningham correction, dimensionless

C_p drag coefficient, dimensionless

C_{in} inlet dust concentration (kg/m^3)

C_{out} outlet dust concentration (kg/m^3)

C_p particle concentration (kg/m^3)

C_1 constant (V)

d dimensionless variable

E electric field (V/m)

E_c corona-starting field (V/m)

E'_c charging electric field (V/m)

E_g radial component of electric field near a particle (V/m)

E_p precipitating electric field

F force (N)

F' force per unit area (N/m^2)

j current density (A/m^2)

j_b areal back-corona current density (A/m^2)

j_f areal back-corona current density (A/m^2)

j_{bl} linear back-corona current density (A/m)

j_{fl} linear forward-corona current density (A/m)

j_i free-ion current density (A/m^2)

j_l linear current density (A/m)

j_p particle current density (A/m^2)

j_s average current density at plate (A/m^2)

j_t total current density (A/m^2)

j_{tl} total linear forward- and back-corona current density (A/m)

k Boltzmann's constant $(J/°K)$

k_A Anderson's precipitator constant

n number of wires

N ion concentration in potential field (m^{-3})

N_0 undisturbed ion concentration (m^{-3})

N_p particle number density (m^{-3})

$(N_p)_{in}$ inlet particle number concentration (m^{-3})

p dimensionless parameter, Eq. 33

p_a pressure (atm)

q ion charge (C)

Q particle charge (C)

Q_{max} limiting particle charge (C)

Q_s particle charge in absence of corona quenching (C)

r	radius (m), radial distance from tube axis (m), radial distance from center of particle (m)
r_0	wire radius (m)
r_1	tube radius (m)
Re	Reynolds number, dimensionless
s	one-half plate-to-plate spacing (m)
S	particle surface per unit volume of gas per meter (m^{-1})
t	time (sec)
T	temperature (°C)
T_A	absolute temperature (°K)
U	potential energy of ion (J)
v_i	ion drift velocity (m/sec)
$\bar{v}_i$	rms ion velocity (m/sec)
V	potential (V)
V_c	corona-starting potential (V)
V_c'	corona-starting potential in presence of dust space charge (V)
V_g	gas flow rate (m^3/sec)
V_0	potential at wire surface (V)
V_{s1}	sparkover voltage for system of one wire (V)
V_{sn}	sparkover voltage for system of n wires (V)
w	particle migration velocity (m/sec)
w_s	migration velocity in absence of corona quenching (m/sec)
α	bipolar charging constant (sec^{-1})
δ	gas density relative to 0°C and 1 atm, dimensionless
δ'	gas density relative to 25°C and 1 atm, dimensionless
ε	permittivity (F/m)
ε_0	permittivity of free space, 8.85×10^{-12} F/m
η	fractional efficiency, dimensionless
θ	polar angle (rad)
κ	relative dielectric constant, dimensionless
κ_g	relative dielectric constant of gas, dimensionless
κ_p	relative dielectric constant of particle, dimensionless
λ	molecular mean free path (m)
μ	viscosity (dP)
ξ	back-corona ratio, dimensionless
ρ_b	ion space charge of back corona (C/m^3)
ρ_d	bulk particle resistivity $(\Omega\text{-m})$
ρ_f	ion space charge of forward corona (C/m^3)
ρ_g	gas density (kg/m^3)
ρ_i	ion space-charge density (C/m^3)
ρ_i'	density of ions available for particle charging (C/m^3)
ρ_p	particle space-charge density (C/m^3)

ρ_v	volume resistivity (Ω-m)
ρ_∞	constant of material (Ω-m)
τ	charging time constant (sec)
ϕ	dimensionless current, Eq. 11

REFERENCES

1. S. Oglesby, and G. B. Nichols, *A Manual of Electrostatic Precipitator Technology*. Part I: Fundamentals, Document PB 196380; Part II; Application Areas, Document PB 196381, National Technical Information Service, Springfield, Va., 1970. The broadest compilation to date of practical design data, operating conditions, and areas of utilization for electrostatic precipitators.

2. M. Robinson, *Bibliography of Electrostatic Precipitator Literature*, Southern Research Institute, Birmingham, Ala., 1969. A listing of 3000 precipitator references, the product of a comprehensive search of the world literature from 1600 to 1969.

3. M. Robinson, and N. Frisch, *Selected Bibliography of Electrostatic Precipitation Literature*, Document PB 196379, National Technical Information Service, Springfield, Va., 1970. One thousand items of the above bibliography are selected for lasting or timely interest, provided with key words, and indexed accordingly to facilitate rapid searching by subject.

4. M. Robinson, "Electrostatic Precipitation," in *Air Pollution Control*, Part I, W. Strauss, Ed., Wiley-Interscience, New York, 1971, pp. 227–235. A more extensive treatment of the present chapter, emphasizing physical principles rather than empirical methods, and giving special attention to current research and new applications. Useful as an updating supplement to H. J. White.[6] Contains 494 references, chiefly to recent literature.

5. H. E. Rose, and A. J. Wood, *An Introduction to Electrostatic Precipitation in Theory and Practice*, 2nd ed, Constable, London, 1966. A survey mostly of established principles and practices, but also including descriptions of the authors' own researches. Handy as a first, rapid overview of the subject.

6. H. J. White, *Industrial Electrostatic Precipitation*, Addison-Wesley, Reading, Mass., 1963. Deservedly recognized as the standard work on the subject in English. Reflects the author's two decades of wide-ranging activities and contributions in the science and art of electrostatic precipitation. Contains 261 references.

7. U.S. Environmental Protection Agency, *Air Pollution Abstracts*, Government Printing Office, Washington, D.C. A monthly journal thoroughly abstracting the international air-pollution control literature. The most complete guide available to current publications in electrostatic precipitation.

CHAPTER 10

Electrostatic Separation

James E. Lawver, *Director*
Mineral Resources Research Center
University of Minnesota
Minneapolis, Minnesota

W. P. Dyrenforth, *Senior Vice President*
Carpco Research and Engineering, Inc.
Jacksonville, Florida

10.1 OVERVIEW

This chapter describes methods of electrically separating from each other two or more solid species in an air ambient. The separation of solids from gases and the separation of solids in a liquid medium are described in Chapters 8 and 14, respectively. Separation engineers frequently refer to the processes described in this chapter as *electrostatic separation*, even though there is always some flow of current involved. Typical applications are:

1. Beneficiation of ores, such as the concentration of the minerals ilmenite, rutile, zircon, apatite, asbestos, hematite, and many others.
2. Purification of foods, such as the removal of trash and rodent excrement from cereal seeds.
3. Sorting of reusable wastes such as separating fiber insulation from copper wire shreds.
4. Electrostatic sizing, namely, the sorting of solid particles according to their size or shape.

Table 10.1 presents a partial list of commercial electrostatic separations.

Dedicated to the memory of J. H. Carpenter for his pioneering work in electrostatic separation. J. E. L. and W. P. D.

Table 10.1. Industrial Separations Made by Electrostatics

Arsenopyrite–feldspathic gangue	Mica–feldspars/silicates
Asbestos–silicates	Molybdenite–silicates
Barite–silicates	Monazite–beach sands
Chromite–ilmenite– magnetite–monazite	Nickel–copper ores–metasilicates
Coal–pyrite	Rutile–beach sands
Coal–shale	Scheelite–pyrite
Cobalt–silver–silicates	Silicon carbide–alumina/silicates
Coke–iron	Spodumene-cassiterite
Copper ore–silicates	Stibnite–silicates
Copper wire–insulation	Wire–thermoplastics/rubber
Diamonds–silica	Wolframite–pyrite
Feldspar–mica gangue	Zircon–beach sands
Feldspar–quartz	Bark–sand
Fluorite–silicates	Barley–rodent excrement
Fly ash–carbon	Cocoa beans–shell
Gold/platinoids–beach sand	Cotton seeds–stems
Gold/platinum–jewelry sweeps	Movie film–paper
Graphite–silicates	Nut meats–shells
Halite–sylvite	Peanuts–shells
Ilmenite–garnet	Plastic–lint
Ilmenite–beach sand	Polyvinals–polyesters
Iron (specular hematite)–silicates	Rice–rodent excrement
Kaolin–iron contamination	Seeds–foreign material
Kyanite–rutile and iron gangues	Soap–detergent
Limestone–silicates	Soybeans–rodent excrement
Magnetite–silicates/rutile	Walnuts–shells
	Wheat–garlic seeds

It is estimated[1] that the free world produces about 14 million tons of products per year by electrostatic separation. Mineral beneficiation accounts for about 95% of the entire production.

The first record of an electrostatic separating machine is a patent[2] issued in 1880 for a device to purify ground cereal. The following year a patent[3] was issued that described a rather crude machine intended to concentrate gold ore, and in 1892 Thomas A. Edison[4] patented an electrostatic separator for concentrating gold ore. These first separators apparently did not meet with industrial success. The first commercial process[5] was a plant at Plattville, Wisconsin, used to electrostatically beneficiate lead and zinc ores. This plant was started in 1908 and marked the initial commercial use of electrostatic beneficiation of minerals. The success of this operation led to numerous applications of electrostatic separation of solids at a variety of locations throughout the world. The process was used both as a single process and in conjunction with gravity processes for concentrating minerals; it also served as an industrial process for purifying foods. The industrial application of the electrostatic process was short-lived, however. In 1912 the new and exciting process of froth flotation[6] was developed. In but a few years the advantages of selectivity and the low capital cost of the flotation process precluded the industrial use of the electrostatic process except in the purification of foods and in the beneficiation of ores in arid regions. The electrostatic process remained little more than a metallurgical curiosity until the titanium shortage of World War II afforded a new opportunity for electrical concentration of titanium-bearing beach sands. The demand for rutile and associated valuable heavy mineral concentrates led to marked improvements in separating machine design and in high-voltage power supplies.

As a result of this impetus, the improved separators (known as "high-tension" separators) are now beginning to supplement and even replace the more conventional gravity and froth flotation processes used in mineral beneficiation. In 1965 the world's largest electrical concentration plant was installed[7] at the Wabush Mines in Canada. This plant is used as a supplement to gravity concentration of iron ore. The "high-tension" circuit has a capacity of more than 6 million tons per year and is used to reduce the silica (SiO_2) content of spiral concentrates from about 8 to about 2% SiO_2. The total high-tension operating cost, exclusive of amortization, is less than $0.05/ton of dry feed.

The future of electrostatic or "high-tension" concentration of minerals is most promising. The success of the Wabush Plant has pioneered the way for similar iron ore beneficiation plants now being designed in several parts of the world. The concept of using metallized pellets as a substitute for scrap in electric steelmaking has sparked an interest in the use of high-tension separators to produce supergrade iron ore concentrates for metallization.[8] The process also offers promise as a method of producing extremely pure iron ore concentrates ($<0.1\%$ SiO_2) for use in making sheet metal directly from iron ore concentrates.[8] Finally, since the process is dry and generally does not require any chemical reagents, the method may replace certain flotation processes that present water pollution problems.

10.2   GENERAL PRINCIPLES

Electrostatic separation is the selective sorting of solid species by means of utilizing forces acting on charged or polarized bodies in an electric field. It follows that the important components of a system designed to make an electrostatic separation of solids (see Fig. 10.1) are:

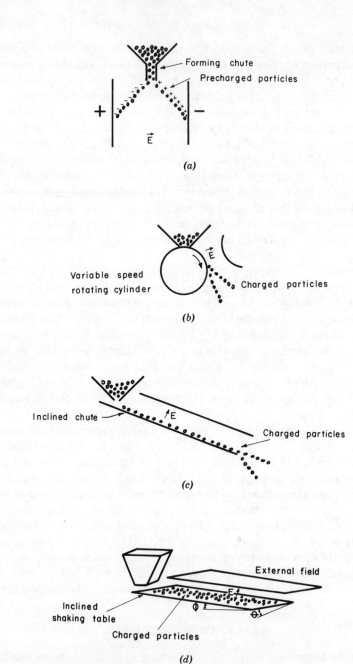

Figure 10.1. System for the electrostatic separation of solids: (*a*) forming chute, (*b*) variable-speed rotating cylinder, (*c*) inclined chute, (*d*) inclined shaking table.

1. **A charging mechanism**. We must either devise a method of selectively charging the species that are to be separated from each other or we must come up with a system that will result in one of the following categories of charge distribution.

a. Particles of two different species enter a separating zone bearing an electric charge of opposite sign. For example, if we are to separate phosphate rock particles from quartz particles, it is rather simple to use a charging mechanism such that the phosphate particles enter an external electric field with a net positive charge while the quartz particles bear a net negative charge.

b. Particles of two different species enter a separating zone where only one type of particle bears a significant electric charge.

c. Solids enter the separating zone such that particles of different species bear the same sign of charge, but the magnitude of the electric charge is significantly different.

d. Particles of different species enter the separating zone with significantly different dipole moments.

Each of these categories of charge distribution has been used successfully in commercial separations.

2. **An external electric field.** An electric field is required that is defined by some configuration of equipotential boundaries and a source of high voltage. Voltage sources are in the range of 10 to 100 kV and are usually unidirectional.

3. **A nonelectrical particle trajectory regulating device.** A physical separation of two types of particles is always made by adjusting the forces and the time forces act on particles, such that different types of particles will have different trajectories at some predetermined time. In addition to electrical forces, it is usually advantageous to utilize forces such as **gravity, centrifugal force**, or **friction** to effect a selective sorting.

a. A forming chute (Fig. 10.1*a*) is used to adjust the direction of the initial velocity due to gravity.

b. A variable-speed rotating cylinder (Fig. 10.1*b*) serves to adjust the centrifugal force acting on particles.

c. The magnitude and direction of the particle velocity vector is adjusted through the use of an inclined chute (Fig. 10.1*c*).

d. The magnitude and velocity of the particle velocity vector is adjusted through the use of an inclined shaking table (Fig. 10.1*d*).

4. **Feeding and product collection system.** All separators must, of course, have some means of conveying feed to the separator and a method of cutting the stream of particles at the desired point so that different species can be collected and conveyed to subsequent stages of separation or to product storage bins.

A cursory review of the literature yields virtually hundreds of patents on electrostatic separators. However, after but a little study of the patents, it becomes apparent that the proposed machines consist of little more than various combinations and permutations of the aforementioned basic ingredients.

10.3 CHARGING MECHANISMS

Although there are many ways to cause solids to acquire an electric charge, only three charging mechanisms are used in commercial electrostatic separation, namely:

1. Charging by contact and frictional electrification.
2. Charging by ion or electron bombardment.
3. Charging by conductive induction.

It should be emphasized at once that these charging mechanisms are by no means mutually exclusive. In fact, charging by contact electrification is always an active mechanism whenever dissimilar particles make and break contact with each other or whenever they slide over a chute or an electrode. Charging by conductive induction is possible whenever particles are touching a grounded conductor in the presence of a high-voltage electric field. And, it is obvious that particles may also acquire a charge (often unwanted) because of corona discharge from a sharp portion of an electrode. Although several charging mechanisms are often active in an electrostatic separation, it is usually possible and desirable to design a separator in which only one charging mechanism predominates. The design of a separating machine usually depends on the type of charging mechanisms used to effect the separation.

10.3.1   Contact Electrification and the Free-Fall Separator

Contact electrification is the charging mechanism most frequently used to selectively charge and make an electrostatic separation of two species of dielectric materials.

Typical examples are the separation of feldspar from quartz, quartz from apatite, and halite from sylvite. This is not to say that conductors will not charge by contact electrification—they usually charge—but since they are conductors, they often lose their charge before they can be separated in an external electric field.

Everyone has had first-hand experience with the phenomenon of contact electrification. When we comb our hair on a dry day we notice that both the comb and our hair have become electrified. People with long hair often amuse themselves by combing their hair in the dark and watching the sparks in a mirror. As one might guess, the comb becomes electrically charged with one polarity and the hair is charged with the opposite polarity. Another common yet somewhat more impressive demonstration of contact electrification is the annoying shock experienced when we touch our fingers to a metal doorknob after walking over a thick carpet. This phenomenon of charging—by making and breaking contact between dissimilar surfaces—is known as *contact electrification* and is the basis of one type of electrostatic separation.*

Man has been aware of this interesting charging mechanism for centuries; in fact, it was recorded by Thales of Miletus 600 years before Christ. Yet, despite our growing knowledge of solid-state physics, the exact mechanism is not fully understood even today.[9] One questionable rule of thumb often used for predicting the sign of the surface charge resulting from contact electrification is Coehn's rule, which states that "when two dielectric materials are contacted and separated, the material with the higher dielectric constant becomes positively charged." In a rather loose way, this rule is in accord with the modern theory of solids. Zwikker[11] pointed out that "the material with the greater number of energy levels will have the higher permittivity and will be more easily polarized so that it will give off electrons to the other contact material."

Coehn's rule was quantitatively formulated by Beach,[12] who stated that the surface charge density is $15 \times 10^{-6}(\varepsilon_1 - \varepsilon_2)$ C/m^2, where ε_1, ε_2 are relative permittivities. It is important to note that this equation indicates the surface charge density that could be obtained at the moment contact is broken. If the electrification is made in air (STP), it is impossible to maintain a surface charge density greater than about 26.6×10^{-6} C/m^2.

Henry[13] remarked that if we consider an ionic surface with interatomic spacing of the order of 3.2 A, then each ion in a surface could be represented by an area of 10 A^2. A surface area of 1 m^2 could represent 1×10^{19} ions or 5×10^{18} ions of each sign. Assuming univalent ions each with a charge of 1.6×10^{-19} C, we have 8×10^{-1} C of charge of one sign per square meter. Thus the maximum charge in air (STP) could be obtained by a transfer on only $(26 \times 10^{-6})/(8 \times 10^{-1}) \times 100 = 0.003\%$ of the available ions. A similar result is obtained by calculating possible electron transfer across a metal surface.

Although Coehn's rule is reported to have been verified for more than 400 substances, it is of limited value to the metallurgist because of the difficulty in determining the effective relative permittivity of the surface layers (which often differ from the substratum) of two substances that are to make contact. Furthermore, there appear to be many exceptions to Coehn's rule,

* The theory of contact electrification is discussed in Chapter 5.

and there is not even agreement on whether the charge is due to the transfer of electrons, ions, or both. It suffices to say that almost any dissimilar surfaces can be charged by contact electrification and that the sign and magnitude of charge can often be controlled by the addition of surface active agents—a subject that is discussed further on.

To illustrate electrostatic separation utilizing contact electrification let us consider the case of two spherical particles of different chemical compositions such as the two mineral particles quartz and apatite. We know from the discussion presented in Chapter 5 that the two species tend to acquire equal and opposite surface charge densities by the mechanism of contact electrification. If two minerals are clean and dry and are put into contact with each other in a temperature range of about −40° to +800°C, the quartz, say, will become charged negative and the apatite positive. If the charged particles are now simply dropped through an electric field, their trajectories will be in opposite directions and it is a simple matter to make a separation. As a practical example, if Florida phosphate ore, composed of small particles of quartz and phosphate rock is (1) water washed to make sure that the mineral surfaces are clean, (2) dried and heated to about 110°C, and (3) vibrated or conveyed so that the minerals make and break contact, the ore can be separated into its components simply by using a uniform electric field (Fig. 10.2).

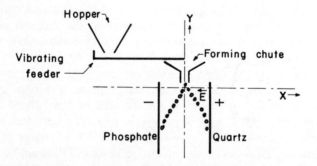

Figure 10.2.   Separation of quartz from phosphate.

For the purpose of this illustration we can calculate the approximate position of the particles at a given time in the electric field if we ignore the Coulombic forces acting on the particles due to the charges on neighboring particles (a catastrophic assumption if the particles are −500 mesh) and also disregard the effect of air friction. Let

E = electric field $(N/C = V/m)$

Q = charge on a particle (C)

$\quad Q = \sum \sigma_s$ where σ_s is the surface charges on the particles due to contact electrification

g = acceleration due to gravity

t = time (sec)

m = mass of particle (kg)

F = force (N)

$$F_{electrical} = Q\mathbf{E} = m\,\frac{d^2x}{dt^2} \tag{1}$$

$$F_{gravity} = mg = -m\,\frac{d^2y}{dt^2} \tag{2}$$

If the initial velocity and displacement are zero, then upon integration the deflection due to the electrical force is

$$X = \frac{1}{2}\frac{QE}{m}t^2 \tag{3}$$

and the vertical displacement due to gravity is

$$y = -\tfrac{1}{2}gt^2 \tag{4}$$

In the case of, say, a -48 to $+65$ Tyler mesh quartz particle (ca. 0.25 mm in diameter) charged by contact electrification with phosphate rock, the numerical value of Q/m is about 9×10^{-6} C/kg.

A typical value of E for a free-fall separator is 4×10^5 V/m $= $ N/C. Then by Eq. 3 we have

$$X = \tfrac{1}{2} \times 9 \times 10^{-6} \times 4 \times 10^5 = 1.8\ t^2 \text{ m} \tag{5}$$

Now according to Eq. 4 the time required for a particle to fall, say, 0.5 m is $[(0.5 \times 2)/9.8]^{1/2} \simeq (0.1)^{1/2}$ sec, so that the X deflection of a particle that has fallen 0.5 meter is, by Eq. 5,

$$X = 1.8 \times 0.1 \simeq 18 \text{ cm}$$

This means that a negative particle will be separated from a positive particle (with the same Q/m ratio) by a distance of 36 cm—ample for a good separation.

Note from Eq. 3 that the deflection of a freely falling charged particle in a uniform electric field varies directly as the charge and intensity of the electric field and inversely with particle mass. The magnitude of the external electric field is limited by the breakdown strength of air, which for practical purposes may be assumed to be about 3×10^6 V/m. Usually, it is the breakdown strength of air that also limits the maximum value of surface charge attainable by contact electrification. Recall that Gauss's law is stated as

$$\oint \mathbf{E} \cdot d\mathbf{A} = \frac{Q}{\varepsilon_0} \qquad (6)$$

Given a sphere with a uniform surface charge σ C/m^2, then by Gauss's law we have

$$EA = \frac{\sigma A}{\varepsilon_0}$$

since ε_0, the permittivity of free space, is 9×10^{-12} C^2/N-m^2 and E_{max} in air, the breakdown strength of air, is 3×10^6 V/m. It follows that

$$\sigma_{max} \simeq E_0 \simeq 26 \times 10^{-6} \frac{C}{m^2}$$

It is interesting to compare the electrical force with the gravitational force acting on a particle in a free-fall electrostatic separation. The electrical force acting on a sphere of radius r (m) with a uniform charge density σ (C/m^2) is

$$F_e = QE = 4\pi r^2 \sigma E$$

The gravitational force is

$$F_g = \tfrac{4}{3}\pi r^3 \rho g$$

where ρ = density (kg/m^3)
$\quad\; g$ = acceleration due to gravity = 9.8 m/sec^2
and

$$\frac{F_e}{F_g} = \frac{4\pi r^2 \sigma E}{\tfrac{4}{3}\pi r^3 \rho g} = \frac{3\sigma E}{r\rho g}$$

It is possible to increase the electrical force acting on charged particles by working under high pressure, since both the limiting value of particle charge density and the external electric field are raised with increasing pressure. This procedure has not proven to be practical because of the high capital and operating costs.

Thus in air at 1 atm and room temperature we have

$$E = 3 \times 10^6 \, \frac{V}{m}$$

$$\sigma = 26.6 \times 10^{-6} \, \frac{C}{m^2}$$

$$g = 9.8 \text{ m/sec}^2$$

$$\frac{F_e}{F_r} \simeq \frac{25}{r\rho}$$

where r = radius (m)

ρ = kg/m^3

More realistic values are

$$\sigma = 5\% \text{ of maximum} \qquad \text{and} \qquad E = 80\% \text{ maximum}$$

which gives the ratio

$$\frac{F_e}{F_g} \simeq \frac{1}{r\rho} \tag{7}$$

The practical upper size limit for the type of separation is about 10 mesh or r = 1 mm, corresponding to F_e/F_g = 1/3 for particles having a specific gravity of 3.0. As the density increases, the upper size limit decreases. The lower size limit is determined by the particle size distribution where fines tend to form clusters by interparticle Coulombic forces and cease to be separated by the external electric field. The lower limit is about 20 μ.

Free-fall separators are best custom designed for a specific application since the feed rate, temperature of separation, and electrode length depend on the properties of the solids that are to be separated.

Figure 10.3 shows a laboratory-scale free-fall–contact-electrification separation of phosphate rock mixed with an equal amount of quartz. The following separation data can be cited for this device:

Feed rate = 200 lb/(hr)(in.) of electrode width
Electrode spacing = 6 in.
Potential difference between electrodes = 60,000 V
Particle size = 0.15 to 0.30 mm diameter

Product	% Wt	% Phosphate	% Quartz
Feed	100	50	50
Concentrate	47	97.1	2.9
Tail	53	8.2	91.8

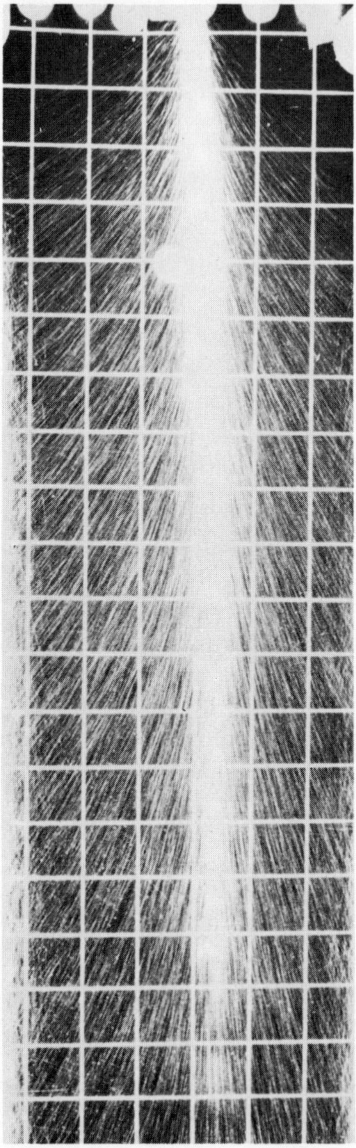

Figure 10.3. Free-fall electrostatic separation of quartz from phosphate rock.

A real-life separation of an ore is, of course, not as easy as this laboratory separation of an artificial mixture. In general, final products are the results of several cleaning or scavenging passes. Flowsheets of commercial separations are illustrated in a later part of this chapter.

10.3.2   Charging by Ion Bombardment and the "High-Tension Machine"

Ion bombardment is the charging mechanism most frequently used to separate good electrical conductors from poor conductors.

Typical among the many possible examples are the separations of ilmenite and rutile from quartz, specular hematite from quartz, and chopped copper wire from insulation.

The usual method of charging solids by electrons or ions is to take advantage of corona discharge. This is accomplished by passing solids through the corona discharge from a fine wire or a series of needle points positioned parallel to a grounded rotor of a separating machine. Corona discharge is produced whenever the wire or needle points are raised to an electrical potential such that the electric field in the immediate vicinity exceeds the electrical breakdown strength of the ambient (usually air). The electric field varies inversely with the radius of a conductor; thus it is practical to use modest potentials and rather fine diameter wires or needle points.

The type of corona that is produced depends on the polarity of the ionizing electrode. If the electrode is positive, negative ions are accelerated toward the electrode, causing the breakdown of air molecules with the result that positive ions are repelled outward from the electrode in the form of a corona glow. Conversely, if the polarity of the ionizing electrode is negative, positive ions are all accelerated toward the electrode and negative-charged oxygen ions are repelled from the electrode to produce a corona discharge. Since ion-bombardment charging does involve a small current flow, at least appreciably more than the so-called electrostatic separators, machines using this type of charging are called electrodynamic separators or, more frequently, the trade name "high-tension separators"[14] is used.

The principle of the separation is simple. Charging is effected by passing the solids over a grounded rotor through an intense corona glow such as is obtained from a fine wire or needle points charged to a potential exceeding the electrical breakdown strength of air. Special electrodes have been developed for preventing sparkover when treating combustible material such as coal and grain.[15] The conducting particles rapidly share their charge with the grounded rotor and are thrown from the rotor in a trajectory determined by centrifugal force, gravity, and air resistance. The dielectrics or poor conductors lose their charge slowly and are thus held to the surface of the grounded rotor by the image force associated with their surface charge. Particles composed of good electrical insulators, such as dry quartz, are at times so tenaciously held to the rotor by the image force that they must be scraped from the back side of the rotor by a fiber brush. On commercial separators it is often desirable to install an ac corona discharge electrode (wiper) on the back side of the rotor to partly discharge the particles that cling to the rotor thus reducing the wear on the brush.

The important design features of a high-tension separator are presented in Fig. 10.4. A granular mixture of conducting and insulating particles is fed

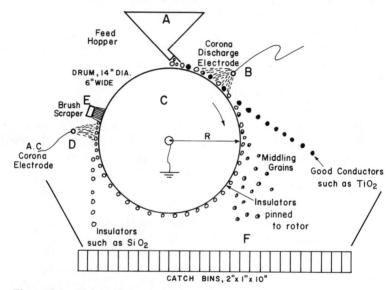

Figure 10.4.   High-tension separator.

through the feed hopper A onto the surface of the grounded rotating cylinder C. All the particles receive a surface charge σ_s as they pass through an intense corona discharge from electrode B. The value of the initial surface charge σ_s depends on particle shape, feed rate, time of exposure, and intensity of corona. The maximum value in air is of course about 27×10^{-6} C/m^2, which is the previously explained maximum charge density that can be sustained in air.

As the particles leave the corona region they lose their surface charge at a rate which is a function of their electrical resistance, the extent of particle contact with the grounded rotor, and the magnitude of their initial surface charge. The good conductors rapidly share their charge with the

grounded rotor and are thrown free of the rotor by a combination of centrifugal force and gravity. This is depicted by the black conducting particles
in Fig. 10.4.

The centrifugal force acting on the particles is

$$F_e = m\omega^2 R \tag{8}$$

where m = particle mass
   ω = angular velocity
   R = radius of the rotor

At very low rotor speeds the conducting particles simply slide off the
rotor under the influence of gravity.

The good insulating particles are attracted to the rotor by their image
force. If we consider a single particle of radius r having a surface charge
σ_s, the image force can be roughly approximated by analogy of the force
acting on a charged particle in front of an infinite plane, as in Fig. 10.5.
It can be shown that the necessary and sufficient condition, Laplace's
equation $\nabla^2 V = 0$, will be satisfied if we imagine a mirror image of the
original charge existing as in Fig. 10.5. It then follows that the force

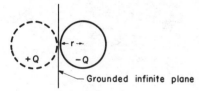

Figure 10.5.   Image force of particle of radius
r acting on a charged particle in front of an
infinite plane.

acting on the particle can be calculated from Coulomb's law, from which
we have

$$F_{\text{image}} = \frac{1}{4\pi\varepsilon_0} \frac{Q_1 Q_2}{(2r)^2} \tag{9}$$

where $Q_1, Q_2 \equiv$ Coulombs

$$\frac{1}{4\pi\varepsilon_0} = 9 \times 10^9 \ \frac{\text{N-m}^2}{\text{C}^2}$$

   r = particle radius (m)

It should be noted that the electric image force of a dielectric particle
bearing a surface charge is actually a very complicated function that has been
derived for only highly idealized conditions. Equation 9 is probably at best
only a crude approximation. For the case of a conducting sphere the

interested reader should consult an article by A. Rusell.[16] By using the simple model of image force, we can gain some insight regarding the importance of the variables associated with a high-tension separator by considering the ratio of the electric image force to the centrifugal force acting on a particle.

It has been determined experimentally that the corona-discharge surface charge on a 0.3-mm diameter quartz particle is about 5.8×10^{-6} C/m^2 after one revolution of a 14-in. diameter drum rotating at 70 rpm.

The ratio of the image force to the centrifugal force can be approximated as follows:

$$\text{image force } F_{\text{image}} = \frac{1}{4\pi\varepsilon_0} \times \frac{(4\pi r^2 \sigma)^2}{(2r)^2} = 36 \times 10^9 \times \pi^2 r^2 \sigma^2$$

and

$$\text{centrifugal force } F_c = \tfrac{4}{3}\pi r^3 \rho \omega^2 R$$

where ρ = density, and the ratio F_i/F_c, called the pinning factor, is

$$\frac{F_i}{F_c} \simeq \frac{8.5 \times 10^{10} \sigma^2}{r \times \rho \times \omega^2 R} \tag{10}$$

For the case in point, we can write

$$\sigma = 5.8 \times 10^{-6} \, \frac{\text{C}}{\text{m}^2}$$

$$\rho = 2.65 \times 10^3 \, \frac{Kg}{m^3}$$

$$r = 1.5 \times 10^{-4} \, \text{m}$$

$$\omega = 7.9 \text{ rad/sec}$$

$$R = 1.78 \times 10^{-1} \, \text{m}$$

Using these values in Eq. 10 gives

$$\frac{F_i}{F_c} \simeq 0.7$$

Equation 10 is important because it points out the significance of the operating variables. We see that the pinning factor increases as particle size, density, and $\omega^2 R$ are decreased. The pinning factor also increases as the square of the surface charge density σ_s. However, unlike the electrical precipitation case, σ_s is not particularly sensitive to time of exposure but is of course very sensitive to the electrical conductivity of the particles. It follows that when trying to separate large insulating particles from small conductors, using closely sized feed will improve the separation.

It remains to discuss at least briefly the rate of discharging a corona-charged particle that is in contact with the grounded rotor, since it is the discharge rate that determines the value of σ_s at any given time in the separating zone.

As a first approximation we note that the total charge $Q = \Sigma\sigma_s$ determines the potential difference V, existing between the particle and the grounded rotor. Furthermore, the discharge current $i = dQ/dt$ is directly proportional to the potential and inversely proportional to the total resistance between the charged particle and the grounded rotor. The resistance R_T is the sum of the particle resistance and the particle–particle and particle–rotor contact resistance. It follows that, since $i = $ voltage/resistance, then

$$\frac{dQ}{dt} = -\frac{K\Sigma\sigma_i}{R_T} = -\frac{K}{R_T}Q \quad \text{or} \quad \ln_\varepsilon Q = -\frac{K}{R_T}t + C$$

when $t = 0$ and $C = \ln_\varepsilon Q_0$.

Thus we have

$$Q = Q_0\,\varepsilon^{-Kt/R_T} \tag{11}$$

where $Q_0 = $ total charge at time $t = 0$

$K = $ constant depending on particle shape and dielectric constant

$R_T = \Sigma$ volume, surface, and contact resistance

It is often useful to define relaxation time T_R as the time required for

$$\frac{Q}{Q_0} = \frac{1}{\varepsilon} \quad \text{so that} \quad T_R = \frac{R_T}{K}$$

It turns out that T_R for pyrite is about 10^{-3} sec and T_R for quartz is about 10^6 sec. It is clear that R_T is the determining variable in separating good conductors from poor conductors. It is sometimes possible to improve separations by altering the value of R_T through the use of surface active agents, humidity control, or temperature control. The Bureau of Mines published a paper in which 95 minerals were classified according to weight percent collected as a conductor (in an electrostatic separator) as a function of the separation temperature.[17]

10.3.3   Charging by Conductive Induction

Charging by conductive induction is a charging mechanism suitable for separating good conductors from good electrical insulators. Under carefully controlled conditions the mechanism can be used to separate two or more semiconductors having large differences in electrical conductivity.

In a classical high school experiment, charging by conduction is demonstrated using an electroscope. When an uncharged electroscope is grounded with the finger in the presence of an external charged body (as in Figure 10.6a), electrons are attracted from the ground to the electroscope. If the ground is removed while the charged rod is still held close to the electroscope, and the charged rod is then removed, the leaves of the electroscope diverge (Fig. 10.6b) because the electroscope now has a negative charge, which has been acquired by conductive induction.

Exactly the same mechanism occurs when conductive particles pass over a grounded rotor in the presence of an external electric field, as in Fig. 10.7.

(a)   (b)

Figure 10.6. Charging experiment utilizing an electroscope: (a) Uncharged electroscope grounded in the presence of an external charged body; (b) charged rod removed, electroscope leaves diverge.

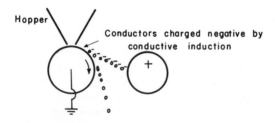

Figure 10.7. Conductive particles passing over a grounded rotor in the presence of an external electric field.

10.3.4   Conduction Separator

Good conductors quickly become highly charged by conductive induction and are attracted toward the external electrode. Poor conductors are only feebly charged by conductive induction but may bear a measurable charge due to particle–particle or particle–rotor contact electrification.

The conduction separator is seldom used for separating good conductors from insulators, since good conductors are easily separated from poor conductors by corona discharge. The principal application has been in the food and drug industry.

10.4 FEED PREPARATION

The success or failure of an electrostatic separation depends as much on feed preparation as it does on the selection of the separating machine.

1. In general, material should be sized to within $\pm$ one diameter of the mean particle size. Difficult separations should be sized even more closely.

2. The surfaces should be free of foreign material. It is often necessary to water or air scrub the feed material prior to making a separation.

Table 10.2. Ten Typical Industrial Separations and the Surface Treatment Required for Each

Mineral Combination	Surface Treatment	Usual Charging Mechanism
Specular hematite–quartz	Drying	Corona discharge
Ilmenite and rutile from poorly conducting, heavy mineral gravity concentrates (zircon, monazite, etc.)	Scrubbing to remove organic slimes; drying (reducing roast at 650°C)	Corona discharge
Zircon (cleaning)– ilmenite, rutile	Drying	Corona discharge
Cassiterite–scheelite	Drying	Corona discharge or conductive induction
Feldspar–quartz	Drying; HF vapors	Contact electrification
Halite–sylvite	Heating to 340°C or drying plus 1 lb/ton of fatty acids	Contact electrification
Pyrite–coal	Drying	Corona discharge or conductive induction
Coal–shale	Humidity control	Corona discharge or conductive induction
Diamonds–silica	Wet scrubbing in NaCl pulp; drying	Conductive induction
Dry foods and drugs from trash and rodent feces	Drying	Conductive induction

3. Unless a separation based on surface conductivity that is controlled by humidity is being attempted, the surface of the particles should be dry and preferably hot enough to assure the investigator that they will stay dry through all the processing steps.

It is sometimes possible to alter the properties of solids to improve an electrical separation. It is far beyond the scope of this chapter to discuss these methods in any detail. However, Table 10.2 covers the major treatments now used industrially.[18]

10.5 LABORATORY TECHNIQUES

One of the more important measurements in any type of electrostatic research is the determination of the charge existing on solids after a given treatment. Perhaps the most straightforward method of making this measurement is the use of a Faraday pail and a good electrometer. The details of a Faraday pail used in the laboratories of the senior author[19] are presented in Fig. 10.8. This instrument has been used with a Keithley

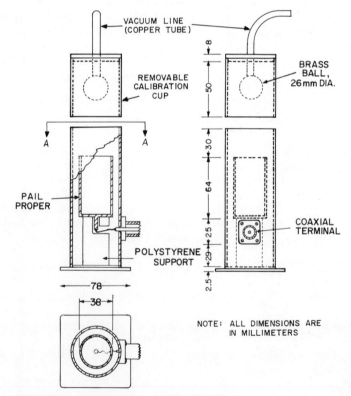

Figure 10.8.   Faraday pail and calibration sphere.

model 610A electrometer having an imput impedance greater than 10^{14} Ω shunted by 30 pF. Two methods of calibration of the pail were given by Lawver.[19]

A second measurement that is of value is that of the apparent particle resistivity of a loosely packed cell of granular solids. As was previously mentioned, the decay time of charged particles resting on a grounded rotor depends in part on the resistance R_T given in Eq. 11.

The numerical value of R_T can be approximated using the conductivity cell and measuring circuit shown in Fig. 10.9.[20]

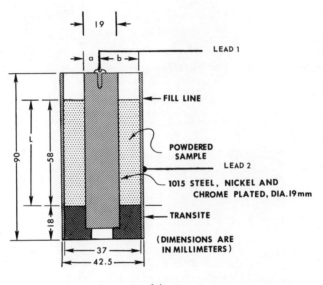

(a)

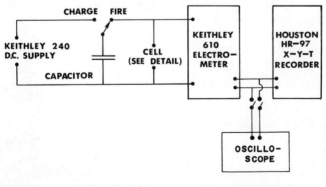

(b)

Figure 10.9. (a) Detail of the conductivity cell; (b) schematic diagram of the discharge circuit for approximating resistance R_T.

Figure 10.10.   Laboratory high-tension research unit.

The voltage of the capacitor is related to time by the well-known equation

$$V = V_0 \, \varepsilon^{-t/RC}$$

where V_0 is the voltage at time $t = 0$, C is the value of the capacitor of the circuit, and the value of C than the capacitance of the conductivity cell.

From the discharge voltage versus time data, we can calculate the value of the resistance of the cell R_c, since a plot of $\ln V_0/V$ versus time is linear and has a slope equal to $1/R_c C$.

Once the value of R_c has been approximated, the resistivity of the powdered sample can be computed by the following well-known expression of the leakage current in a sheathed cable:

$$\rho = \frac{2\pi LR}{\ln b/a} \; \Omega\text{-m}$$

where L, b, and a are the dimensions in Fig. 10.9.

In addition to charge and resistance measurements, it should be possible to make laboratory tests utilizing the various charging mechanisms. A typical laboratory research test unit appears in Fig. 10.10.

10.6   INDUSTRIAL APPLICATIONS

In conclusion of this review of electrostatic separation of solids, it seems desirable to cite a few examples of industrial separations. As has been previously mentioned, the vast majority of industrial separations are based on utilizing large differences in electrical conductivity to separate solid species using "high tension"—that is, corona-discharge machines. The examples cited are limited to high-tension applications.

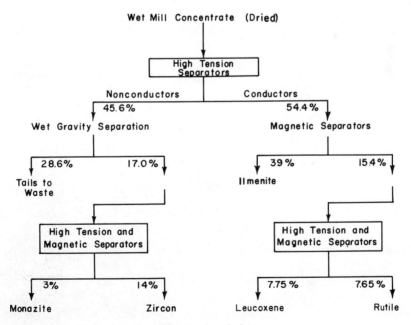

Figure 10.11.   Dry mill portion of heavy mineral plant.

Figure 10.12. High-tension plant for heavy minerals.

10.6.1   Concentration of Heavy Minerals

The term heavy minerals refers to certain minerals heavier than quartz which occur as valuable deposits in beaches, dunes, and streams. These deposits have been enriched by gravity segregation. The valuable heavy minerals include ilmenite, rutile, zircon, leucoxene, sillimanite, kyanite, garnet, and rare earth minerals such as monazite. These deposits are beneficiated by making a wet gravity concentrate using such devices as sluices or Humphrey spirals and then making final concentrates by a combination of electrotatic (high-tension), dry-magnetic, and gravity separation methods. Figures 10.11 and 10.12 show a typical flowsheet for the dry portion of a heavy mineral concentrating plant[21] and interior views of a plant, respectively.

10.6.2   Concentration of Iron Ore

Another important application of high tension is its use in the beneficiation of iron ore. The largest iron ore high-tension plant is at the Wabush Mines in Labrador (Fig. 10.13). The iron ore is crushed and ground to produce a gravity feed sized to about -0.6 mm. This material is then concentrated wet by gravity to produce a product that is dried and further concentrated by high tension. At Wabush it was desired to produce a concentrate containing less than 2.0% silica (SiO_2) so that a pellet made from this concentrate would contain, after the necessary bentonite had been added for cohesion, less than 2.5% SiO_2. A 2.5% SiO_2 pellet sells for a premium. In the beginning it was planned that such a high-grade concentrate could be produced on equipment that used the basic differences in specific gravity of Fe_2O_3 and SiO_2 as the means of separation. This method worked, but the iron losses were high, particularly in the second and third (cleaner and recleaner) passes. A flowsheet was finally designed and tested using gravity equipment for the initial concentration and high tension for the final finishing of the product. It was highly successful, and Wabush Mines is now producing more than 6 million tons per year of high-grade premium pellets for blast furnace consumption. The Wabush high tension flowsheet appears in Fig. 10.14.

The capital cost for the high-tension equipment at Wabush to handle 1000^+ tons/hr of total feed is approximately $1.7 million (1971 dollars). The cost of the conveying equipment to distribute the feed and remove the products usually runs about 35% of the cost of the separators when using

Figure 10.13. High-tension section of Wabush plant.

belt conveyors only. When using a combination of belt conveyors and bucket elevators, this cost drops to about 25%. The operating cost for the high-tension separators including power, operating labor, maintenance materials, and labor, but not including drying is approximately $0.04/ton of new feed. High tension for iron ore has proved to be a low-cost but highly efficient means of concentration.

The first all dry electrostatic or high-tension iron ore concentration plant in the world has recently been developed and tested in northern Sweden. The government-owned Luossavaara Kiirunavaara Ab (LKAB) Company began construction of a plant at Malmberget, Sweden, early in 1972 to produce high-grade hematite concentrate at an annual rate of 1 million tons. The ore is crushed and ground dry, with heat applied during grinding. After removal of a small amount of magnetite by the use of dry magnetic separators, the ore is fed to the high-tension process to remove silica and apatite (phosphorus). Test work showed that high tension could reduce the

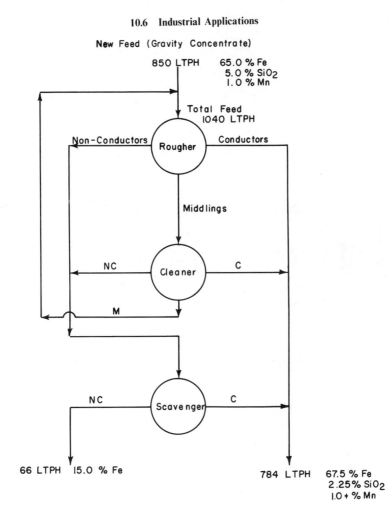

Figure 10.14.   Wabush high-tension flowsheet.

phosphorus content to a lower value than any other method available and yet maintain a higher efficiency in upgrading the iron by removal of silica. Extensive pilot plant testing has proved that the process is as efficient as laboratory results predicted; the flowsheet is shown in Fig. 10.15.

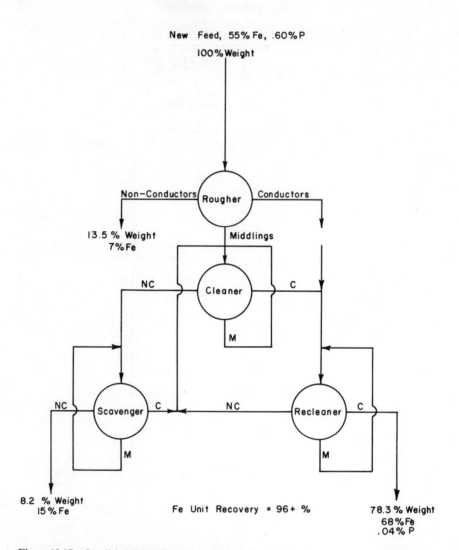

Figure 10.15. Swedish LKAB high-tension flowsheet.

REFERENCES

1. J. H. Carpenter, private communication, Carpco Research and Engineering, Inc., Jacksonville, Fla.
2. T. B. Osborne, U.S. Patent 224,719,1880.
3. C. H. Hill and E. H. Whited, U.S. Patent 245,299,1881.
4. T. A. Edison, U.S. Patent 476,991,1892.
5. H. A. Wentworth, *Trans. AIME*, **43**, 411 (1912).
6. AIME, *Froth Flotation*, 50th Anniversary Volume, 1962.
7. *Skillings' Mining Rev.*, **56**, No. 38 (September 1967).
8. R. M. Funk and J. E. Lawver, *Trans. AIME–SME*, **247**, 23 (March 1970).
9. W. R. Harper, *Contact and Frictional Electrification*, Oxford Press, 1967.
10. J. Lawver, "Fundamentals of electrical concentration of minerals," *Mines Mag.*, January 1960.
11. C. Zwikker, *Physical Properties of Solid Materials*, Interscience, New York, 1954, p. 253.
12. R. Beach, *Elect. Eng.* **66**, 325 (1947).
13. P. Henry, *Brit. J. Appl. Phys.*, Supplement No. 2 (1953).
14. Trade name coined by J. H. Carpenter, Carpco Research and Engineering, Inc., Jacksonville, Fla.
15. W. Grogg, U.S. Patent 2,860,276.
16. A. Rusell, *Proc. Roy. Soc.* (*London*), *A*, 524–531 (1909).
17. U.S. Bureau of Mines Bulletin 603, "Electrostatic Separation of Granular Materials," p. 72, 1962.
18. J. Lawver, *J. Electrochem. Soc.*, **116**, No. 2 (February 1969).
19. ——— and J. Wright, *AIME–SME Trans.*, **241**, 445 (December 1968).
20. ——— and ———, *AIME–SME Trans.*, **224**, 243 (March 1969).
21. A. G. Naguib, private communication, Carpco Research & Engineering, Inc., Jacksonville, Fla.

CHAPTER 11

Electrostatic Coating

EMERY P. MILLER, *Vice President*
Technical Operations
Ransburg Corporation
Indianapolis, Indiana

11.1 OVERVIEW

From the very early experiments involving Coulomb's law, it became apparent that electrically charged bodies would be attracted to or repelled from each other. If the bodies were free to respond, this attraction or repulsion was made apparent by the movement of the bodies. The experimental work done in the field of electrostatic air cleaning in the early 1800s is ample evidence of the effort that was put forth, even at that early date, to find commercial uses for these phenomena.[1] As is often true, these early efforts failed only because the collateral developments had not advanced sufficiently to provide suitable auxiliary equipment. Modern developments in these related areas have resulted in equipment that has allowed wide industrial application of these fundamental laws. One of these areas is the electrostatic application of coatings.

The term coating is a broad one, encompassing all processes in which any material, dry or wet, is applied over a surface to produce thereon a layer or coat. Electrostatic coating includes all processes of the coating type in which electrostatic forces are employed, either directly or indirectly, to bring about the deposition of the material on the surface.

Considering that almost every article or product manufactured and sold in today's industrial market is coated at least once during its production, it is readily apparent that the coating field is a large and varied one. The paper sheets that form the pages of this book have been coated to produce the surface quality needed for good print reproduction; the brass hardware used in our homes has been coated to protect its original luster from chemical corrosion; automobiles have at least three successive coatings applied to the surface to protect the base metal against corrosion and to provide a beautiful final appearance; even the transparent plastic sheet in which meat products are wrapped in the supermarket has perhaps received several coatings to increase its physical strength, to render it resistant to moisture penetration and ultraviolet light transmission, and possibly to allow it to be heat sealed to itself to form a tight package.

In all these coating processes, the applied materials represent an item of cost in the final product. It is highly desirable, therefore, to have these materials applied in all cases with the smallest possible loss. In many cases it is further desirable to have the coatings applied automatically and with the utmost simplicity. In still other cases, it may be that the material must be applied on the surface in a special condition, state, or orientation, to satisfactorily perform its intended purpose. The use of electrostatic forces in the coating process may make it possible to meet all these specifications.

In its most simple form, an electrostatic coating operation is visualized as taking place in the following manner. The article to be coated is supported so that its surface can be approached without obstruction. The coating material to be applied is distributed in the space about the surface in the form of finely divided particles. The suspended particles are then charged to one electrical sign (negative) and the surface to the other (positive). There is an attraction of the particles to the surface, and as a result they move toward and accumulate on the surface to form the coating. The various electrostatic coating applications are somewhat sophisticated modifications of this simple situation. They differ from one another in the manner in which the particles are formed, the means by which they are charged, the peculiar aspects of the methods by which the particles are distributed about the surface, and perhaps in the way in which they collect upon it. These differences, as well as possible similarities, become apparent as we discuss the various individual methods and applications in present use.

11.2   LIQUID ELECTROSTATIC COATING

11.2.1   Automatic No. 1 Processes

In the early 1930s a great number of the industrial coating operations were related to the painting or finishing of objects to produce on their exposed surfaces a protective or decorative coating. Such operations generally involved the use of liquid plastic solutions which were applied wet and then cured to develop the desired solid plastic surface film. It was normal procedure to apply these materials with an ordinary air spray gun. The process was fast, but very inefficient. Depending on the shape of the surface, between 50 and 80% of the material sprayed from the gun might escape deposition. Considerable incentive existed to devise a method that would save this oversprayed material and eliminate the labor and expense associated with its collection. It was conceived that charges might be placed on the sprayed particles so that they would be attracted to the articles and not blown past to become waste material. Two methods were developed, and patents were issued on both processes.[2,3]

The idea of Ransburg and Green[3] was reduced to commercial practicality in the early 1940s; many such installations were made throughout the world, and some are still in use today. The arrangement of the components of such a unit is shown in Fig. 11.1.

In this automatic process, the articles to be coated are supported on a conveyor and carried into a coating enclosure or booth. Inside the enclosure, extending along the path of article travel, a suitable charging-deposition electrode is located. The electrode has distributed over its surface a series of elements such as sharp points or fine wires. The electrode is insulated from ground so that it can be held at a potential of approximately 100,000 V with respect to the articles. It is spaced approximately a foot away from the articles to avoid the occurrence of disruptive discharges.

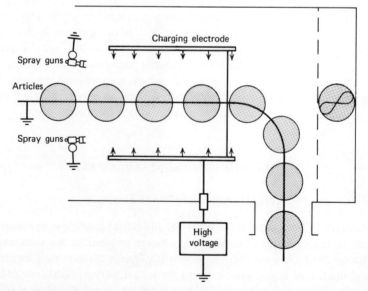

Figure 11.1. Schematic layout of a No. 1 automatic electrostatic spray-coating installation, parallel spray introduction.

At the entrance end of the booth, there is located an exhaust system which creates a general airflow from the open exit end of the booth toward the exhaust. The airflow is parallel to the path of the parts and through the space between the parts and the electrode.

One or more grounded air spray guns are placed at the exit end of the booth. These are arranged to form and project small droplets of the liquid coating material into the space between the electrode and articles so as to generally fog the material into the moving exhaust stream and generally around the article.

When the voltage is applied to the electrode, an electric field is established to the article. This means that the article must be electrically conducting, if it is to serve as the other electrode in this field. Several techniques have been developed whereby an article that is not electrically conducting can be made adequately conductive, at least on its surface. For example, phenolic molded parts can have a surface coating applied which has a high graphite content. Such a coating, when used on phenolic–wood dust, molded toilet seats not only gives them the needed electrical conductivity, but also acts as a primer–sealer for the surface. Rubber steering wheels, although normally good non-conductors, become conducting when they are heated to approximately 250°F. Heating the wheel as it is sprayed is further advantageous because a heavier, more durable coating film can be applied under these conditions. Glass objects that resist being coated at room temperature readily accept coating in an electrostatic field when the glass is heated above about 400°F. Chemical formulations have also been developed which leave a very thin conducting layer when they are applied to the surface of an object. Quaternary ammonia salts dissolved in low concentrations in organic solvents are applied to wooden furniture parts. The evaporation of the solvent leaves a gel-like residual layer of the salt on the surface. This layer is adequately conducting and does not interfere with the properties of the coatings that are applied over it as protection or decoration for the wood surface.

With an adequately conducting article and the arrangement of elements shown in Fig. 11.1, the application of a voltage to the electrode creates a high electrical gradient at the points or wires on the electrode surface. A corona discharge is created about the points, and air molecules (ions) charged to the same sign as the electrode move through the field from the electrode to the article. A small electric current flows from the high-voltage supply to the electrode, across the space to the article, through the article and its support to the conveyor, from the conveyor to ground, and thus back to the grounded side of the supply. A coating material particle sprayed into the field of charged air particles itself becomes charged to the same sign as the electrode by air ion bombardment.

We now have in this arrangement all the elements of the simplified electrostatic coating process. The charged particles experience a force toward the article. If the velocity imparted to them is not directed toward the article, the two forces compete for control of the particle. The particle responds to the resultant of these two forces and eventually is deposited on the article. If it passes the article, it is possible for the electrical force to reverse its flight and return it to the rear side of the article. In other cases, it is possible for the particle to escape deposition because the electrical force cannot dominate its motion. Such material particles are carried into the exhaust plenum by the exhaust air in the enclosure. By extending the electrode along the path of the article, the charging and deposition forces have longer to act and so deposit a greater percentage of the particles.

If the article is a single item supported in the electric field, coating material is collected on all its surfaces by the "wrap-around" action of the electric field deposition forces. Complete uniformity of coating is not obtained, however, since the deposited liquid continues to be electrically conducting, and coated and uncoated surfaces exhibit the same attraction for new material. To achieve uniformity, material must be distributed equally about the article and must experience equal attraction.

In actual production, the parts are usually supported from the conveyor on a rotatable hook and are rotated by a friction drag as they are moved along by the conveyor. All surfaces are thus comparably presented to the field and to the spray, and uniform coatings are readily obtained.

With this arrangement, it is possible to capture on the object at least half the material that would be lost if a normal air spray gun were used. In the spraying of items like refrigerators, a material transfer efficiency of 40% might be obtained with a hand-held air spray gun, whereas with electrostatics, the transfer efficiency can be increased to 70%. The wrapping effect of the electrostatic field offers an added advantage over straight air spray methods. Edges and surfaces not directly in the mechanical path of the spray are effectively coated with the electrostatic process. The surfaces on the back of panel members immediately adjacent the edges likewise receive protective coating with this method without special effort.

This process has served industrially to coat all types of production items, including refrigerators, automotive parts, military hardware components, waste baskets, oil cans, powder containers, and cannister sets. It has been used to apply extremely thin oil coatings to electrolytic tin plate in continuous sheet form. Starch solutions have been applied to paper webs as they are formed at the wet end of the paper-making machine, to enhance the surface properties of the sheet. In the latter cases, where flat sheets are being coated, the electrode system is usually an extended plane electrode located parallel to and above the moving sheet. A series of guns introduces the material into the field space essentially parallel to the sheet. The particles are deflected out of their flight paths to be deposited on the web.

Several modifications of this original automatic, air atomizing, electrostatic charging, and depositing process have been introduced. In one of these, the spray guns spray the material through the electrode system directly toward the articles on the conveyor.[4] In another, the electrode is actually mounted directly on the front of the atomizer and the entire gun–electrode assembly is charged to high voltage.[5] In this case also the material is sprayed at the object across the conveyor path. In both techniques, the particles are produced by air atomization, charged by ion bombardment, carried into the field by the air stream, and deposited by the combined action of the electrostatic attraction and the mechanical velocity imparted to them by the atomizer.

The second modification described previously is being used to apply liquid ceramic slurry to washing machine parts. The applied slurry is then dried and placed in a furnace where the frit is fused to form a porcelain enamel coating on the object.

11.2.2 Automatic No. 2 Process

The original or No. 1 type of electrostatic processes indicated the potentialities of electrostatics in the coating field. Although considerable material was lost when such methods were used, this inefficiency was the incentive which spurred individuals to look for still better methods.

It was quickly realized that the material lost with the No. 1 system escaped deposition because the air stream of the atomizing process blew the particles beyond the control of the electric field. A means was needed which avoided the use of air atomization. A method was developed in which the particles were formed at a point of high electrical field strength by the action of the field itself and charged at the instant of their formation. No adverse force then existed which would prevent the particle from following the lines of field force directly to the article. This method is now referred to as the No. 2 process. As applied to automatic coating operations, this process is used in three modifications; the blade, the bell, and the disc.

11.2.3 No. 2 Blade Modification

The blade modification, although used less than the bell and disc modifications, can be best used to illustrate the electrostatic aspects and method of operation of the No. 2 type units (Fig. 11.2). In this modification the atomizing head is a wedge-shaped electrically conducting member having a length comparable to the length of the spray pattern desired. The head is placed

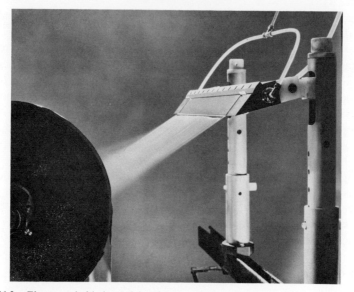

Figure 11.2. Electrostatic blade applying liquid coating to flat sheet stock.

adjacent the surface to be coated at a spacing of about 1 ft. It is oriented so that its length is parallel to the surface and so that the sharpened edge of its wedge-shaped cross section is directed toward the surface. It is supported on insulators and held at a potential of approximately 100 kV with respect to the surface.

When the voltage is applied, an electric field is created between the head and the object which has a fairly low gradient at the object surface but which has a high gradient at the sharpened front edge of the head. Coating material is pumped to the head so that it flows as a uniform exposed liquid film at this forward edge. At this position, it is exposed to the action of the high gradient electric field. The liquid is strongly attracted toward the article. In addition, however, the field forms the extended film into a series of liquid cusps or pointed liquid extensions spaced uniformly along the exposed length of the film. The extensions reach out into the field because they are attracted to the article. The small liquid element at the end of each of these extensions becomes a point of charge concentration. That small portion is attracted to the object or, if you prefer, repelled from the body of liquid on the atomizer. When the liquid tip is repelled by a force which is equal to the surface tension forces holding it to the liquid body, the liquid column necks down and a small, highly charged liquid particle leaves the cusp. These particles leave the head and travel along the electrical lines of force extending

Figure 11.3.   Liquid cusps formed at blade edge by high-gradient electrostatic field. Charged particles from cusps' tips are visible.

from the head to the object surface. This attraction and deposition can be seen in Fig. 11.2. The particles collect on the surface, the charge which they transport flows through the object to ground, and the particles flow and coalesce to form a coating film.[6]

It should be noted that in this process, the particles are formed without the presence of any second, possibly disturbing agent. They are charged by contact since, in fact, they actually carry away with them an appropriate portion of the charge on the edge. Being highly charged, they are deposited by strong deposition forces without any outside interference. The process potentially can produce 100% deposition of all material leaving the head.

The condition existing at the blade edge at any specific instant is a stable one resulting from the combined action of the high voltage field, the quantity of material available, the electrical characteristics of the material, and the rheology of the material. For any specific quantity of material delivered at a constant rate to the edge, a stable condition of uniformly spaced liquid extensions (see Fig. 11.3) will be established along the edge and charged particles of a rather narrow size distribution will be obtained. If the voltage is raised, the spacing between the cusps decreases and they become shorter. Smaller particles are formed. If the voltage is lowered, the cusps spread apart and become longer, and the particle size increases. When voltage is kept constant and the rate of material delivery is increased, the cusps separate and become longer. As would be expected, the rate at which material can be atomized from such a device to produce a satisfactory coating is limited. In practice, this is about 4 cm³ of material per minute per centimeter of edge length.

Such a blade type atomizer is best suited to coating operations requiring a long, extended, uniform pattern. The coating of continuous flat sheet material is an ideal application so long as the coating rate required by sheet speed or coating thickness is not too large. The blade is finding some industrial application in the finishing of appliances and appliance parts which have large flat areas to be coated.

11.2.4   No. 2 Bell Modification

The No. 2 electrostatic atomization and charging principles used in the blade device just described have been utilized in the No. 2 bell coating modification illustrated schematically in Fig. 11.4. The atomizer takes the form of a bell or funnel mounted on a motor for rotation about its axis of symmetry. The outer edge at the open end of the bell is sharpened. This open end is directed toward the surface to be coated. The bell and its rotating mechanism are mounted on an insulated support so that they can be held at high voltage with respect to the surface to be coated.[7] The bell is rotated at a rate of about 900 to 3600 rpm. Coating material is pumped to the internal surface of the bell at its small end. The rotation causes this material to spread uniformly over the inside surface of the bell and to flow to the forward larger end. At the sharpened forward edge of the bell or funnel, the material becomes exposed as an extended thin film to the action of the high-gradient electric field which is established there by virtue of the charge on the atomizer.

As at the edge of the blade device, the liquid film at the bell edge is formed into a series of liquid cusps. These are established on equal spacing about the circumference of the bell. From these cusps, as before, there is produced a spray of small, lightly charged liquid particles. These proceed to the target along a path which is the resultant of the centrifugal and electrical forces. The envelope of the sprayed particles looks very much like an umbrella extending from the head to the target.

If the surface being coated is a flat sheet, the material is deposited in a

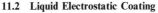

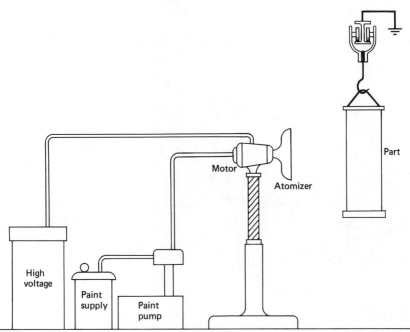

Figure 11.4.   Schematic elevation of automatic No. 2 electrostatic bell coating installation.

doughnut-shaped pattern. Movement of the flat surface relative to the atomizer produces an extended coated area on the sheet. In case the article being coated has a multiplicity of surfaces, the charged particles follow the lines of force from the atomizer to the head and land on these various surfaces. Surfaces hidden from the possibility of direct impingement of particles from the atomizer become coated by the electrostatic wrapping action. The part can also be rotated as it passes through the spray pattern to more easily obtain equal treatment of the various surfaces.

11.2.5   No. 2 Disc Modification

The No. 2 disc modification employs the same electrical principles of operation as the blade and the bell, but the atomizer has a different configuration and the parts being coated are handled in a different manner. As in the schematization of Fig. 11.5, a disc-shaped element is mounted on a drive so that it can be rotated about a vertically disposed axis. The outer edge of the disc is sharpened and the entire assembly of disc and drive motor is mounted on an insulating support. This assembly is positioned at the center of a loop in the article-carrying conveyor, which makes an arc of about

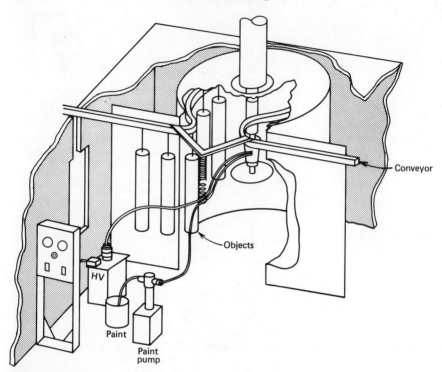

Figure 11.5. Schematic elevation of automatic No. 2 electrostatic disc coating installation.

300° about the disc. The disc can be rotated at 900 to 3600 rpm, although higher rates can be used if desirable.

A high-voltage direct current is applied to the disc, creating an electrical field that diverges from the disc edge to the surface of the articles on the conveyor. Coating material is pumped to the surface of the rotating disc at its center. The rotation spreads this material over the disc surface and causes it to flow to the outer edge of the disc, where it appears as a thin extended 360° film.[8]

When no voltage is applied to the disc this material is thrown off the disc edge by pure mechanical forces. The cusp formation is very irregular and each has a bulbous head which is eventually thrown off as a large particle. The attachment thread shatters giving many small particles (Fig. 11.6a).

When the high-voltage field is applied under the same conditions, the film at the edge is exposed to the high gradient existing there. It is shaped, as before, into a series of liquid cusps equally spaced around the disc perimeter (Fig. 11.6b). The cusps extend out into the field toward the articles, and from the tips of the cusps, a cloud of highly charged particles is

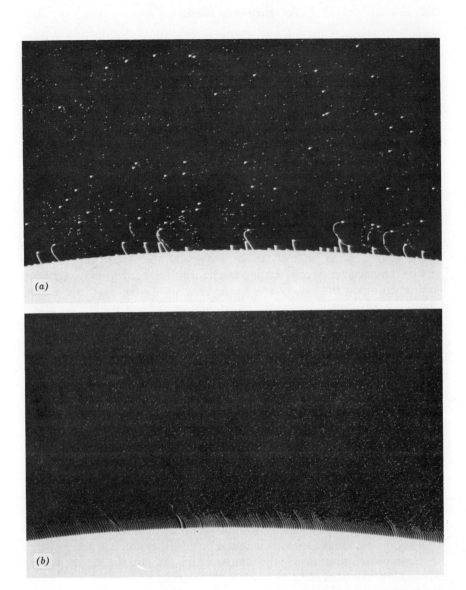

Figure 11.6. Particle formation at rotating disc edge: (*a*) without electrostatic field applied; (*b*) with electrostatic field applied.

atomized. The particles move out into the space between the disc and the articles on the conveyor and follow a path to the article dictated by the resultant of the centrifugal and electrical forces. They disperse in space because they are all charged to the same electrical sign. The overall attraction of the particles to the articles causes the deposition of those particles which would miss the articles in the absence of the electric field.

Since the material is emitted from cusps located around the entire circumference of the disc, the parts receive coating from the time they enter until they leave the loop. The entire emission of the disc is effectively used.

Because the particles are charged and disperse, they do not land on the articles as a narrow band of material (as would be the case of purely mechanical deposition) but are deposited in a relatively wide band. With the disc edge to article set at about 12 in. this dispersion will cover an article height of about 8 in. Where the surface to be coated is higher than the 8 in. obtainable from a stationary disc, the disc and rotating motor can be mounted on a reciprocating mechanism arranged to move them up and down along the axis of the conveyor loop. As the parts are moved about the conveyor loop, the disc is stroked up and down over the article height. Where the article surface to be coated is complex, the parts on the conveyor can be rotated or mechanically indexed at appropriate spots about the loop in order that the various surfaces be suitably exposed to the disc to ensure equivalent coating.

Numerous industrial applications of the disc and bell modifications of the No. 2 process have been made for the coating of a wide variety of items. Golf balls, toilet seats, automotive trim parts, oil filter containers, acoustical panels, bicycles, and folding metal chairs are representative of this variety. Throughout the world, almost all refrigerators, washing machine and dryer jackets, water heater cases, and similar coated appliance items are now done by one or another of these processes. When these techniques are employed, the overall attraction of the particles to the ware makes it possible to obtain about 2.5 times as many parts per gallon of material as can be obtained with ordinary nonelectrostatic air spray methods. This increased production per unit of sprayed material obviously represents great savings in the manufacturing costs of the item and explains in part the great success which electrostatic coating methods have achieved.

11.3   ELECTROSTATIC HAND GUNS

11.3.1   Automatic Deposition

The No. 1 and No. 2 processes just described are automatic deposition arrangements. In such equipment installations, the parts to be coated are carried to and through the coating station from some remote point, moved to the curing station, and eventually conveyed to the unload or finished stock position. In the coating station the atomizers are in specific positions about the articles so that the charged material particles will be appropriately distributed about the articles to produce a satisfactory coating. Usually there is no occasion for any plant personnel to be near the atomizing equipment when it is in operation. Starting and stopping of the operation, as well as any adjustment, can be accomplished from remote stations. In this sense these systems are somewhat inflexible. A unit set up to coat flat sheet could not effectively be used to coat appliances.

To introduce greater flexibility into the operation and still retain the advantages of electrostatic attraction, hand-held devices have been developed which can be manipulated by an operator to control the distribution of the charged particles about the article. These devices combine the deposition efficiency of electrostatics with the flexibility of a hand-held unit. Since high voltages are required and the spacing of the device from the surface can freely be altered by the operator, the device must be designed to be free from electrical discharge under these conditions. In addition, the operator must likewise be protected against the possibility of electric shock. Such devices have been successfully designed—one variety uses the concepts of the No. 2 type electrostatic automatic units; others feature air atomization or hydrostatic atomization in combination with ion-bombardment charging.

11.3.2   No. 2 Process Hand Gun

In the No. 2 process hand gun, the liquid material is presented to the action of the high voltage field as a thin film; it is atomized under the influence of the high gradient at the film, charged by contact with a high-voltage electrode, and deposited by the action of the electric field between the atomizer and the surface being coated. The device (Fig. 11.7) consists of an insulated elongated body with a handle attached at one end and a rotating bell-type arrangement at the other.[9] The bell is made of an electrically insulating material but has on its outside surface a coating of a material which is a poor conductor. The bell is rotated at about 600 rpm by an air motor or a small electric motor in the gun body near the handle. A shaft of insulating material allows the bell surface to be charged without also placing a charge on the drive motor or the handle.

Coating material is delivered to the rear inside surface of the bell by way of a valve in the handle and a tube running lengthwise through the gun body. A high-voltage cable extending from a suitable supply to the gun carries approximately 90 kV dc potential to the conducting surface of the bell by means of a sliding brush contact. The coating on the bell conducts this potential to the sharpened forward edge.

Figure 11.7. No. 2 electrostatic hand gun being used to finish office furniture.

In use, the atomizing bell is set into rotation, the voltage applied, and the coating material is pressure fed to the rear of the bell. The rotation causes the material to flow across the inside surface of the bell and form a thin extended liquid film coextensive with the forward sharpened bell edge. The material film comes under the influence of the electric field at this edge and is formed into the cusps from which the charged particles are created. By manipulation of the gun over the surface of the article to be coated, the material is deposited by the field on the surfaces of the article selected by the operator.

The voltage supply is designed to have a high internal impedance, and its output is delivered to the atomizing bell through a high resistance. The exposed brush contacting element is kept extremely small so that its electrical capacity will be low. The distributed resistance on the surface of the bell makes it impossible to discharge any sizable portion of its stored charge by contacting it at a single point. The combination of all these features makes it possible to have a practical, operable, safe electrostatic hand spray gun.[10]

Guns of this type, since they apply material essentially without loss, are being widely used in the on-site refinishing field. Metal office furniture can be cleaned, scuff sanded, and refinished with a self-curing epoxy material without removing it from its place of use. Only the floor area immediately adjacent the desk pedestals need to be protected from stray material. In a matter of only a few hours, an entire office can be reconditioned in matching colors and be ready for continued use, thanks to the effective depositing action of electrostatic fields.

Although the No. 2 hand gun allows material to be applied at very high efficiency, its operation is subject to several limitations. Because atomization results from the interaction of the high-voltage field with the liquid film, the degree to which this occurs is related to the electrical characteristics of the material. Not all materials respond equally well. Some are too conducting and actually ground or short out the applied voltage so that the bell cannot be held at atomizing potential. With materials that are too insulating, the particle forming and charging functions are disturbed. Still other materials appear to be too highly polar and for this reason do not atomize with full effectiveness. In most cases, the material can be adjusted so that it will work well. Nevertheless, it would be highly desirable to eliminate this required adjustment.

In addition, successful operation of the No. 2 process depends entirely on the electrical deposition forces of the field. This means that the field distribution about the part controls the amount of material deposited on various portions of the article. Little or no material can be deposited in deep recesses or in areas that are shielded by the Faraday cage type of field shielding. Here again, on certain objects it would be desirable to have an added force that could be used to project the charged material into such cavities or shielded areas.

To meet these two desires, electrostatic hand guns have been developed which use air or hydraulic pressure to atomize or create the particles and then charge them by corona charging. The particles are deposited under the resultant action of the electrostatic forces and the mechanical impulse imparted to them by the forming agent.[11,12]

11.3.3 Air Atomizing Electrostatic Hand Guns

Guns that employ air atomization to form the particles are referred to as air-electrostatic guns. A typical gun of this type appears in Fig. 11.8. Generally, these guns have a body of insulating material with a normal grip-type handle on one end and an air-cap, fluid-tip assembly on the other. The latter components are made of insulating material; they generally accomplish the same functions they serve in an ordinary spray gun and are of similar construction. The liquid coating material exits from the fluid tip

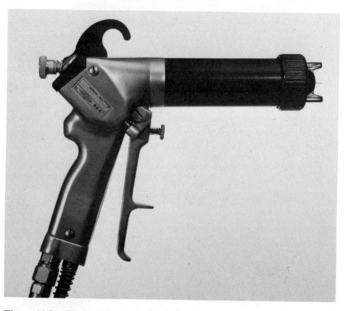

Figure 11.8. Typical electrostatic air hand gun.

at relatively low pressure as a liquid stream and is atomized by jets of air from the various air outlets in the air cap impinging on it. It is then projected outward away from the gun in the form of a spray. Some guns of this type have modified forward ends which accomplish the liquid breakup in a different manner. As a rule, however, the fluid–air relationship is one that produces an expanding spray of small particles.

At or near the position of particle formation in all guns of this type there is located an electrode which is maintained at corona-producing potential. This can be a small pointed wire element within the body of the gun or it can be the liquid jet itself. Regardless of the exact configuration, the existence of this corona at or near the site of particle formation charges the produced particles by ion bombardment. In a sense, these guns are portable No. 1 type systems in which the charging electrode is carried by the gun and moved with it over and about the article surface.

The charged spray particles are projected away from the gun toward the surface. They possess the velocity and direction given to them by the atomizing air. For some particles this velocity is so high that they would miss the article surface if no other forces were present. However, since the particles pass through the corona-formed charged air particles, they are charged and become attracted to the article. This electrical attraction attempts to overcome the mechanical movement, altering the particle's flight so that it lands on the surface. In the actual use of such a gun, about half of the particles that normally would escape are redirected to the surface as useful material. The overall transfer efficiency can be as high as 80% under ideal conditions.

At the present time, eight or ten manufacturers produce air-atomizing electrostatic spray guns in this country.[13] The voltage applied to the charging electrode in these guns varies from 30 to 60 kV, depending on the manufacturer. In most guns the voltage is supplied from a cascade or ladder-type circuit energized from 110-V ac circuits. In several of the guns a voltage of 4 to 6 kV is produced by the primary supply, and this is delivered to the gun where it is multiplied by an EGD device to a value of approximately 60 kV, which is applied to the charging electrode in the gun. In all guns intended for manual use, the shock and discharge hazard to which an operator is exposed is kept low by using voltage supplies having high impedance values and gun-charging electrode systems having low electrical capacities. An operator who has the grounded handle of such a gun in one hand can touch the charging electrode of the gun with the other without experiencing any adverse discharge or shock.

11.3.4   Hydraulic Atomizing Electrostatic Guns

In some electrostatic hand guns, hydraulic pressure instead of air is used to atomize the liquid materials. Such guns, referred to as hydraulic electrostatic guns, are made by numerous manufacturers, with various modifications. All these devices, such as the one in Fig. 11.9, have a main elongated body portion of insulating material. The handle is at one end of this body member. On the forward end of the body, there is an extremely small orifice—equivalent diameter about 0.007 in. This is the outlet orifice from which the liquid is atomized. The coating material, which is pumped from a container to the gun at a pressure of about 1000 psi, is valved to the orifice by motion of a trigger on the gun. As the material is released from the orifice, it atomizes into a fine spray. By correctly shaping the orifice this material can be spread into a fan-shaped spray.

A sharp-pointed, small-wire electrode charged to high voltage terminates adjacent the point of particle formation. A corona is formed at the electrode termination, and the material particles formed at approximately this same position become charged as they pass through this ion cloud. Once charged, the particles move out away from the gun under the combined influence of the mechanical momentum and the electrical attraction due to their charge. Since there is no air blast associated with particle formation in this type of gun, the mechanical momentum quickly dissipates leaving the

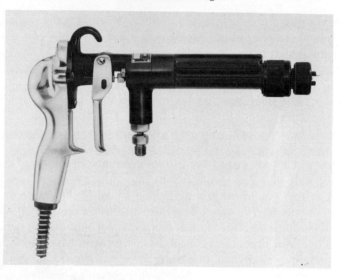

Figure 11.9.   Typical electrostatic hydraulic hand gun.

particle to be controlled by the electrical forces. Therefore, the transfer efficiency with such guns is usually slightly higher than that obtainable with air-electrostatic guns. Because hydraulic guns are somewhat inflexible insofar as pattern adjustment, delivery control, and degree of atomization are concerned, they have not been as widely accepted in industry as the air-type electrostatic guns.

Electrostatic guns of the air and hydraulic atomizing types have been modified to be attached to mechanical motion devices and are being used in automatic coating systems.

In addition to being used for the normal application of paint to all types of manufactured articles, these various types of liquid electrostatic application equipment also serve in many unusual processes. The air atomizing electrostatic guns can apply a spray of metallic organic compounds to the surface of hot glass objects. The applied compound decomposes on the hot surface, imparting to the object a metallic luster with a color that is characteristic of the particular metal in the compound. The high cost per gallon of these metallic compounds makes the high deposition efficiency of electrostatic application almost a necessity for the economic production of such items.

The electrostatic bell-type equipment is used in the formation of a type of nonwoven sheet. Polyurethane materials having some elastomeric characteristics are sprayed onto a flat surface. As the material comes under the influence of the high-voltage field at the edge of the bell, it is elongated into filament-shaped elements. As the material is attracted by the electrical forces of the field, the elongated elements become unstable and break off. These filaments are charged and are attracted to the oppositely disposed flat surface. The rotation of the bell causes the filaments to be deposited in jackstraw fashion on the surface. The elongated filaments cross over one another, adhering at their intersecting positions. The random deposition imparts a nondirectional strength to the porous sheet formed as the applied elongated fibers solidify. This composite porous sheet is then stripped from the backing and used as a sheet of nonwoven synthetic cloth. Such materials are also applied to a rigid backing to form a suede or pilelike finish on such backings. These are used as liners for the underdash compartment and trunk walls in some automobiles.

All types of wooden and plastic items are being coated using electrostatic methods. In these cases the nonconducting parts are rendered conducting by one of the techniques described earlier.

Liquid electrostatic deposition techniques have been used to apply liquid insecticides to growing plants. The liquid particles are charged and dispersed from suitable electrostatic guns carried by tractors or by air-borne vehicles as they move across the planted fields. A charged cloud of the insecticide accumulates over the (grounded) plants. The space-charge field created by the cloud of charged particles deposits the sprayed materials on the exposed plant foliage. It is particularly effective in causing deposition on the underside of the leaves, which normally are surfaces that are hidden to straight mechanical application. Better overall application and consequently better protection from insect attack results.

Wherever liquid materials are to be applied to surfaces for whatever reason, electrostatic deposition methods should be considered for possible use. One of the available methods is sure to meet the required conditions of application. Many times an electrostatic method actually makes a process practical which under other conditions would be uneconomical.

11.4   SOLID ELECTROSTATIC COATING

Many industrial coating applications do not involve the deposition of liquid particles but are concerned with depositing dry particulate materials which are powdered or dustlike. Again, in this case the basic concepts of Coulomb's law can be applied. If the particles can be distributed about the surface to be coated and given an electrical charge which is opposite to the surface, they will be urged toward the surface and will be deposited there as an effective coating. This elementary concept has been elaborated on in many respects to fit the demands of many industrial operations. Electrostatic deposition of many kinds of dry particulate material is being used to help solve many industrial coating problems.

11.4.1   Electrostatic Formation of Sandpaper

One of the earliest large-scale applications of electrostatic forces in the manufacture of a product that involved the deposition of a dry powderlike material was the manufacture of sandpaper or similar abrasive items.

In this process[14] two extended flat electrodes, illustrated schematically in Fig. 11.10, are positioned parallel to each other at a separation of about 4 in. and are insulated so they can be charged with respect to each other in the manner of two plates of an air dielectric condenser. A continuous belt of semiconducting material passes over a group of rolls and is in contact with the upper surface of the lower electrode plate. The paper or textile web, which is to become the backing for the abrasive product, is fed from a roll through a glue applicator which places a layer of adhesive on its one face. The web then passes through the spacing between the

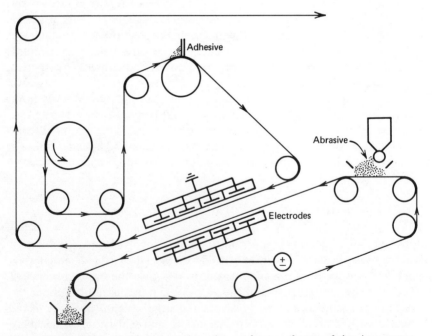

Figure 11.10. Schematic of electrostatic equipment for manufacture of abrasive papers.

electrodes with its glued surface facing downward and with its other surface in contact with the upper electrode. The electrodes may be of special design, depending on the imagination of the builder; usually they are metal sheets encapsulated in a semiconducting material.

A dc voltage in the neighborhood of 100 kV is applied across the two electrodes. The ground frit—agate, sand, or other dry material forming the abrasive component of the product—is distributed over the surface of the lower moving belt by a hopper feed mechanism at a point outside the electrodes. The belt carries this material on its surface into the space between the electrodes. Simultaneously, the glued surface web is also moving through this space.

When the material on the lower belt arrives over the surface of the electrode, it becomes charged by contact to the same sign as the electrode. It is repelled from the lower electrode and attracted toward the upper one. If the forces are sufficiently large, each particle will be moved upward. At the top electrode, the glued paper backing intercepts the flight of the particles and they are imbedded in the glue. If an arriving particle lands on top of an already adhered particle, it will eventually become charged to the sign of the top electrode and will be repelled toward the bottom belt. Upon hitting the belt it again has its sign reversed and makes a second trip to the glue layer.

Each particle executes this up and down motion until it is either affixed to the glue or is carried out of the field between the electrodes by the motion of the web and belt. As the upper web leaves the electrode space, it is mechanically beat or vibrated so that all unattached material returns to the lower belt for collection in a suitable hopper. The coated web is carred into an oven where the glue is set to attach the grit firmly to its surface.

If this deposition of the grit onto the paper was the only action that occurred, no advantage would exist for using electrostatics instead of pure mechanical distribution of the material into the glue. However, another phenomenon takes place. Most of the abrasive material particles are ground and have an asymmetric shape. When such particles become charged on the lower belt, they first stand on end to align their long axes with the lines of the electric field. When the force becomes sufficient, they travel to the upper electrode in the same oriented position and, so to speak, dive head first into the adhesive. They are oriented on the paper in this peculiar but very beneficial position—sharp tails out. The glue is set and the particle held there to form a better, faster-cutting, electrocoated paper.

Obviously, many factors contribute to the optimum operation of such a system. The conductivity of the particle, its weight, the rate at which the frit is fed into the electric field, the magnitude of the field between the electrodes, the ambient humidity, the nature of the adhesive, and many such factors must be correlated. The operation depends on the particles being satisfactorily charged and on their accepting and retaining this charge for a sufficiently long time. Any factor that can in any way influence this requisite becomes important. Proof that this correlation can be effectively accomplished is seen in the many miles of abrasive paper and cloth that have been manufactured in this manner.

11.4.2   Electrostatic Flock Coating

Another practical development of solid particle electrostatic deposition which has many features in common with sandpaper making is that of electro-static flock coating.[15] Flock is a generic term used to describe a product made by cutting fibers or filaments to a relatively short length. A quantity of nylon flock, for example, might be made by taking a bundle of continuous nylon filaments and progressively cutting across the end of the bundle to cut off the ends of all the filaments. Each cut-off portion then becomes a small solid particle of material which is cylindrical, with a length larger than its diameter. Flock coating consists of applying such a material onto a glue-coated surface and allowing the glue to dry in order to adhere the flock. Such coated surfaces have a felt or suede appearance and are readily recognized on many toys.

Electrostatic flock coating is a method of applying these materials to the glue with the use of electrostatic forces. Here, as in the sandpaper case, the principal advantage of using electrostatics resides in the ability of the charged flock particles, as they obey Coulomb's law, to align themselves with the electric field and so to land on the glue in an oriented position. Since the particles also make repeated attempts to be attached to the glued backing, the coated surface becomes completely covered with an upright standing pile. After the glue or adhesive dries and the attached fibers are well secured, the excess unattached flock is removed by a vacuuming process to leave the finished surface. Since the best flocks are fibers all cut to the same length, the finished surface is a very uniform, highly attractive one, having the appearance of rich suede or velvet. It likewise has a resiliency that can be sensed when the surface is brushed or felt.

Equipment for electrostatic flocking has been developed for various types of applications. Large extensive units for applying flock to continuous paper or textile webs have been in industrial operation for years. As in the electrostatic manufacture of sandpaper, the backing web is glued and carried into an electric field established between two plate electrodes. The glued surface is directed downward away from the upper of the two electrodes. The flock is distributed over the surface of a belt which carries it into the electric field in contact with the lower electrode. As the flock enters the field, it is charged. The filaments oscillate in the field, aligned with their length parallel to the field lines. The fibers are attached to the glue covering the entire glued area with a suedelike nap. Excess fibers are removed and the glue is allowed to dry to form the final product.

If the field lines are perpendicular to the glued surface, the fibers will be oriented to the surface in a perpendicular fashion. If, however, the field lines are otherwise disposed to the surface, the fibers can be affixed to the surface at an angle. A pattern can thus be created on the surface. If, for example, the top plate electrode in the deposition equipment is replaced by a series of rods extending along the direction of web movement, a pattern of lines in this direction will be recognizable in the final coating.

Patterning can also be accomplished by printing the glue onto the surface in the desired pattern so that fibers adhere only to these selected areas. Lettering on surfaces is done in this manner. Athletic uniforms, novelty sweaters, and similar knit garments are printed by this method. In most of these applications the garment is stretched over a metal form, the desired lettering is silk screened in glue on the surface, and the flock is deposited on the surface, using the form as one of the deposition electrodes.

Electrostatic methods are used to produce flocked wall paper. The backing paper is uniformly glued and flocked in the normal process with perpendicularly deposited fibers over all. Before the glue is set, the surface is rolled with an embossing roll that flats the flock in selected areas to produce the desired pattern contrast. This type of product has found use in providing the red "gay nineties" decoration in many bars and lounges.

Curtain material with patterned adhered dots of tufted flock, artificial suede for play shoe tops, and many other articles are now produced in this fashion. Carpeting with adhered pile instead of the normal "woven-in" pile is also manufactured. A much longer flock fiber is needed for this application.

Small hand-held electrostatic flocking units are available which utilize the same deposition and orientation forces for the application of flock to structural surfaces. Such equipment consists of a hopper for the fibers, suitably arranged with respect to an electrode system that can be charged to a high dc voltage with respect to the grounded surface to be coated. When in use, the fibers are electrically agitated in the hopper and are propelled by this agitation into the field between the electrode and the surface. As the operator moves the unit about over the previously glued surface, the flock fibers adhere and the surface is coated. Excess material is removed after the glue cures. Entire walls of a room can be coated in this manner. Recently some novelty coatings have been applied to automobiles with portable electrostatic equipment.

Since the movement of the flock is entirely caused by the electric field forces, the electrical characteristics of the flock fibers are highly important for good operation. Fibers are usually conditioned after cutting to be given the correct surface conductivity. Large-scale continuous operations are done in temperature- and humidity-controlled environments. There are flocks manufactured and sold which are specifically for electrostatic applications.

11.4.3 Electrostatic Powder Coating

Painting or the application of decorative or protective coatings to the surface of articles has always been a widely used industrial process. Such processes generally have utilized liquid materials. In the early 1950s a method was developed in Germany by Erwin Gemmer of Knapscak-Griesheim, whereby coatings or finishes could be applied to metal surfaces using dry, powdered-resin coating materials. In this method the part to be coated was heated and brought into contact with the resin powder. The powder fused upon hitting the hot surface and adhered there, eventually flowing and leveling to form the desired coating. In most of these cases the powder was placed in a porous-bottomed container and air was blown up through it so that the powder was fluidized. The hot part was then dipped into this air–powder mixture. Coating application could not be too well controlled.

Nevertheless the fluidized bed method has had wide industrial use. All sorts of wire goods (dishwasher baskets, fencing, etc.) are being vinyl coated by this purely mechanical method.

The basic capabilities of this method as a coating technique and the ability of electrostatic forces to move small particulate material have been combined to produce a number of electrostatic powder coating methods. These differ in the manner in which the powder is presented, charged, and brought into contact with the article surface.

11.4.4 Electrostatic Fluidized Bed

The electrostatic fluidized bed was the earliest method featuring electrostatic deposition of resin powders on articles to produce commerically acceptable finishes.[16] In this method, schematized in Fig. 11.11, the powdered resin is fluidized in a normal mechanical fluidizing bed. A series of charging electrodes is immersed in the fluidized powder. When these electrodes are connected to a high voltage, a corona is produced at their tips or points. The corona creates air ions which charge the powder particles in the bed by ion bombardment. Under the action of this charge they repel one another and are attracted to any grounded surface in the vicinity. A conducting object

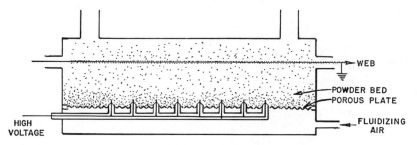

Figure 11.11. Electrostatic fluidized bed.

held above the charged powder in the bed becomes such a surface and quickly accumulates a powder layer on its surface. If the object surface is heated, the attraction still exists. Under these conditions the particles hitting the surface are fused and melted onto the surface. Since the melt is electrically conducting, the charge on the accumulated material leaks off to the underlying metal surface and flows to ground. Additional powder continues to be attracted and accumulated on the surface as long as there is sufficient heat at the exposed surface to sinter the arriving particles.

If, on the other hand, the surface is not heated, the particles are still attracted. Under these conditions the particles are held to the surface by electrostatic attraction. Since the particulate material is normally a non-conductor, only that charge effectively in contact with the surface can leak off to ground. A charge is accumulated and retained on the surface as the powder layer accumulates. Soon the charged layer no longer accepts additional material, and the coating process stops. It is, so to speak, a self-limiting procedure. The electrostatic attraction of the retained charge for the surface causes the powder to stay on the surface. The article is then carried into an oven where heat is applied to melt and flow the accumulated powder to a solid continuous film.

The electrostatic fluidized bed method has been successfully applied for coating continuous wire webs, sheets, and small parts. A bed is constructed having a width equal to the width of the web to be coated and a length along the travel of the web commensurate with the film to be accumulated and the speed of the web. The web is then simply moved across the bed. The powder jumps out of the bed and adheres to the web surface. Individual articles are similarly handled. The deposition forces of the field are largest at that portion of the object which is closest to the bed, diminishing rather rapidly as the distance above the bed increases. For this reason the technique is best applied to sheet-type surfaces or to discrete objects which are not too tall.

11.4.5   Electrostatic Powder Hand Guns

The limitations of the electrostatic fluidized bed, insofar as it is a fixed-position applicator, have been overcome by another recently introduced method. Hand-held devices, such as the one in Fig. 11.12, have been developed which receive the powder from a suitable supply, charge it, and dispense it as a charged powder cloud over and about the object being coated at the discretion of the holder. This material is then attracted to the surface to form the coating.

The object being coated can be heated or unheated, as desired. As in the case of the fluidized bed, a heated article continues to accumulate sprayed material as long as heat is available to produce sintering. In the case of the unheated article, the charge accumulation creates a self-limiting film acceptance. It is possible, therefore, to obtain coatings with this method which are of quite uniform thickness. This method of applying the powder to unheated objects is receiving the greatest acceptance of all the various powder application methods.

In hand spraying powder electrostatically for the coating of surfaces, the powder material is aerated in a fluidized bed, picked from this bed by a venturi-type pump, and delivered in a supporting air stream to the gun. The powder leaves the gun in the form of a dispersed fan that is directed toward the surface to be coated. Adjacent to the dispersal point in the gun there is located a sharpened charging element, which is connected to a source of high voltage. As the powder passes from the gun it is charged by ion bombardment. This charge urges the powder toward deposition on the article. The charging elements in the gun, which have low electrical capacity, are energized by high-impedance circuits; thus no discharge of high energy can occur when the elements are contacted by grounded objects.

The charging elements are usually held at about 60 kV direct potential with respect to the oppositely disposed surface. These voltage supplies usually have negative output and are adjustable from 0 to 60 kV. There is some evidence, although it is not fully conclusive, that some powder materials are more effectively applied by using one polarity than with the other. Some manufacturers provide units capable of being switched from one polarity to the other at the discretion of the operator. With adjustable output with respect to both polarity and voltage magnitude, a completely flexible arrangement is available, allowing the operator to match the application conditions with the powder characteristics to obtain optimum functioning.

When the object is covered with powder it is carried into an oven, where it is heated to melt the powder and to allow it to flow. The

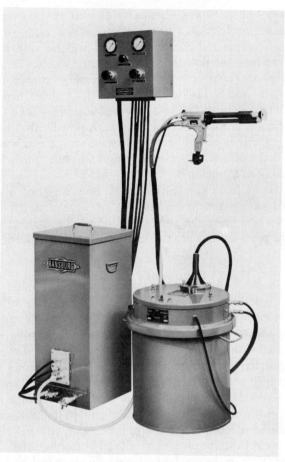

Figure 11.12.   Typical electrostatic powder hand gun.

coatings formed by this technique can be equivalent in all respects to those obtained with liquid materials, and they are obtained without the use of the usual solvents and diluents that contribute to hazard and pollution. Coatings of materials like Teflon and nylon, which are not readily soluble in normal finishing solvents, can be applied by powder methods.

Using hand electrostatic powder methods it is possible to accumulate about 80% of the powder dispensed from the gun on the surface. The other 20% escapes deposition in any one application, but it can be recaptured in a cyclone- or bag-type filter. If the material collected is kept in an uncontaminated condition, it can be reused in the equipment, thus obtaining overall efficiencies close to 98 to 99%.

The ultimate film thickness obtainable as well as the overall efficiency of deposition is closely related to the electrical properties of the powder. If the powder accepts the charge readily, it will be very effectively deposited. If, however, the powder loses its charge easily, the film that can be accumulated on a cold surface may be extremely variable. If the powder is highly resistive and holds its charge for a long time, then only a very thin film may be obtained. The best powders, since they need to meet these various conditions, are carefully formulated. A powder that would deposit at near-100% efficiency would be ideal because such a rate would avoid all the troubles connected with the reclaiming and reuse. Such ideal powder has not as yet been formulated.

The electrostatic hand gun features have been designed into automatic powder dispersion equipment devices. The units are mounted on automatic manipulators which move the guns to distribute the charged powder over and about the parts as they are carried through the application enclosure.

11.4.6 Electrostatic Powder Disc Unit

A special disc applicator has recently been introduced to supplement the capabilities of the other automatic methods in this rapidly expanding area. The device consists of a disc element made of insulating material. Its surface is coated with a material which is a poor electrical conductor. The coating is connected to a high-voltage source at the disc center to produce a high-gradient field around the outer sharpened disc edge. Powder is introduced into a swirling air stream at the center of the disc and is spun centrifugally outward across the disc face and past the disc edge. As it passes the edge in a 360° plane pattern, it becomes charged electrically by ion bombardment from the air ions produced at the disc edge. The charged material moves toward the articles under electrical attraction. The grounded articles to which the powder is attracted are carried on a conveyor about the disc. The powder emission from the entire circumference of the disc is thus utilized.

The powder particles as they move away from the disc all carry the same electrical sign; thus they tend to disperse on their way to the articles. This means that a coating band of a width greater than the mechanical projection of the disc thickness is produced on the articles.

When it is desired to coat longer parts on the conveyor, the entire disc unit is supported on a reciprocator arranged to move it up and down along the axis of the loop. The entire unit is placed within a cylindrical enclosure. Undeposited material is collected in an exhaust system arranged to withdraw air centrally from the bottom of the enclosure. This material is carried by the exhaust air through a cyclone collector system. The cylindrical symmetry of the system and the long deposition time over which the part accumulates its coating gives this arrangement even greater promise than other techniques.

The disc in this system can be energized at 90 to 100 kV to obtain added efficiency.

11.4.7   Other Applications Using Dry Powder

An unusual arrangement for applying powder recently introduced in Europe is illustrative of the manner in which various approaches are being taken to use the very basic attraction laws to meet the demands of a particular production need. In this arrangement, which might be referred to as a powder curtain, an effort is made to direct a relatively broad uniform pattern of charged powder at the object in order to achieve added uniformity of the film collected on the part. The objects are supported on a conveyor. Adjacent to the path of the conveyor in a coating enclosure there is located a wide porous belt which has considerable width in the direction of travel of the parts. This belt is arranged to move about a top and bottom cylindrical pulley. The movement of the belt portion adjacent to the articles on the conveyor is upward past the articles. It then passes over the top pulley and returns to the bottom one. As the belt passes about the bottom pulley, it moves through an enclosure where a layer of powder is deposited on its surface. As this layer is carried upward by the belt, it passes an air jet being blown through the belt toward the articles from a pipe member behind the belt. The pipe member is sloped from top to bottom across the belt width. A powder spray is thus progressively directed toward the article's surface as the belt is blown clean of the carried powder.

A fine wire is stretched across the belt on the side opposite the air jet, which is the same side as the article. This wire is held at high voltage with respect to the grounded articles so that a corona charging atmosphere exists about its length. The powder being blown from the belt surface travels past the wire, becomes charged by ion bombardment, and is attracted to the article surface. This technique is too new to be adequately evaluated. It appears to have some merit over other proposed methods.

Electrostatic powder application is also being used in commercial areas other than coating, flocking, and sandpaper manufacture. Systems have been developed specifically to apply powdered fluxes to surfaces that are to be subsequently welded or otherwise processed. On steel manufacturing lines, metal modification fluxes are being applied ahead of the annealing step. These convert when the annealing heat is applied and react with the steel surface to produce on the surface an alloy that is more durable and resistant to corrosives than the base metal.

Powder systems are also in use which electrostatically apply talc to raw rubber belting or sheet to dust its surface so that it can be subsequently handled. The overall attraction brought about by the use of electrostatic fields keeps the talc from escaping from the coating enclosure into adjacent manufacturing areas.

Powdered adhesives are being applied to flat surfaces which later are laminated together at elevated temperatures and pressures.

Solid starches are deposited onto freshly printed sheets just ahead of the stacker in order to prevent the offset which normally occurs when such pages are stacked face to face at the end of the press.

Salts and seasonings are being applied to food products as these products, spread out on the surface of a carrying belt, are carried from one processing operation to another.

There is an ever-broadening field of interest in the use of electrostatic deposition forces in various industrial fields. Hopefully, greater ingenuity will be brought to bear on these newer problems as the result of the older applications having been reviewed in this chapter.

REFERENCES

1. J. Hohlfeld, " Das Niederschlagen des Rauches durch Elektrizität," *Arch. Ges. Naturlehre*, **2**, 205 (1824).

2. E. Pugh, U.S. Patent 1,855,869, April 26, 1932.

3. H. P. Ransburg and H. J. Green, U.S. Patent 2,247,963, July 1, 1941.

4. K. Sittel, *Elect. Eng.*, **79**, 288 (April 1960).

5. J. Sedlacsik, U.S. Patent 2,710,773, June 14, 1955.

6. W. A. Starkey and E. M. Ransburg, U.S. Patent 2,685,536, Aug. 3, 1954.

7. E. M. Ransburg, U.S. Patent 2,893,894, July 7, 1959.

8. C. C. Simmons, U.S. Patent 2,808,343, Oct. 1, 1957.

9. E. P. Miller and L. L. Spiller, *Paint Varn. Prod.*, June 1964.

10. J. W. Juvinall and J. C. Marsh, U.S. Patent 3,048,498, Aug. 7, 1962.

11. J. W. Juvinall, E. Kock, and J. C. Marsh, U.S. Patent 3,169,882, Feb. 16, 1965.

12. Staff, *Furniture Des. Manuf.*, September 1966.

13. A. C. Walberg, *Prod Finish.*, **35**, No. 11, 66, (August 1971).

14. R. L. Melton, R. C. Benner, and A. P. Kirchner, U.S. Patent 2,187,624, Jan. 16, 1940.

15. J. O. Amstuz, *Rubber Age*, **49**, No. 1 (April 1941).

16. G. Nicolas, "La Samesation," *Surfaces* No. 17, June 1965.

CHAPTER 12

Electrostatic Imaging

CHARLES D. HENDRICKS, *Director*
Charged Particle Research Laboratory
Electrical Engineering Department
University of Illinois
Urbana, Illinois

12.1 OVERVIEW—BACKGROUND AND HISTORY

We should, perhaps, begin this section by a definition of electrostatic imaging. As a very unsophisticated definition, but one which carries the pertinent ideas, let us use the following:

Electrostatic imaging is a process or processes by which an ordered arrangement of electric charges is deposited on a surface or in a volume of a material as a latent electrostatic image and subsequently "developed" to convey visual information to an observer.

The deposition and development of the latent electrostatic image may be accomplished by any of a number of methods. Several techniques of charge deposition and development are discussed in this chapter.

The history of electrostatic imaging in a very strict sense extends back to the times of the ancient Greeks and others who noted that bits of paper, straw, and feathers were attracted to amber that had been rubbed. Such behavior was noted by Thales of Miletus[1] in about 600 B.C. as well as by many others at later times.* William Gilbert in *De Magnete*[3] noted that many materials in addition to amber possess the property of attracting bits of material when they are rubbed.

Although the effects of electric fields on small objects have been known

---

* For a discussion of electrostatics before the time of Gilbert see *The DE MAGNETE of William Gilbert*, by Duane H. D. Roller.[2] Dr. Roller points out that we have only a second-hand reference to Thales' statements on amber and lodestone through a discussion by Aristotle.

for many centuries, the production of a visible image from a deposition of charge on a surface was not noted in recent times until 1777. In that year G. C. Lichtenberg[4] directed electrical sparks over the surface of a good insulator and found that dust settled into a pattern following the spark track. This experiment was outlined in many elementary physics texts[5] of the 1920s and 1930s. The student was directed to initiate a spark from a Wimshurst machine to the surface of a glass plate which has been cleaned and dried very carefully. He then had to dust the surface with finely powdered resin or lycopodium powder. The resulting pattern was called a Lichtenberg figure. In subsequent times various mixtures of powders were used for developing Lichtenberg figures, including vapor deposition.

There are examples of Lichtenberg figures which are primarily nuisances. In very dry weather we often find on processed 35-mm photographic film exposed areas that appear very much like miniature lightning flashes. These exposed areas are the result of somewhat the same processes that produce ordinary Lichtenberg patterns. As the film is pulled or rolled through the camera (in haste, perhaps, to obtain "just one more good shot"), the film becomes highly charged by sliding out of the cartridge or over the roller or other surfaces. Electric fields may become high enough to produce a discharge or spark along the surface. The local area of the film is discharged, and the luminosity from the discharge exposes the film and is subsequently developed, showing up as treelike patterns on the processed fil:.ı.

For many years an accepted method of cleaning glass surfaces was bombardment of the surface by the discharge from a Tesla coil.[6] Following such bombardment, the instructions for the technique suggested breathing on the surface through the mouth. If the vapor from the breath condensed into droplets on the surface, the surface was presumed to be still dirty and further cleaning was recommended. If the vapor deposited in a thin, "nonreflecting" film, the surface was considered to be clean and ready for use. This was a so-called black-clean surface, named because the continuous film of water appeared black in comparison with the light scattering areas composed of small droplets on the "dirty" surfaces.

Early examples of "breath figures" are those of Riess (1838)[7] and Karstens (1842).[8]

Karstens was able to produce visible images of real objects—probably the first true examples of electrostatic imaging. To accomplish this feat, Karstens placed a coin on an insulating plate mounted on a grounded metal surface. A high electric potential was applied to the coin, the coin was removed, and the insulating plate was breathed upon. The breath figure formed on the insulating plate was an image of the face of the coin that had rested against the plate.

Recently a method of electrostatic image formation has been incorporated into several commerical machines. This is known by the general term of electrography. A conducting point or edge is dragged along the surface of an insulating material backed with a grounded metal plate. If an electrical potential is applied to the point, a charge pattern is deposited on the insulating surface. This process is discussed at length in a later section, but we should note here that this technique was employed by Ronalds before 1842 in a recorder of atmospheric electrical effects. A lightning rod was connected to the point being moved over the surface of an insulating disc in a spiral pattern. After the disc had been "exposed" sufficiently it was dusted with a powder to develop the track of the point. This process is remarkably similar to that used in modern electrographic printers and recorders.

Another modern technique which had its origin before the turn of the century is known as "frost" or "thermoplastic" recording. The basis of this method of image formation is the propensity of a charged fluid surface to deform under the influence of the electric forces present at the surface.[8] In 1897 Swan[9] reported that charge patterns on a thermoplastic material caused surface deformations when the plastic material was softened by heat. More recently, the same principle of surface deformation of a fluid was used in the Eidophore[10] image-projection system. In this system an electron beam is used to deposit a charge pattern on the surface of an oil film. The depth of deformation is dependent on the electron beam current and hence can be modulated. By using reflection techniques, a bright image of the surface can be projected to large screens.

Until the end of the nineteenth century very little practical use was made of electrostatic imaging and recording techniques. In fact, prior to the early 1930s little work of a practical nature ever came to fruition as useful commerical devices.

In the period beginning about 1920 the techniques of generation and control of electron beams developed very rapidly. The vacuum devices which were the forerunners of present television camera tubes were invented. These devices generally used a beam or beams of electrons to transform an optical image focused on a light-sensitive screen into a modulated signal that could be amplified, modified, transmitted, and received as could any other information-carrying signal. The received signal could be used to synchronize and modulate an electron beam in a cathode ray tube (CRT), thus reproducing the original image as a "picture" on the CRT screen. This is, of course, a simplified description of a television system.

It was a short step from this early system to the desire to produce "hard copy" from CRT displays. Recording of high-frequency transient signals was accomplished in some devices by inserting a photographic film or paper inside the vacuum envelope. The "writing" on the film was done by the electron beam itself rather than by the light output from a phosphor, as is usual today. However, even though the writing speed of the film-in-vacuum device was very high, its inconvenience outweighed its advantages for all but laboratory uses.

A desire to use the convenient techniques available for electron beam deflection in printing and other imaging contexts led to attempts to bring the electron beam out into the atmosphere or to use its electrostatic effects on the atmospheric side of the glass vacuum envelope of the CRT. Between 1928 and 1940 Selenyi[11] invented a number of processes for writing with electron or ion beams on insulating surfaces. The latent images were subsequently developed by dusting with powders of various types. Some of Selenyi's devices used the effects found outside the glass face of the CRT due to the charge provided inside the tube by the electron beam. However, as a result of the thick curved glass face of the tube, the effects outside the tube were quite diffuse, and Selenyi's developed images lacked sharpness and resolution.

In later efforts, Selenyi used a thin membrane of cellulose as a window in the CRT face through which the electron beam could pass. An insulating sheet backed by a conducting plate held at a positive potential provided a surface to which electrons were to be attracted. The electrostatic image formed on the insulating sheet was developed by powder dusting. It is almost certain that the image was formed not by electrons but ions since the electrons "attach" to air molecules upon emerging into the air and form negative ions. The negative ions would be deposited on the insulating sheet to form the image. It should be noted here, however, that the mean free path of high-energy electrons may be long enough to actually reach the sheet if it is very close to the thin film window in the CRT face.

The problems associated with electron beam transmission through thin windows were severe, and other ideas arose for accomplishing the same result with less difficulty. The most useful from a commercial standpoint has been the so-called pin tube. In this device a relatively standard CRT is constructed. Embedded in the face of this tube, however, and extending from inside to outside, is a row of very fine wires—all parallel to and insulated from one another. As an insulating paper or film is moved past the row of pins of the pin tube and with the paper surface very close to the tube face, the intensity-modulated electron beam is scanned across the row of pins. As the beam impinges on each pin, a discharge occurs in the air at the outer end of the pin. A positively biased plate behind the paper may be employed to assist in moving to the paper the ions produced by the discharge. The quantity of charge deposited opposite each pin is dependent on the electron beam current to the pin. Development of the charge image thus produced is accomplished by powder dusting or other methods. A commerical version of the pin-tube system is currently in use for printing large quantities of address labels.

Other schemes for generating electrostatic images were proposed by many inventors, but until Carlson invented a system utilizing a photoconductive layer, very little work of commerical interest was accomplished. Carlson's invention gave rise to an entire industry, and because of its importance, a short summary of events leading to the original copying machine and its subsequent development into a commercial product is now presented.

Chester F. Carlson, who was trained as a physicist and in law, had been interested in graphic arts since boyhood. In his work in the field of patents he saw the need for a means of copying documents which would be inexpensive and easily operated by office personnel. He studied the published literature as thoroughly as possible and concluded that a combination of electrostatic charging and a photoconductive layer would produce an invention of value. In 1938 Carlson employed Otto Kornei to conduct some developmental work on the invention and on October 22, 1938, produced a copy of an image on a small sulfur-coated zinc plate. After attempting to interest many companies and governmental agencies in his invention, Carlson finally entered into an agreement with Battelle Memorial Institute in Columbus, Ohio, to develop the invention. In 1947 the Haloid Company (now Xerox Corporation) licensed the invention and began commerical development of "xerography." After considerable research and development, the first commerical equipment was introduced in 1950, although the process had been announced and demonstrated publicly in 1948.

A detailed description of the history of this important invention written by Mr. Carlson is to be found in Ref. 13, together with an extensive bibliography of early patents and published works on electrostatic printing.

Inherently, there is much in common between this chapter and the next on nonimpact printing.

12.2   IMAGE PRODUCTION

Electrostatic imaging and optical photography require the same essential steps through the complete process. We might label these as:

1. Sensitization.
2. Latent image production.
3. Development.
4. Fixing.

Most electrostatic methods are similar, but sometimes the four steps just listed must be broken into substeps or combined into fewer steps. To clarify the sequence of steps, we now discuss a number of methods of accomplishing the general steps in electrostatic imaging, which are, generally:

1. Surface charging.
2. Latent image formation.
3. Image development.
4. Image transfer.
5. Fixing.

In our discussion we treat steps 1 and 2 together although they are usually separate physical processes. Figure 12.1 illustrates graphically the processes which several commerical systems have in common.

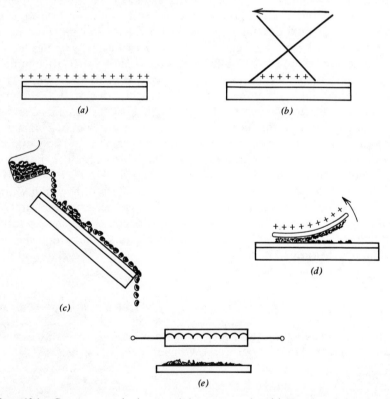

Figure 12.1. Common steps in electrostatic image processing: (*a*) Charging of photoconductor plate. (*b*) Exposure—latent image formation. Note that the surface charge remains in the dark image areas and leaves from the light areas. (*c*) Development with negatively charged toner. (*d*) Image transfer (not necessary if photoconductor is part of final copy, e.g., zinc oxide layer on paper). (*e*) Fixing of image—heat fusing in some cases, drying or vapor fusing in others.

To form an image of charge on a surface, we can either initially deposit the charge in a pattern or uniformly cover a surface by a charge layer and then remove the charge from selected areas to form a pattern.

The following few paragraphs assume a system of a conducting plate covered by a very thin layer of insulating material, which is in turn covered by a layer of photoconducting material. In this case, latent electrostatic images are formed by charging the surface of the photoconductor in the absence of illumination and optically exposing the charged surface to an appropriate image. The photoconductor permits charge migration (discharge of surface) in illuminated areas and maintains a surface charge in dark areas. Thus a "latent" charge image is produced on the surface of the photo-conductor. This charge image can be developed subsequently to obtain a visible image for transfer or observation.

The dark resistivity of amorphous selenium is 10^{11} to 10^{14} Ω-m and the permittivity is about 5.4×10^{-11} F/m. This corresponds to a charge transfer time constant T_E of 5.4 sec to 5.4×10^3 sec. Of course, for use as an imaging material the time constant must be large enough to allow a charge image to be formed and developed before the charge leaks off the surface. Thus the charging of the surface is accomplished in darkness and the surface charge remains until it is discharged by the optical image.

The process for removing the surface charge in illuminated areas is complex and has been studied at length by many. However, for our purposes, we discuss the process in general, letting the more interested reader examine the references for more detailed explanations.

When light falls on the surface of a photoconductor such as selenium, anthracene, or zinc oxide powder in a plastic binder, the photons produce electron–hole pairs in the material. If the layer of photoconductor is sufficiently thin and if there is an externally applied electric field present, the charge carriers (holes and electrons) can be separated and collected at an electrode structure. If the field is small and/or if the layer is thick, the holes and electrons may recombine; or they may be trapped and fail to be collected on an electrode as a current. If the thickness of the layer is less than the distance a carrier can travel before recombining with a carrier of opposite sign or without being trapped locally in the material, the electric field applied will cause the carrier to move through the material and appear as a current at an electrode in contact with the photoconductor. The carriers may also move to a surface and provide neutralization of a charge at that surface.

Let us assume an imaging system as described to be a photoconducting layer (e.g., amorphous selenium), a very thin, highly insulating layer (e.g., alumina), and a conducting plate (e.g., aluminum). Let a uniform positive surface charge be applied to the photoconductor surface in the dark. This surface charge will have a negative image charge on the conducting plate, thus an electric field will exist in the alumina layer and in the photoconductor. Now let half the surface be illuminated with light of appropriate wavelength. Electron–hole pairs will be produced in the photoconductor and will migrate under the influence of the electric field. The electrons will travel to the surface, neutralizing the positive surface charge, and the holes will travel toward the conducting plate to neutralize the image-charge electrons on the plate. On the nonilluminated half-surface, the positive surface charge will remain unchanged except for a gradual decay due to the normal dark current in the photoconductor material. At the edge of the charged area, there will be a strong fringing field, and over the uniformly charged area there will be a weak external electric field. This field configuration is discussed in more detail in the section on development.

We should note that essentially the same process occurs when paper coated with zinc oxide is used in place of the selenium–alumina–aluminum system. The zinc oxide is the photoconductor, and the paper (which must have a suitable conductivity) acts as the conducting electrode.

The most widely used and perhaps the most satisfactory method of charging a surface is by the use of a corona source.[14] If a small-diameter (0.005 in.) wire is maintained at a potential·of about 5 kV, the air in the near vicinity of the wire becomes ionized. An arrangement such as that in Fig. 12.2 can provide a deposition of charge on the surface of the grounded plate and the insulating layer on its surface. The surface of the insulating material (photoconductor) becomes charged until the fields from the deposited surface charge are strong enough to turn other

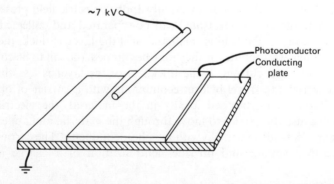

Figure 12.2. Schematic arrangement of the corona wire of the charging system and the sandwich plate consisting of the photoconductor layer and the conducting backing layer. The very thin insulating blocking layer (sometimes alumina) between the photoconductor and the conducting plate is not shown.

approaching ions away from the surface. By moving the wire along parallel to the surface, a relatively uniform charge density can be applied to the insulating surface. If the upper part of the wire is surrounded by a portion of a grounded metallic cylinder, the effects of the corona will be limited to the region of the plate. A positive wire potential leads to a positive charge on the plate. The magnitude of the surface charge density produced on the plate is dependent on the potential applied to the wire, the spacing of the elements of the system, and the rate at which the corona source is moved over the surface.

Other methods have been proposed to charge the surfaces, but these have not proved to be useful commerically except for special purposes. In Carlson's initial efforts to produce a copying system, the photoconductive layers (sulfur) were charged by rubbing, that is, by triboelectric effects. Some of the other methods proposed and studied include charge induction during exposure to the appropriate image, contact charging by a blade or roller held at a high potential during contact with the surface, and charge transfer from another surface. The most interesting of these, perhaps, is the charge induction method. A transparent electrode (e.g., tin-oxide-coated glass) is brought close to but kept separate from the photoconducting surface. The transparent electrode is held at a high potential and the photoconductor is exposed through the electrode. In the illuminated areas the photoconductor permits charge migration and a surface charge appears on those areas. The nonilluminated areas, on the other hand, remain uncharged since the photoconductor maintains a higher resistance in those dark areas. Thus the processes of charging (sensitization) and latent image formation occur simultaneously.

If we assume the photoconductor layer to be charged uniformly over its surface, the next step in the electrostatic image formation process is exposure to an optical image of appropriate wavelength. This step is typical of photographic processes in that the spectrum of the light employed to form the image should match as closely as possible the spectral response of the receptor. Photoconductive selenium, for example, is most sensitive to light in the short-wavelength end of the spectrum and almost totally insensitive to yellow light. Zinc oxide, when suitably dyed and used as a photoconductor for copying work, has a spectral response curve similar to that of the human eye. By appropriate additives, the spectral response and sensitivity of most photoconducting materials can be shifted and improved as compared with the pure material. The requirement that the spectral response of the imaging system match the incoming light spectrum is very important and also very difficult to satisfy.

It is not always necessary to employ photoconductors in the image formation process. As in the case of the "breath images" of coins formed by Riess in 1838, a latent electrostatic image can be formed on an insulating surface by contact with a suitably shaped conductor at a high potential. This circumstance has been utilized in printing schemes in which a paper web has been contacted by a conducting "letter," such as a typewriter character with a high potential between the character and a backing plate on the opposite side of the paper. A charge pattern of the character is left on the paper and can be developed subsequently.

Similarly, if the potential of a conducting contact sliding on a paper web is controlled, a surface charge of controlled magnitude is deposited on the paper. Several such sliding contacts whose potentials are controlled by means of a digital computer have been developed into a computer printout device. Other devices are used for signal recording and facsimile systems.

The charging process in electrographic systems is rather complex and is probably a combination of several mechanisms. A conducting stylus rubs on a surface but is only partly in actual contact with the surface. The surface areas in direct contact with the stylus will be charged by conduction. The factors determining the magnitude of surface charge density are the electric potential of the stylus, the conductivity of the surface material, and the time constant for charge transfer in the surface material. The areas not in direct contact with the stylus may become charged by gas breakdown processes, by field emission processes, or by surface charge migration from nearby charged areas.*

The behavior of the charge pattern after it leaves the stylus is determined primarily by the conductivity and permittivity of the surface material. Low-conductivity materials maintain distinct charge patterns for several weeks under some conditions. Other materials which are suitable for imaging process may retain the charge pattern for only a few seconds or minutes. An ideal material for direct imaging would retain the charge only long enough for the pattern to be developed, whereupon it would immediately discharge. Many commercial paper materials are quite satisfactory for use with these techniques.

It may be desirable to produce a charge image by electrographic techniques on a highly insulating surface. The image is then developed and transferred in the normal fashion. A possible surface for this purpose could be made up of a very highly insulating material by itself or by such a surface covered by many conducting areas insulated from one another. These conducting "dots" could be highly charged by a stylus contact and developed, and the image transferred to a paper web. The dots could be in

---

* For a discussion of the charging processes, see Chapter 4. The processes discussed in Chapter 4 can also be used in accomplishing the charging of surfaces for imaging.

a pattern and all dots charged equally, or the dots could be uniformly distributed on the surface and charged according to a signal voltage applied to the stylus.

Electronic control of the behavior of an electron beam in a pin-tube system provides a very satisfactory means of printing for some purposes, as has been mentioned.

For electrostatically printing material that is to be repreated over and over again, as for books, newspapers, other forms of multicopy information, it is convenient to form an image plate that has charge-retentive areas built in where characters are to be printed. Charge is deposited on the character areas, and the resulting latent image is treated as any other charge image.

12.3   DEVELOPMENT OF ELECTROSTATIC IMAGES

The electrostatic image has a number of characteristics that are important in the development processes. To understand the character of the observable image after development, we should discuss briefly the most important of the latent image properties.

Figure 12.3 presents an expanded view of a typical photoconductor-insulating layer-conducting electrode system with positive charge on the photoconductor surface. In the surface of the conductor next to the photoconductor is a layer of negative charges. If the entire system were isolated and had zero charge before the positive surface charge was applied, the net charge of the system would have to be positive, and there would be positive charges left on the back of the conducting surface. However, let us assume that the conductor is connected to ground during the charging process. The situation would resemble Fig. 12.3a, with a net total charge of nearly zero on the system. There would be a very weak electric field outside the photoconductor layer and a much more intense field inside the layer.[15] The situation is similar to a parallel-plate capacitor that has been charged and the negative plate connected to ground. There is some external field around the capacitor but very little compared with that between plates. As we mentioned in the last section, the photo-conductor layer must be thin enough to allow photo-generated charge carriers from the layer to reach the surface and the backing plate during exposure without too much recombination or trapping. The applied surface charge must give rise to an electric field that is intense enough to cause separation and transport of the charge carriers through the layer in a time short enough to avoid recombination and trapping. However, the field must not be so strong that the material breaks down or that injection of carriers into the photoconductor under dark conditions

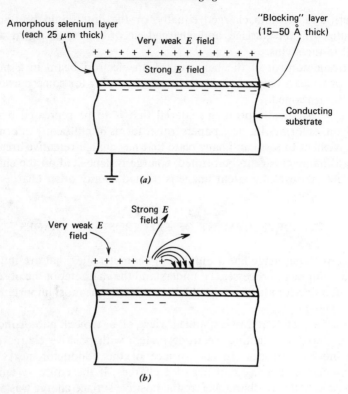

Figure 12.3. Schematic representation of an imaging plate made up of a conducting substrate, a thin "blocking layer," and a photoconductor, in this case, amorphous selenium: (a) The plate has been charged uniformly over its entire surface. (b) To the right of the centerline, the plate has been exposed to light of an appropriate wavelength and hence has been discharged in that region. To the left of the centerline the surface charge is unchanged. The fringing fields appear in the diagram.

is caused. If the necessary electric field is E_{max}, then the potential difference between the photoconductor surface and the backing electrode is

$$V = E_{max} d$$

where d is the thickness of the photoconductor layer (the insulating layer between the conducting plate and the photoconductor is assumed to be very thin compared with the photoconductor thickness).

In Fig. 12.3b we see the charge configuration and electric field distribution after half the plate has been exposed to light. A notable aspect of the field distribution is the fringing at the edge of the charge pattern. Not only is there a strong gradient of the field* at the edge, but the magnitude of the field external to the photoconductive layer is large along the edge compared with the field external to the uniformly charged area.

Development of the electrostatic image is the process of supplying material, usually in small particle form, to the charged areas to provide visibility of the image. The small particles or "toner" are charged appropriately to be attracted to the charge image by reason of the external electric field generated by the charge image.

Several forces act on free toner particles in the vicinity of the charge image, namely, Coulomb forces, dielectrophoretic forces, gravitational forces, and aerodynamic forces. There may be mechanical forces, also, if there are collisions with other bodies by the toner particles. For the time being, however, we ignore collisions and aerodynamic forces† and discuss the other three.

The force on a small charged object owing to the presence of a uniform electric field is

$$F = qE$$

where q is the charge on the object (C) and E is the electric field intensity (V/m). The force will be in the direction of the electric field.

A 10-μ diameter toner particle typical of those used in many systems carries a charge of about 0.5×10^{-14} C. If we assume an electric field intensity near the photoconductor of about one-third the breakdown strength of air (which is approximately 3×10^6 V/m), we have $E = 10^6$ V/m. The Coulomb force on the toner particle for these conditions will be 0.5×10^{-8} N.

If we assume a toner particle to be approximated by a 5-μ radius sphere whose density ρ is 10^3 kg/m^3 (the same as water), we find the gravitational force on the particle is

$$F_g = \tfrac{4}{3}\pi r^3 \rho g$$

or about 0.5×10^{-11} N. Thus we see that the gravitational force is three orders of magnitude less than the Coulomb force and can, therefore, be considered insignificant in most instances of toner deposition. Of course we

---

* The gradient of the field (or any other quantity) is the maximum rate of change of the quantity with distance evaluated at the point in question. Another way of stating this is to say that the gradient of $\Phi(x, y, z)$ is the directional derivative $d\Phi/ds$, where $ds = \sqrt{(dx)^2 + (dy)^2 + (dz)^2}$ and ds is taken in a direction that makes $d\Phi/ds$ a maximum. Thus the gradient has a direction associated with it as well as a magnitude (i.e., it is a vector quantity). The expression is usually written grad Φ or $\nabla\Phi$. For more detail see any good text on vector analysis.[16]

† These forces are ignored for now, not because they are unimportant, but because they are discussed in another section of this chapter. In aerosol and carrier development processes, these forces may be very important.

can envision situations in which the force owing to gravity is a very important aspect of a particular process. However, we must remember that the Coulomb force on *toner* particles near (within a few microns) a charge image will be orders of magnitude larger than the gravitational force on the particle.

The particle, in addition to being charged, will also become a dipole in the electric field by reason of induced polarization. If we assume the object to be a small sphere of radius a (which is a sufficiently good approximation for most work and particularly for the purposes of our discussions) then the strength of the dipole will be*

$$M = \frac{\varepsilon - \varepsilon_0}{\varepsilon + 2\varepsilon_0} a^3 E$$

where ε is the permittivity of the particle material, ε_0 is the permittivity of air, and E is the electric field intensity in the absence of the sphere.[17]

The force on a dipole whose moment is M in an electric field E is†

$$F = M \left(\frac{\partial E}{\partial S} \right)_{\text{max}} \cos \theta$$

where θ is the angle between the dipole axis and the field gradient. The force expression can be written in terms of components in the x, y, and z directions as

$$F_x = M_x \frac{\partial E_x}{\partial x} + M_y \frac{\partial E_x}{\partial y} + M_z \frac{\partial E_x}{\partial z}$$

$$F_y = M_x \frac{\partial E_y}{\partial x} + M_y \frac{\partial E_y}{\partial y} + M_z \frac{\partial E_y}{\partial z}$$

$$F_z = M_x \frac{\partial E_z}{\partial x} + M_y \frac{\partial E_z}{\partial y} + M_z \frac{\partial E_z}{\partial z}$$

where M_x, M_y, and M_z, and E_x, E_y, and E_z are the dipole moment components and the electric field components in the x, y, and z directions, respectively. Since M is a direct function of E, the components of the force on the dipole can be written as

$$F = \frac{\varepsilon - \varepsilon_0}{\varepsilon + 2\varepsilon_0} a^3 E \left(\frac{\partial E^2}{\partial S} \right)_{\text{max}} \cos \theta \quad \text{or} \quad E = \frac{1}{2} \frac{\varepsilon - \varepsilon_0}{\varepsilon + 2\varepsilon_0} a^3 \left(\frac{\partial E^2}{\partial S} \right)_{\text{max}}$$

Since the force is a direct function of the gradient of the electric field intensity E, in a *uniform* field there will be no translational force on the particles as a result of dipole forces.

---

* The dipole moment is perhaps a better term. This expression contains both the magnitude and direction of the dipole and hence is a vector quantity.
† The expression $(\partial E/\partial S)_{\text{max}}$ is the gradient of the electric field.

In a nonuniform electric field, the dipole will experience a translational force. The force will be in the direction of increasing electric field and will be very small compared with the Coulomb force we expect on a toner particle. For a typical toner particle of 10-μ diameter and a relative permittivity (dielectric constant) of $\varepsilon_r = 4$ in an electric field of 10^6 V/m whose gradient is 10^{10} V/(m)(m), we find a force on the induced dipole of about 3×10^{-12} N. This very rough estimate is still about three orders of magnitude smaller than the Coulomb force.

From these considerations, we see that free toner particles carrying a negative charge in a field produced by a positive charge image will follow (approximately) the field lines to the charge image. In the case of a uniformly charged thin photoconductor layer on a conducting back plate, we have noted almost no external electric field is generated by the charge except at edges. Thus to obtain toner coverage in these areas, a grounded conducting surface is brought very near the charged surface, but not touching it. The presence of the grounded electrode (called a "development electrode") gives rise to an electric field between the photoconductor surface and the electrode which is proportional to the surface charge density. If toner is introduced between the charged surface and the development electrode, the toner is attracted to the surface. The toner, being charged oppositely to the surface, will neutralize the surface charge. The quantity of toner per unit area attracted to the surface before the surface charge is reduced to a noneffective level will be proportional to the original surface charge density, which in turn was inversely proportional to the intensity of illumination on the surface during optical exposure. The density of the developed image thus tends to reproduce the continuous tones of the original optical image. The spacing between the development electrode and the photoconductor surface should be as small as possible consistent with the requirements that there be some surface roughness and also that toner particles pass into space without hindrance to accomplish development.

12.4 DEVELOPMENT PROCESSES

There are a number of methods by which toner particles can be charged and presented to the electrostatic image. These include: (1) cascade, (2) powder cloud, (3) liquid aerosol, (4) fur brush, (5) magnetic brush, (6) powder tray, (7) liquid immersion, (8) toner transfer, and (9) other variations and combinations of the first eight methods.*

* The Eidophore, thermoplastic, and frost development systems are discussed separately.

12.4.1   Cascade Process

The cascade is most widely used in commercial systems for copying documents. The cascade process, developed by workers at Battelle Memorial Institute, employs a large carrier particle (500–750 μ diameter) coated with a very thin (less than a monolayer) layer of toner particles. The toner and carrier materials are chosen such that they have very high bulk resistivity (of the order of 10^{14}–10^{15} Ω-m) and such that the toner particles will be charged by triboelectric effects by the carrier. For most positive-to-positive reproducing processes utilizing selenium photoconductors, the toner is charged negatively and the carrier positively. If the carrier and toner combination is poured or "cascaded" over the surface on which the charge image has been formed, toner particles leave the carrier surface and are deposited on the image. (As has been mentioned previously, without a development electrode there is edge development but little or no toner coverage in large, uniformly charged areas or in areas where surface charge varies only slowly with lateral distance.) As the carrier and toner roll, slide, and bounce across the image, the strong fields at the edges of the charge pattern collect toner from the carrier and build a toner layer on the charge image. The toner density is greatest near the image edge and less toward the more uniformly charged areas.[15]

When a development electrode is present, the toner is deposited in the more uniformly charged areas, also.

It is important in most copying processes that toner not be deposited in areas where the copy is supposed to be white. Therefore, the carrier should maintain the toner particles attached to it except in the proper external field. Alternatively, the passage of carrier beads over uncharged areas should pick up or "scavenge" toner particles left there by previous carriers. If the carriers are not coated too thickly with toner, this scavenging action is very effective and is a strong advantage for cascade development processes.

During the agitation of the carrier–toner combination, toner particles are moved, turned over, and rubbed on carrier surfaces. Since the resistivities of the toner and carrier materials are high (10^{14}–10^{15} Ω-m), the time constants for charge transfer are long, and nearly all surfaces of the toner and carrier particles are likely to become charged by triboelectric effects, remaining charged for a long period. Measurements made on standard carrier–toner combinations from commercial machines give 10^{-12} to 10^{-14} C charge for carrier beads and 5×10^{-15} to 10^{-16} C for toner particles.

Many materials have been tested and used for carrier and toner particles. Some of the most successful appear to be glass- or plastic-coated glass beads for carrier particles and carbon-impregnated resin or plastic for toner particles. As has been mentioned, the materials are selected for resistivity and triboelectric interaction properties as well as for other quantities such as availability, low cost, and ease of manufacturing.

12.4.2   Powder Cloud

The process of introducing charged toner particles to an electrostatic image is often easiest accomplished by simply blowing dust over the image. Finely divided carbon particles blown pneumatically through a nozzle become charged and are a very satisfactory developer for electrostatic images. The transfer of the image to a sheet of paper presents some problems, but it can be accomplished by making one surface of the paper tacky or sticky. The carbon then adheres to the paper and the image is transferred.

The powder cloud development process is capable of providing very high resolution development and can be employed with a closely spaced development electrode. The particle sizes may be very small (of the order of 1μ) and so can be projected on the development electrode and the photoconductor without giving rise to scratches on the photoconductor or clogging in the narrow space. The small size of the particles is also advantageous in providing high-resolution images.

12.4.3   Liquid Aerosol

A development similar to the powder cloud process involves a liquid aerosol. In this process a very fine mist is produced by spraying a liquid through an extremely small orifice with or without a nearby charging electrode. The droplets in the mist become charged in the spraying process as described in Chapter 4. In principle, this method of development is very much like the electrostatic spray painting described in Chapter 11. The droplets are charged and are attracted by the action of electric fields to their proper positions on the electrostatic image. The possibilities of very high resolution development and some of the advantages and disadvantages of the method are immediately apparent. The problems of cleaning and maintenance of a trouble-free system are more severe with liquid aerosols than they are with dry materials.

In both dry powder cloud and liquid aerosol development, there are many ways of producing the charged particles. Almost any means that can be devised is satisfactory for development of an image. However, not all methods are satisfactory for *commercial* processes. It must be recognized that many schemes that work quite well in the laboratory fail in operating systems because of rather mundane reasons such as high cost, difficult maintenance, or low reliability. The prospective inventor would do well to think about the practical aspects of a process after conceiving his invention.

12.4.4   Fur Brush Process

Toner of appropriate composition dusted onto a fur brush (of real or artificial fibers) becomes charged by triboelectric effects in much the same manner as the toner in the cascade developer. The choice of fur and toner composition determine the relative polarities of brush fiber and the toner particles. Agitation of the brush by rubbing on a surface or stirring of toner produces better toner charging and more satisfactory developing action. There is a force holding the toner to the fur bristles which must be exceeded in order to remove the toner and deposit it on the image. The existence of this minimum force requires a minimum or threshold field strength from the image, below which development will not take place. This is quite satisfactory and avoids the deposition of toner in regions where it is not desired.

If the brush with attached toner is moved across an electrostatic image, the fields of the image will cause some of the toner to leave the bristles and be deposited on the image. The bristles are left with a net charge, and if there were no leakage, the brush would become so highly charged that no more toner would be removed by the fields of the image. This would result in poor development. Thus a resistivity of the order of 10^{11} Ω-m or thereabouts is required in order for the fibers to discharge in a few seconds. Most natural fur proves to be satisfactory if the base ends of the bristles are grounded and the relative humidity is high. Note that the same discharging of carrier material occurs rather naturally in the cascade process by contact of carrier beads with metallic walls of the storage containers and manipulative apparatus.

12.4.5   Magnetic Brush Process

It is possible to form an artificial "brush" of iron particles clinging to the pole of a permanent bar magnet. If the filings have been mixed with toner particles of proper triboelectric properties, toner particles may be charged and carried to an image by means of this "magnetic brush." In its developing action, the magnetic brush accomplishes results like those of the fur brush, but with some differences. Since the magnetic brush is a conductor, it serves as a development electrode and is able to provide solid-area development to a much greater extent than is the fur brush.

The long chains of the iron particles clinging to the pole of the magnet are very flexible and generally cause little abrasion of the photoconductive surface. This behavior is not of consequence for processes in which the photoconductor is only used once (e.g., zinc oxide paper) but it is of considerable importance when the photoconductive surface is a selenium-coated drum or plate that is to be used many times.

12.4.6   Powder Tray Development

A flexible photoconductor may be caused to slide into and out of a U-shaped trough in which a mixture of two kinds of particles has been placed. If the mixture is composed of toner particles and other particles of about the same size (let us call them antitoner particles) chosen so that the toner will become appropriately charged by triboelectric action against the antitoner, an image on the photoconductor can be developed with toner in the charged areas and antitoner in the uncharged areas. Some toner is surely present in the noncharged areas, but it should be relatively a small amount compared with the antitoner. If the developed image is transferred to a sheet of paper and then fused on the paper, the antitoner can be made nonfusing and can be wiped off, while the toner will remain fused to the paper. Any toner in the noncharged region will fuse to the antitoner and can be wiped off with the antitoner. A number of toner–antitoner combinations have been proposed, but a thermoplastic (methacrylate toner and nonfusible particles of calcium carbonate) has proved to be quite sucessful. Particle sizes most appropriate seem to be in the range of 3 to 10 μ diameter.

12.4.7   Liquid Immersion Development

Liquid development of electrostatic images has achieved considerable commercial success. Although it has not enjoyed the popularity of the cascade process, nevertheless, it is used in many successful copier systems. The liquid development process is basically similar to many others. Charged particles are caused to migrate to the charge image on a surface by the electric fields external to the image. In this particular case the charged toner particles are carried by a liquid medium rather than by a gas, as in aerosol development, or by carrier beads, as in cascade development. The properties of the toner particles and the liquid are so chosen that the toner particles become appropriately charged owing to the formation of a charge double layer at the interface between the solid and liquid phases.

The surface of a particle placed in the liquid may acquire a charge by several possible processes. There may be an acquisition or a loss of electrons as a result of a chemical reaction at the surface, an orienting and attachment of local dipolar molecules, an acquisition of ions by adsorption, or a formation of various other bonds (covalent, etc.). The particle surface then has a net charge, and immediately outside the surface (in the liquid) there is a layer of charge of the opposite sign. The layer of charge in the liquid is composed of two regions: the Sterm layer and the Gouy layer. The Sterm layer is the region immediately adjacent to the particle surface and the potential gradient in this region is very large. The Gouy layer, which extends from the Sterm layer some distance into the liquid, is actually a lower gradient transition region in which the potential drops to the potential of the ambient liquid. If an externally produced electric field is applied to the liquid–particle mixture, the particles will move under the influence of the electric field. This is the phenomenon known as electrophoresis. In the development process, the external electric field is produced by the charge image and causes the toner particles to migrate and be deposited on the image.

Most commercial imaging processes that employ liquid development produce the charge image on a paper coated with the photoconductor (e.g., zinc oxide in a suitable binder). The electrostatic image is produced by charging the paper surface uniformly and then exposing it to the optical image. The paper is passed through a tray or trough containing the developer (toner–liquid mixture); after the image is developed, the paper is squeegeed or passed between rollers to remove the excess liquid and dried. It is very convenient, therefore, to use in the developer a liquid with high vapor pressure which can be evaporated easily. Kerosenelike liquids often serve this purpose.

By the use of a development electrode or development rollers it is possible to obtain good area development in liquid systems. Without roller or development electrodes the image will be toned in the fringe regions in the same manner, as has been discussed for cascade developers.

12.4.8   Other Methods

Many other methods of development are possible but none has enjoyed great commercial success. One method that appears to have some promise is roller development; in this process, a roller or other applicator is coated with a "paste" or a thick, ink type of material. The roller is drawn over the electrostatic image and the paste or ink is transferred to the image from the roller. This process is capable of solid-area development and can, if employed with care, produce development images of good quality and high resolution.

Vapor development was mentioned earlier as a method providing high-resolution images that is also of historical interest. " Breath figures " may be considered to be made by vapor deposition. Enhanced condensation in areas of high electric field intensity as compared with that in low-field areas provides the image development.

12.5 IMAGE TRANSFER

Once an electrostatic image has been developed, it often must be transferred to another surface. This is particularly true of the processes that employ reusable photoconductor surfaces and of those which use a conducting plate on which a nonconducting surface is produced in the shape of the desired image, which is then charged and developed (electroprinting). In the cases mentioned, and in many others, the final image is to be transferred to a sheet of paper or other flexible material. The transfer of the toner or other developer material may be accomplished by several means.

12.5.1 Electrostatic Transfer

The electrostatic transfer method is perhaps the most common and in many ways the most useful. When a charge image (positive charge, e.g.) is developed by negative toner whose bulk resistivity is high, the toner may retain a negative charge on the surfaces not actually in contact with the image. If a sheet of paper or other nonconducting material is placed over the developed image and sprayed with a positive charge, the electric field penetrating the sheet will act on the negative toner and cause it to transfer to the paper. Of course, all the toner will not transfer; indeed, it is possible to spray the paper first with positive and then with negative charge and transfer only a small amount of toner. When the sheet is stripped away from the imaging surface, a well-toned image is still present. A second sheet of paper can then be placed over the image and the transfer process repeated. A well-toned image can quite easily be made to yield seven or more readable copies of printed material.

The electrostatic transfer process described here is rather more complex than might at first be thought. When the charged paper is stripped from the toned image, air breakdown is almost certain to occur between the image area and the paper. This phenomenon can affect the production of further transfer images, the cleaning of photoconductor surfaces, and so on. The conductivity of the paper is only of minor importance; but if it is not low, special provisions must be made in the transfer process. Electrostatic transfer of developed images can be made to metallic plates, provided the plate is maintained at an appropriate potential with respect to the photoconductor layer. The potential would normally be positive and sufficiently high to effect the toner transfer when the plate is in virtual contact with the toner layer. There is some danger of sparking through the photoconductor layer with attendant damage to the photoconductor.

12.5.2   Adhesive Transfer

A very convenient method of image transfer involves making the toner or the paper surface of adhesive material. The paper is rolled or squeezed against the imaging surface on which the toner is present. Then the paper is stripped from the imaging surface—the toner sticks to the paper surface, and the image transfer to the adhesive surface is accomplished.

12.6   SURFACE DEFORMATION IMAGING

In 1940 a system of theater television was proposed and developed by Fritz Fischer at the Swiss Federal Polytechnical Institute in Zürich. In a sense this was a revival of some of the techniques discovered much earlier by J. W. Swan.[9] The system invented by Dr. Fischer was known as the Eidophore system. An electron beam was used to deposit charge on a thin layer of oil on a conducting backing surface. The charge per unit area deposited was controlled by modulating the intensity of the electron beam. Because of the presence of the charge layer on the liquid surface, the oil deformed into a pattern determined by the electron deposition. The oil surface was illuminated by an optical system that allows only light scattered from the surface of the deformed oil surface to pass. This system makes possible the projection of very bright images on screens as large as those used for theater projection of movies.

In 1959 W. D. Glen of General Electric reported a method of recording and projecting images formed by electron beams on plastic surfaces. In Glen's system, an electron beam writes an electrostatic charge image onto a film of thermoplastic material. After the charge image is deposited, the plastic film (on a conducting backing) is heated until it becomes soft. The charge image causes the surface to deform, and the deformed surface is observed by optical reflection techniques. An electron beam of about 0.5 A/cm^2 current density is sufficient to "write" video bandwidth images on this system.

The charging of the plastic surface can be accomplished by means of a conducting stylus which is moved across the surface along successive lines very close together. A voltage corresponding to the desired charge pattern is added to a dc bias voltage and applied to the stylus. We might visualize a cylinder much like that on an old Edison recorder-playback system. On the cylinder, as the stylus follows its closely spaced helical path, a charge image would be deposited. The recording surface would be a thermoplastic whose resistivity is very high (say, 10^{13}–10^{14} Ω-cm). Thus a charge pattern would be retained for a long period of time. After deposition of the charge, the cylinder would be heated to soften the plastic and cooled to preserve the resulting deformations. The plastic sheet could then be removed from the cylinder and flattened for use as a projection image.

The Xerox Corporation has devoted considerable effort to the development of a process called "frost" imaging. A plastic layer is deposited on a photoconducting layer backed by a conducting plate. The surface of the thin plastic layer is flooded with charge to form a uniform surface charge density. This step is executed with the materials in darkness. The surface is then exposed to an optical image, which leads to a redistribution of charge due to activation of the photoconductor. After the optical exposure, the plastic is softened either by heating or by exposure to vapors of a suitable solvent. In the illuminated regions, charge carriers have migrated through the photoconductor to the base of the plastic layer. Thus, owing to the original surface charge and the charge brought through the photoconductor, the plastic layer is subjected to strong electrostatic fields in the illuminated regions and will deform when softened. The surface deforms and presents an appearance much like that of frosted glass. Hence the term "frost" imaging. The deformation of the surface forms cells about 5 μ across, surrounded by ridges. The depth of the depression is a fraction of the lateral dimensions of the cell. Thickness of plastic layers are from about 1.0 to perhaps 10.0 μ.

It is possible to use plastic organic photoconductors, in which case there is only a layer of the plastic on a conducting backing plate.

The observation of the frost image is best done with an optical system similar to a Schlieren projection system. Since the frost image is only made visible because of scattering processes, high contrast is possible with systems that reject light scattered outside very small angles. It is, however, quite easy to view frost images with ordinary projection systems because such systems have limiting apertures.

Resolutions of 100 line pairs per millimeter are quite readily obtainable in frost systems. The method thus lends itself well to high density storage of printed images. However, the feeling should not remain that only high-contrast line copy can be recorded by this technique. On the contrary, very good continuous-tone rendition is possible using frost techniques (true also for the Eidophore and thermoplastic recording systems).

12.7   COLOR PROCESSES

There is always a strong enticement to try to find a multicolor imaging process similar to the single-color processes we have. It is obviously possible to use repeated charging–exposure–development steps to produce multicolor images. The primary requirements are accuracy in the registration of successive images and exposure through filters of appropriate color. This assumes that toner particles or other developing material can be made in the various colors. This type of imaging has been accomplished and is used for multicolor map reproduction, drafting and architectural layouts, and so on.

There is, however, always the lure of a single-exposure, single-development color process. To say that there has been considerable research and development effort devoted to this topic would be an understatement. The most promising process appears to be the selective control of toner deposition on the substrate by electric fields in the presence of the optical image. Selective charging or influence of the behavior of various colors of toner particles by light of chosen wavelengths is possible. This influence of toner particles is a technique that can be used to control the migration of multicolor particles to charged surfaces.

12.8   SUMMARY

Electrostatic imaging processes are many and varied in the details of their operation. However, the common feature seems to be charge or field control of deposition of a developing medium onto a surface. This may be achieved by a controlled charge pattern and similarly charged developer particles or by a relatively uniform charge on the surface and controlled developer characteristics. The surfaces may be photoconductors of selenium, zinc oxide, or various other materials; they may be preformed patterns of conductors on insulators, or they may be insulators with a charge pattern deposited by an electron beam, by a conducting stylus, or by a flood of ions through a mask onto the surface. The developer may be a vapor, a liquid, an aerosol (solid or liquid), a liquid containing solid or liquid particles in suspension, a paste, a sheet on which particles are lightly held, or carrier beads to which the developing particles attached by triboelectric charges. The surface itself may serve as its own image developing material by migration or by deformation. The image may be transferred after being developed, or we may observe it or use it where it is deposited initially. It may be used for printing by ink processes, observed visually, projected on a screen, used for electronic circuit production (or it may *be* an electronic circuit), or it may be used for information storage. The image may even be observed by tactile methods in the case of Braille printing.

The uses of electrostatic imaging are numerous and include office-style copiers, computer output, high-speed reproduction of maps, document storage and retrieval, color printing, book printing and reproduction, facsimile printing, and many others.

There are numerous published papers in which much of the research and technical information on electrostatic imaging appears. Two published works of great use in this field are *Xerography and Related Processes*,[13] and *Electrophotography*.[18] Both books contain extensive bibliographical information on almost all the topics covered in this short exposition. I particularly call the interested reader's attention to the chapter by the late Chester F. Carlson in *Xerography and Related Processes*.[13] Mr. Carlson was the inventor of the process most people think of as xerography, and his own record of the background and his development of the process is necessary reading for those involved in electrostatics. Also to be recommended is the extensive patent listing found in Schaffert's book, *Electrophotography*. This is, I think, the most complete published listing of patents available in this field. The list is very extensive and can be most useful to anyone interested in electrostatic imaging or related fields.

REFERENCES

1. Encyclopedia Britannica, Vol. 8, Chicago, 1955, p. 169.

2. Duane H. D. Roller, *The DE MAGNETE of William Gilbert*, Manno Hertzberger Publisher, Amsterdam, 1959.

3. W. Gilbert, *De Magnete*, Chiswick Press, London, 1900. This book was reissued in 1958 in a collector's series by Basic Books, New York.

4. G. C. Lichtenberg, *Novi Comment. Gottingen*, **8**, 168 (1777).

5. D. C. Miller, *Sparks, Lightning, Cosmic Rays*, Macmillan Co., New York, 1939, pp. 17, 18.

6. J. Strong, *Procedures in Experimental Physics*, Prentice-Hall, Englewood Cliffs, N.J., 1938, p. 165.

7. P. T. Riess, *Die Lehre von der Reibungs Elektrizität*, Berlin, 1853.

8. T. J. Baker, "Breath Figures," *Phil. Mag.*, **44**, 752 (1922).

9. J. W. Swan, "Stress and other effects produced in resin and in a viscid compound of resin and oil by electrification," *Proc. Roy. Soc. (London)*, **62**, 38 (1897).

10. J. C. Mol. "The Eidophor system of large-screen television projection," *Photogr. J.* **102**, 128 (1962); see also W. E. Glenn, "Thermoplastic recording," *J. Appl. Phys.*, **30**, 1870 (1959).

11. P. Selenyi, "A method of and apparatus for recording by means of cathode rays, rapidly occurring phenomena," British Patent 305,168, 1929; "On the electrographic recording of fast electrical phenomena," *J. Appl. Phys.*, **9**, 637 (1938).

12. C. F. Carlson, "Electrophotography," U.S. Patent 2,297,691 1942.

13. J. H. Dessauer and H. E. Clark, *Xerography and Related Processes*, Focal Press, London, 1965.

14. J. D. Cobine, *Gaseous Conductors*, Dover New York, 1958; also see L. Walkup, U.S. Patent 2,777,957, 1957.

15. H. E. Neugebauer, "Electrostatic fields in xerography," *J. Appl. Opt.*, **3**, 385 (1964); see also R. M. Schaffert, "The nature and behavior of electrostatic images," *Photogr. Sci. Eng.*, **6**, 197 (1962).

16. An excellent treatment of the gradient is presented C. R. Wylie, *Advanced Engineering Mathematics*, 2nd ed., McGraw-Hill, New York, 1960, p. 480.

17. L. Page and N. I. Adams, *Principles of Electricity*, 2nd ed. Van Nostrand, New York, 1949, pp. 86–98.

18. R. M. Schaffert, *Electrophotography*, Focal Press, London, 1966.

CHAPTER 13

Nonimpact Printing

DONALD S. SWATIK, *Manager, Advanced Development Engineering*
Printer Products Division
Computer Peripherals, Inc.
Rochester, Michigan
(A joint venture of Control Data Corporation
and The National Cash Register Company)

13.1 INTRODUCTION

13.1.1 Overview

The technologies employed in nonimpact printing are multidisciplinary and encompass many areas of endeavor. A comprehensive treatise on the processes utilized in nonimpact printing would include: electrooptic, magnetic, thermal, electrolytic, photographic, and electrostatic concepts. An analysis of viable nonimpact printing techniques reveals, however, that the majority are based on electrostatic phenomena, either solely or in conjunction with other technologies.

Electrostatic forces are used to disrupt liquid surfaces and produce beams of charged ink droplets that are electrostatically deflected into readable images. Electrical discharges from fine-metal styli are utilized to place programmable patterns of charge on dielectric surfaces; the charge patterns are then developed with electroscopic or electrophoretic toners. Electrostatic induction is employed to charge uniform-sized ink droplets precisely and thus control their electrodynamics. In these and sundry other applications, electrostatics serves to generate and/or control the charged particles employed in the formation of intelligible images.

Nonimpact printers are presently being utilized in many computer printing applications. The computer printer market, however, is dominated by electromechanical impact printers. Present impact printers range from serial character printers operating at speeds from 10 to 100 characters per second (CPS) to line printers with speeds of 100 to 2000 lines per minute (LPM). There are commercial nonimpact printers that operate from 30 to 750 CPS in the serial character mode and up to 5K (5000) LPM in the line at a time mode. These nonimpact printers have not displaced their impact printer counterparts for the following reasons: (1) many nonimpact printers require special forms,* (2) simultaneous multiple copies cannot be produced, and (3) print quality is inferior to that of impact printing.

Nonimpact printers have served primarily to perform functions unobtainable with impact printers. Electrostatic matrix printers have been utilized in applications where alpha-numerics and graphics are both required and for the printing of extended character fonts, such as the Katakana or Kanji alphabets. Address labels for the major publishing houses are printed by pin-CRT electrostatic printers at rates of 135,000/hr and at a usage of 3 billion labels annually. The high-speed capability of nonimpact printers (units have been constructed capable of more than 30K LPM) permits a single printer to be used where otherwise multiple impact printers would be required. Nonimpact printers are also being used in terminal applications, where their low noise level and reliability are important factors.

The potential applications for nonimpact printing are unlimited. The development of future nonimpact printers that are viable on standard forms and are capable of producing simultaneous multiple copies will expand the spectrum of nonimpact printing to the entire computer printer market— a market that is projected to surpass $1 billion annually in the mid-1970s.

13.1.2   Nonimpact versus Impact Printing

In order to ascertain the utility of the various electrostatic nonimpact printing technologies, it is instructive to outline the requirements for computer printers in general. The parameters listed in Table 13.1 can be used to compare nonimpact and impact printing.[1]

13.1.2.1   *Print Quality*

Most nonimpact printers employ a dot-matrix character format. Unless a high-order matrix is used, the characters produced are not as legible as the solid characters typical of impact printing. Impact printers, however, are subject to vertical or horizontal registration errors due to imperfect timing or alignment of their complicated electromechanical printing systems. Hence, in general, nonimpact printing exhibits superior alignment, but individual impact characters are more legible. In applications where print quality is of the utmost importance as in optical character recognition

---

* The term "forms" is used to denote the recording medium.

Table 13.1. Comparison of Impact and Nonimpact Printing.

NON-IMPACT VS. IMPACT PRINTING

	NON-IMPACT	IMPACT
PRINT QUALITY	DOT MATRIX CHARACTERS	FULL CHARACTERS (IMPERFECT ALIGNMENT)
MULTIPLE COPIES	NO	UP TO 6
SPECIAL FORMS	DEPENDS ON TECHNOLOGY	NO
PRINT SPEED	5000 LPM (30,000 LPM)	2000 LPM WITH 48 CHARACTERS
NOISE LEVEL	INHERENTLY LOW	HIGH
FONT SIZE	LIMITED ONLY BY CONTROL ELECTRONICS	MECHANICALLY LIMITED (EFFECTS PRINT SPEED)
FONT SELECTION	LOGIC CONTROLLED	REQUIRES MECHANICAL CHANGE
RELIABILITY	POTENTIALLY BETTER THAN IMPACT	

(OCR) and magnetic ink character recognition (MICR), impact printing possesses a definite advantage.

13.1.2.2 *Multiple Copy Capability*

No nonimpact printing technologies that have been commercialized produce simultaneous multiple copies. The production of multiple copies by nonimpact printing is still in the research and development stage.[2,3] Additional copies must be produced sequentially at the expense of print speed. If six copies are desired, for example, a nonimpact printer would have to function at 12K LPM to equal the throughput of an impact printer operating at 2K LPM. Thus, if simultaneous multiple copies are desired, present nonimpact printers are not viable.

13.1.2.3 *Special Forms Requirement*

Many nonimpact printers require special dielectric-coated forms; however, recent advances in nonimpact printing technology have produced systems that work with standard forms.* If special forms are used, their availability, storage characteristics, and cost must be considered. At high print speeds, the cost of these forms may exceed the initial cost of the printer in less than a year of operation.

* "Standard forms" denotes conventional papers typical of those used in impact printers.

13.1.2.4 *Print Speed*

The fastest impact printers are capable of 2K LPM. Present commercially available nonimpact printers are capable of 5K LPM, and units have been built that printed in excess of 30K LPM. Hence, if high print speed is mandatory, and especially if only a single copy is needed, nonimpact offers increased performance relative to impact printing.

13.1.2.5 *Emitted Noise Level*

In most nonimpact printers, mechanical motion is limited to the forms-handling system; hence they are inherently quiet. Impact printers are intrinsically noisy and require specially designed acoustic cabinets to lower their operating noise level. Accordingly, if a printer is to be used in a quiet environment, the nonimpact variety offers better acoustic characteristics.

13.1.2.6 *Font Size and Selection*

In impact printers, font size is limited by mechanical restraints, and increases in font size can be obtained only at the expense of reduced print speed. Also, in order to change the font size or style, the type elements must be physically replaced. In nonimpact printers, utilizing dot-matrix characters, both font size and style can be controlled logically to within the limitations of the matrix format used. In addition, the dot-matrix approach allows for the simultaneous printing of graphics with alpha-numerics

13.1.2.7 *Reliability*

By utilizing electrostatic forces in place of mechanical forces, the number of moving parts in nonimpact printers is dramatically reduced relative to mechanical impact printers. Mechanical motion is limited to the forms-handling system, and literally hundreds of mechanical components can be eliminated. By reducing the complexity of the printer system, the potential for improved reliability is increased.

13.1.3 Non-Impact Printing Technologies

Nonimpact printing technologies can be classified according to the basic imaging process employed. Three generic types of nonimpact printers that utilize electrostatic phenomena are: electrodynamic, electrographic, and electrophotographic. The fundamental distinctions between these three technologies can be seen by comparing the means by which the writing medium (ink or toner) is imaged on the recording medium (forms).

1. In electrodynamic printers, the forms are passive. The printing ink is formed into macroscopic droplets which are imaged by electric fields external

to the forms. The forms serve only as a receptor for the imaged ink droplets, and upon contact, a visible image is instantly produced. This technique is viable on standard forms, and no image development or fixing is required.

2. Electrographic printers employing electrostatic technologies* produce images by placing controlled patterns of charge on dielectric forms.† The electric field of the charge pattern on the forms attracts charged toner particles and produces a visible image. Therefore, the forms are active and their properties influence the imaging process. Most electrostatic printing systems require the use of special dielectric-coated forms to ensure that the charge image remains immobile until it is developed.

3. As in electrographic printing, charge images are utilized in electro-photographic printing to attract charged toner particles and to produce a visible image. Unlike electrographic printers, however, this process does not occur on the forms but on an intermediate photoconducting surface that is exposed by a programmable light source. After the image is formed and developed, it is transferred to the forms and fixed. The forms are passive and do not participate in the imaging process; accordingly, standard forms can be utilized.

13.2 ELECTRODYNAMIC PRINTING

13.2.1 Ink Jet Printers

The ink jet printing technique produces instantly visible images on standard forms by the electrostatic deflection of charged ink droplets into electrically programmable dot-matrix patterns. This principle is analogous to the deflection of electrons in a cathode ray tube. The operation of the ink jet printer can be subdivided into three sections: droplet production, droplet charging, and droplet deflection.

The operation of the ink jet printer is based on the capability to produce a colinear stream of uniformly sized, equally spaced ink droplets. Using droplets with uniform size means that the dynamics of the droplets can be electrostatically controlled by varying only their charge. The droplet stream is created by taking advantage of the natural dynamic instability of cylindrical liquid jets.[4] Whenever a jet is perturbed at a wavelength exceeding the initial jet circumference, the action of surface tension renders the jet unstable and leads to the production of droplets. If the perturbation

---

* Electrographics also includes other printing concepts such as electrolytic and electro-sensitive recording.

† A reusable dielectric surface can also be used, as in ovography.

is created by friction or other random source, a plurality of unstable wavelengths will be present and will lead to the production of randomly sized and spaced droplets.

From an energy standpoint, the most rapid instability occurs when the wavelength of the perturbation is equal to 9.016 times the initial jet radius. Hence, if a jet is externally excited at this optimum wavelength, it will dominate the disruption process and a stream of uniformly sized, equally spaced droplets will be produced.[5] One droplet will be created for each cycle of the perturbing waveform. The disruption of a liquid jet into 100-μm diameter droplets at a frequency of 100 KHz is shown in Fig. 13.1.

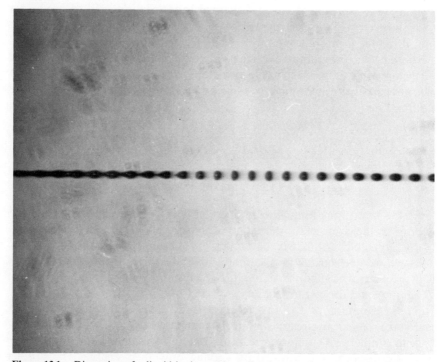

Figure 13.1.   Disruption of a liquid jet into 100-μm droplets at a frequency of 100 KHz.

The components of an ink jet printer are schematically depicted in Fig. 13.2. The ink jet itself is produced by a pressurized ink supply that feeds a small-diameter nozzle. Uniform droplets can be produced by either vibration or pressure modulation of the jet at a single frequency. This is most usually accomplished by means of either piezoelectric or electromagnetic trans-

INK JET PRINTER

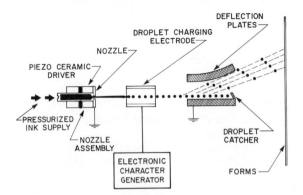

Figure 13.2.   Basic elements of an ink jet printer.

ducers. For usable droplet sizes, excitation frequencies exceeding 100 KHz can be utilized, yielding more than 10^5 droplets/sec.

The radius of the ink droplets r is primarily determined by the initial radius of the jet a and is approximately equal to twice the jet radius. For example, to produce 0.254-cm high characters, nozzles in the 25 to 125 μm diameter range are required, depending on the dot-matrix format used. The nozzles can be of various designs including watch jewels, holes etched, or lased in thin flat plates, and short capillary tubes.

The droplets produced by the disruption of a liquid jet are uncharged. Individual droplet charging is accomplished by passing the jet through a cylindrical charging electrode. The jet and the charging electrode form a coaxial capacitor. When a voltage V_p is applied to the charging electrode, a charge is induced on the jet. The jet is longitudinally positioned so that it disrupts inside the charging electrode; thus the induced charges on the jet are trapped on the droplets as they break off.

The time available to charge an arbitrary droplet is determined by the droplet production frequency f. To allow for the independent charging of each droplet, the pulse width of the voltage applied to the charging electrode must be less than the interdroplet period $1/f$. In addition, the conductivity σ of the ink must be high enough to ensure that the droplets will become fully charged within the duration of the charging electrode pulse. That is to say, the electric Reynolds number $\mathbf{R_e}$ for the system must be much less than unity.

$$\mathbf{R_e} = \frac{\varepsilon/\sigma}{1/f} \ll 1 \tag{1}$$

where ε is the permittivity of the ink. If a conductive ink is used, the charge per droplet q can be determined from the capacitance of the coaxial geometry as follows:

$$q = \frac{8\pi V_p \varepsilon_g r^3}{3a^2 \ln (b/a)} \qquad (2)$$

where b is the inner radius of the charging electrode and ε_g is the permittivity of interelectrode gap. The droplet specific charge q_s is found by dividing Eq. 2 by the mass per droplet:

$$q_s = \frac{2V_p \varepsilon_g}{a^2 \rho_m \ln (b/a)} \qquad (3)$$

where ρ_m is the mass density of the ink.

Thus for a given geometry, the specific charge and hence the dynamics of the droplets are determined solely by the charging electrode voltage. For 100-μm diameter droplets, charging voltages in the 50 to 300 V range are required. The charge on each droplet is controlled by a programmable character generator that is synchronized with the droplet production frequency.

The charged droplets are then deflected into the desired matrix patterns by passing the stream through a nontime-varying electrostatic field. This field is produced by means of a pair of planar electrodes, one held at ground potential and the other maintained as a potential in the range of 1 to 5 kV. If the electrodes are parallel and fringing is neglected, the electric field between them will be of constant amplitude E_0 and the force F exerted on a droplet by the field will be

$$F = q_s E_0 \qquad (4)$$

The total transverse displacement of a droplet from the stream X is then

$$X = \frac{1}{2} q_s E_0 t^2 \qquad (5)$$

where t is the transit time of the droplet thru the field.

The deflection of charged 100-μm ink droplets in an electrostatic field is represented in Fig. 13.3. Alternate* droplets were charged in sequence to

---

* If all droplets were charged there would be coupling between adjacent droplets. To avoid this, only every other droplet is charged.

MODEL 3000 MICRODIAL®
MOUNTING INSTRUCTIONS

The following instructions apply to the installation of the Amphenol Model 3000 MICRODIAL to an Amphenol Potentiometer. It may, however, be used with other multi-turn devices, ten turns or less, requiring accurate position indication in turns and decimal parts of turns.

1. Drill panel as shown under suggested panel holes. The angular position of the .062 dia. hole is determined by the desired position of the Potentiometer Terminal Lugs.

2. Secure mounting plate to Potentiometer shaft and panel. Use adapter ring for 3/16 thick panels and spacer for 1/16 thick panels. Use special nut for 1/4" shafts.

3. Set Potentiometer against its counter-clockwise stop.

4. Slide MICRODIAL onto Potentiometer shaft until it rests lightly on the mounting plate face. Apply **Light** finger-tip pressure and rotate knob **counter-clockwise** to index keys in their ways. **DO NOT ROTATE CLOCKWISE.**

5. Press knob section into mounting plate to seat assembly. A slight snap will indicate proper operating position.

6. Tighten Allen set screw locking the MICRODIAL to the Potentiometer shaft.

7. Remove the #4-40 holding screw and replace with the #4-40 Allen set screw.

8. Check MICRODIAL for zero-setting and apply set-screw hole seals. **Note:** Should it be necessary to remove the MICRODIAL, rotate to zero-point. Remove **one** #4-40 Allen set-screw and replace it with a #4-40 holding screw. Loosen the other set-screw and pull MICRODIAL knob forward to disengage snap-on assembly.

9. Parts Included:

3020 Series	3040 Series
(1) Assembled and tested 3020 MICRODIAL	(1) Assembled and tested 3040 MICRODIAL
(1) Adapter Ring	(1) Adapter Ring
(2) Set-Screw Seals	(2) Set-Screw Seals
(1) #4-40 Set-Screw	(1) #4-40 Set-Screw
(1) #4 Allen Wrench	(1) #4 Allen Wrench
	(1) Spanner Wrench
	(1) Spacer
	(1) Hex Nut

BUNKER RAMO AMPHENOL

AMPHENOL CONTROLS DIVISION

Box 1368 • 120 South Main Street • Janesville, Wis. 53545

Printed in U.S.A.   62945

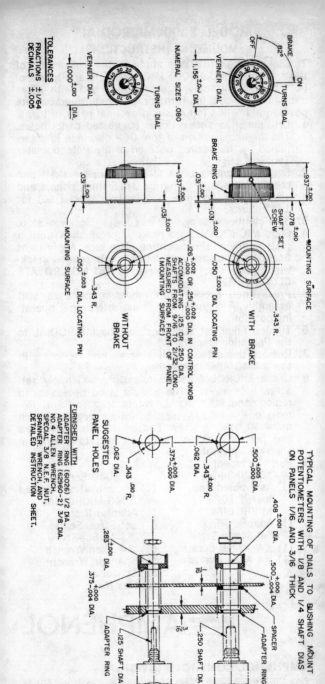

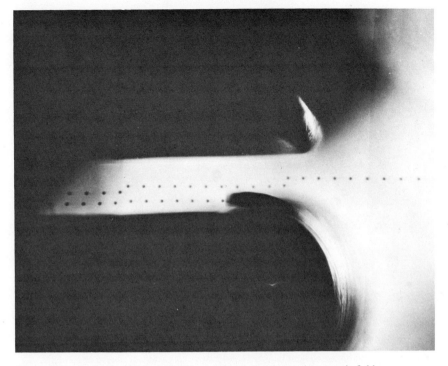

Figure 13.3. Deflection of charged 100-μm ink droplets in an electrostatic field.

11 distinct levels by varying the charging electrode voltage from 50 to
150 V. The variation in droplet deflection is evidenced by the "saw tooth"
droplet pattern produced. Uncharged droplets are not affected by the field
and remain in the initial stream, where they are mechanically trapped and
recirculated.

The ink jet concept is capable of droplet deflection in only one plane.
Two-dimensional characters are formed by utilizing relative motion between
the ink jet and the forms. This is accomplished either by scanning a
single jet past stationary forms or by continuously moving the forms past a
linear array of ink jets. A print sample depicting 0.254-cm high 9 × 11 dot-
matrix characters produced at 500 CPS from a single jet appears in
Fig. 13.4.

Because of the large number of droplets produced per second, the ink
jet technique is capable of extremely high print speeds. Besides the droplet
production rate, speed is dependent on the size of the matrix format used
and the number of columns served by a single jet, as in Table 13.2.

Figure 13.4. Ink jet print sample: 9 × 11 dot-matrix characters produced at 500 CPS.

Ink jet printers possess the distinct advantages of being viable on standard forms and not needing postrecording image-fixing systems. Also, since the ink jet printer is electrically controlled, it is easily interfaced to computer systems; typical drive voltages of 100 V are within the range of inexpensive solid-state devices.

Table 13.2. Speed of Ink Jet Printers for a Droplet Production Frequency of 50 KHz

	Speed (LPM)		
Dot-Matrix Format	1 Column/jet	2 Columns/jet	3 Columns/jet
5 × 7	61K	30K	20K
7 × 9	37K	18K	12K
9 × 11	25K	12K	8K
11 × 15	13K	7K	4K

13.2.2 Ink Spray Printers

Ink spray printing technology is similar to ink jet printing in that imaging is accomplished by the electrostatic deflection of a stream of charged ink droplets. The two techniques differ fundamentally, however, in the manner by which the ink droplets are generated. In ink spray printers, the droplets are created by the electrohydrodynamic (EHD) spraying of a liquid surface.

The EHD spraying of liquids has been studied for many years as a source of multimolecular particle beams.[6] When a liquid surface is subjected to an external electric field, it experiences an outward stress due to the interaction of induced surface charges with the field. If this outward stress exceeds the internal surface tension forces of the liquid, the surface becomes unstable and disrupts by ejecting one or more charged droplets.

For printing applications, the liquid surface consists of a hemispherical meniscus of ink held at the tip of a small-diameter nozzle by the force of surface tension. An intense electric field is impressed on the meniscus by applying a high voltage between the liquid and a nearby accelerating electrode. As this voltage is increased, the hemispherical meniscus is elongated into a prolate spheroidal geometry; then, at a well-defined voltage, the surface becomes unstable and disrupts. The minimum voltage V_{min} needed to initiate EHD spraying is found by equating the surface tension forces of the liquid hemisphere to the outward stress acting on the induced surface charge.[7]

$$V_{min} = 300(20\pi c\gamma)^{1/2} \tag{6}$$

where c is the radius of the hemisphere and γ is the surface tension of the liquid.

For example, a 0.025-cm radius hemisphere of liquid with a surface tension of 30 dynes/cm will become unstable at a voltage of 2.1 kV.

At voltages greater than V_{min}, the liquid hemisphere will be drawn out into a long slender filament with a radius 10 to 100 times smaller than the initial radius. The action of the electric field is greatest at the tip of the filament, and it disrupts into a stream of highly charged droplets. These droplets then react with the field and are accelerated away from the filament. This process is continuous only if the conductivity of the liquid is sufficiently high to maintain the surface of the filament as an equipotential. If the surface does not remain as an equipotential, it will relax back to its initial hemispherical geometry after each emission and the stream of droplets will be produced in periodic pulses corresponding to the oscillation of the meniscus. A photomicrograph of EHD spraying from an electrified capillary is presented in Fig. 13.5.

As a droplet breaks off from the filament, it carries with it part of the induced surface charge. Therefore, the droplets generated by EHD spraying

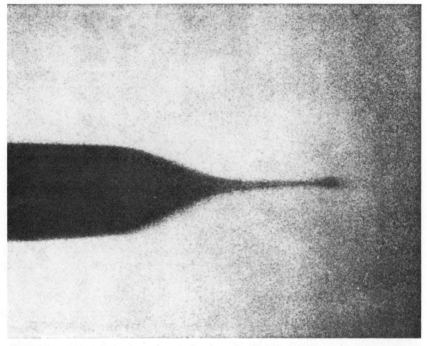

Figure 13.5. Electrohydrodynamic spraying from an electrified capillary. (Courtesy of C. D. Hendricks.[8])

are simultaneously charged as they are produced. The maximum specific charge $q_{s, \max}$ of a droplet of radius r is limited by the Rayleigh criterion for the stability of an electrified spherical droplet.

$$q_{s, \max} \leq \frac{3\gamma^{1/2}}{\pi^{1/2} \rho_m r^{3/2}} \tag{7}$$

The droplets produced by EHD spraying possess specific charges near but not exceeding the Rayleigh limit.[9]

Figure 13.6 is a schematic diagram of an ink spray printer. The ink is supplied to a nozzle from a reservoir maintained at a slight positive hydrostatic pressure. This pressure is sufficient to produce a meniscus at the tip of the nozzle, but not large enough to trigger a flow of ink. The nozzle itself is preferably but not necessarily a conductor and can be easily fabricated from small-diameter hypodermic tubing. The outside diameter of the nozzle is limited by the voltage available to produce the EHD spray and is typically in the range of 100 to 500 μm.

INK SPRAY PRINTER

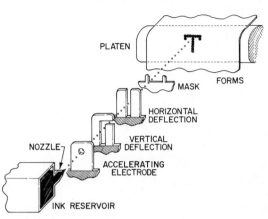

Figure 13.6.   Schematic diagram of an ink spray printer. (Courtesy of Teletype Corporation.)

Immediately adjacent to the nozzle is an accelerating electrode that serves to modulate the production of ink droplets. It consists of a flat plate containing a hole concentric with the nozzle to allow for passage of the droplets. When a voltage of sufficient amplitude is placed between the nozzle and accelerating electrode, the meniscus disrupts and a stream of droplets is produced. Typically voltages in the 1.5 to 5 kV range are required. Since no droplets are created until the voltage exceeds the minimum given by Eq. 6, the accelerating electrode can be biased to a value near this minimum and modulation can be achieved by relatively small variations (0.5–1 kV) in the accelerating electrode potential.

The stream of ink droplets is then imaged into the desired dot-matrix format by electrostatic deflection. Since the droplets are produced with a nearly uniform specific charge, programmable deflection is obtained by varying the magnitude of the deflection fields. This is accomplished by electronically controlling the voltage applied to the deflection electrodes. Two sets of electrodes can be employed to obtain both horizontal and vertical deflection to allow for printing on stationary forms, or a single set can be used in synchronization with either forms or printhead motion.

The displacement of an ink droplet from the stream is governed by its specific charge and the amplitude of the deflection field as given by Eq. 5. For constant specific-charge droplets, variations in the deflection potential in the range of 500 to 1000 V are required to produce 0.25-cm high characters. Unwanted droplets are deflected into a mask and are recirculated. The physical components used in an ink spray printhead are shown in Fig. 13.7.

Figure 13.7. Ink spray printhead used in Teletype Corporation Inktronic printer. (Courtesy of Teletype Corporation.)

NON-IMPACT
PRINTING

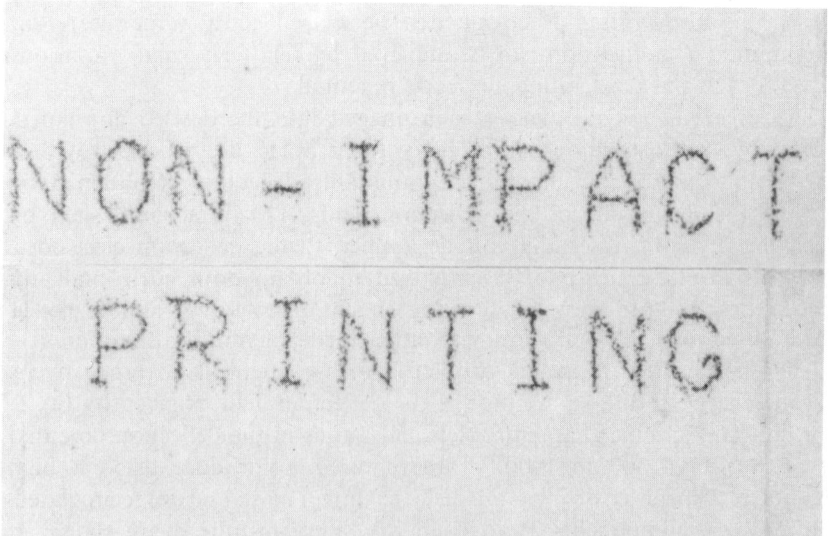

Figure 13.8. Ink spray print sample. (Courtesy of Teletype Corporation.)

The droplets produced by the ink spray technique are very small, typically less than 0.0025 cm in diameter. To achieve readable print densities, each dot in the required matrix pattern must be composed of many ink droplets. The number used is controlled by the pulsewidth of the deflection voltage. In the print sample appearing in Fig. 13.8, each dot was produced by deflecting the jet to that position for 250 μsec. A character composed of 32 dots would be formed in 8 msec or at a rate of 120 CPS. A line printer using this technique with one nozzle for every two columns would have a print speed of 3600 LPM.

13.2.3   Modulated-Aperture Printer

In modulated-aperture printing, dot-matrix characters are formed by modulating ion beams with electric fields and utilizing these ions to selectively charge an aerosol ink cloud. The charged ink droplets are subsequently directed to standard forms by an external electric field, and a visible image is produced.[10] The function of the modulated-aperture printer can be divided into four parts: (1) ion beam production, (2) ion beam modulation, (3) ink droplet charging, and (4) transfer of the ink droplets to the forms.

A schematic diagram of a modulated-aperture printer is given as Fig. 13.9. A corona source typical of those employed in xerography is used to produce a uniform line source of ions. The corona discharge is produced by maintaining a relatively large potential difference between the corona source and the platen. The electric field produced between the source and platen (propulsion field) also serves to accelerate the ions toward the platen.

The ion beam is controlled by an aperture board placed between the source and platen; it is biased to correspond to an equipotential of the propulsion field. The aperture board resembles a printed circuit card; it has electrodes on both sides of an insulating substrate, and a linear array of equally spaced apertures is placed in it. For typical printing applications, an aperture diameter of 0.025 cm is used.

The flow of ions through an aperture is modulated by the voltages applied to the electrodes surrounding that aperture. When there is a potential difference placed between the electrodes of an aperture, the insulating substrate appears as if it had a double-layer charge on it. In the aperture itself, the fringing fields from this double-layer charge control the flow of the ions.[11] If the electric field of the aperture aids the propulsion field, the lines of force of the propulsion field pass directly through the aperture and the ion beam passes through unattenuated. However, if the aperture field opposes the propulsion field, the lines of force emanating at the corona source terminate on the aperture board, and the ion beam is

MODULATED APERTURE PRINTER

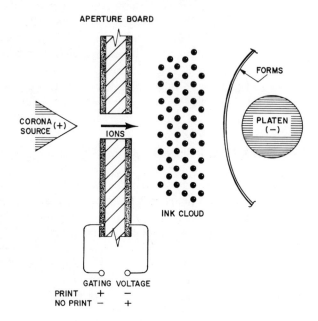

Figure 13.9. Modulated-aperature printer. (Courtesy of ElectroPrint, Inc.)

effectively blocked. The flow of ions can be modulated by control voltages in the 100 to 200 V range, depending on the ion energy and the aspect ratio of the apertures.

Next the modulated ion beams pass through an aerosol ink cloud that is directed between the aperture board and the platen. The ink cloud consists of uncharged droplets approximately 10 μm in diameter. When an aperture is activated, the cylindrical ion beam produced collides with the aerosol and the ink droplets are charged until their surface potential equals the potential through which the ion beam has been accelerated.

The charged droplets immediately react with the propulsion field and are accelerated to the forms placed in front of the platen. If the propulsion field is uniform, the droplets follow straight-line paths and reproduce the dot image of the aperture board.

Printing is accomplished by synchronizing the control of the aperture board with the forms motion. Figure 13.10 is a sample of Japanese Kanji characters printed in a 14 × 20 dot-matrix format. The aperture board

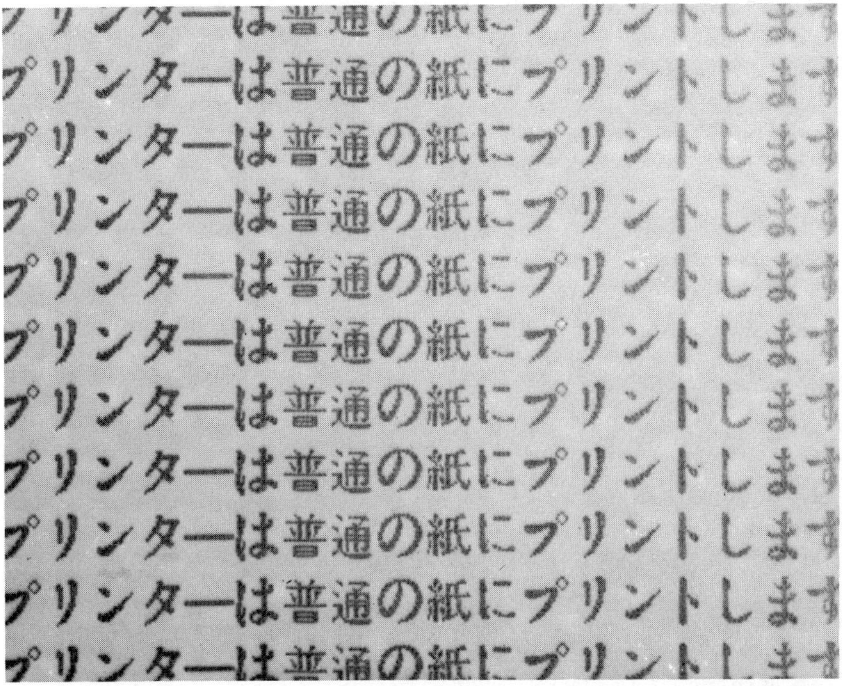

Figure 13.10. Printing of Japanese Kanji characters at 22 IPS by modulated-aperture printer. (Courtesy of Electroprint, Inc.)

used contained 100 apertures/in. Print speeds of up to 20K LPM have been achieved in modulated-aperture printing.

13.3   ELECTROGRAPHIC PRINTING

13.3.1   Electrostatic Printers

In conventional electrostatic printers, imaging is accomplished by placing electrostatic charge patterns on dielectric-coated forms. The charge is usually transferred to the forms by means of electrical discharges from electrically programmable electrodes.* The latent electrostatic image is subsequently made visible by development with either liquid or solid toner.

* Imaging can also be accomplished by electron beams, either directly or by use of a CRT with a Lenard window.

Basic to the operation of electrostatic printers is the phenomenon of electrical discharges. An electrical discharge occurs when there is an electrical breakdown of the air gap between two electrodes. The potential difference needed to initiate a discharge is a function of both the ambient air pressure P, and the gap length g. For uniform fields, the breakdown potential V_B is given by Paschen's law:[12]

$$V_B = BPg/\ln\left(\frac{APg}{\ln 1/C}\right) \tag{8}$$

where A and B are constants of the ambient gas and C is the coefficient of electron emission by positive ion bombardment of the electrodes.

For air gaps in the 8 to 100 μm range at atmospheric pressure, Paschen's law can be written as[13]

$$V_B = 312 + 6.2 \times 10^6 g \tag{9}$$

The minimum breakdown potential in air occurs for gaps in the 8-μm range and is equal to 360 V. At smaller gap lengths, the breakdown potential actually increases. In practice, gap lengths of 8 μm cannot be accurately maintained; therefore, voltages greater than the minimum breakdown potential are used. Typically potentials in the 500 to 1000 V range are required for efficient printer operation.

To ensure that the latent electrostatic images remain immobile until development, the relaxation time ε/σ of the dielectric forms must be long compared with the time interval Δt between imaging and developing. That is to say, the electric Reynolds number R_e must be much greater than unity:

$$R_e = \frac{\varepsilon/\sigma}{\Delta t} \gg 1 \tag{10}$$

It is necessary to use special high-resistivity multilayer surface coatings that are insensitive to large changes in ambient relative humidity.

Various configurations are possible for electrostatic printers, depending on the electrode geometry employed. Three geometries that have been successfully utilized in electrostatic printers are: (1) linear array of metal styli, (2) character-shaped electrodes, and (3) pin-matrix cathode ray tube. These are presented schematically in Fig. 13.11.

The simplest electrostatic printing system employs a linear array of metal styli which span the entire width of the print station. Each stylus is electrically isolated from its neighbors and can be independently driven electronically. When a stylus is pulsed "on," a dot of charge is placed on the forms. Accordingly, printing is in a dot-matrix format and two-dimensional characters are produced by synchronizing the stylus-array drive circuitry with the motion of the forms.

ELECTROSTATIC PRINTING

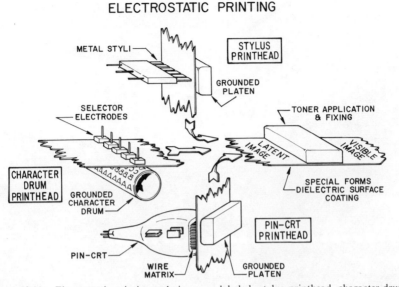

Figure 13.11. Electrostatic printing techniques, as labeled: stylus printhead, character-drum printhead, pin–CRT.

The styli are usually produced in modules by printed circuit techniques, and all must be of uniform tip geometry with equal interstylus spacing. Typical arrays contain between 20 and 50 styli per centimeter; an individual stylus diameter would be in the 100 to 250 μm range. In line printer applications, more than 1000 styli are used. Figure 13.12 is a photomicrograph of a stylus array.

Sufficient charge can be deposited on the forms to produce developable images in less than 1 μsec; accordingly, an array of 1000 styli can be sequentially addressed in approximately 1 msec. These extremely high imaging rates allow for the use of a single high-voltage drive circuit that can be time-shared among all the styli in a printer. Imaging is then accomplished in a raster scan format.[14] For 5×7 dot-matrix characters with a three-dot line spacing between lines, a line can be printed in 10 msec. Thus, a print speed of 6000 LPM can be easily obtained with a single high-voltage driver. It is interesting to note that if each stylus had an independent drive circuit, a theoretical print speed of 6 million LPM would be obtained. A sample of electrostatic printing in a 5×7 matrix format appears in Figure 13.13.

Solid characters can be produced in place of dot-matrix characters by utilizing a rotating character drum resembling the print drum used in

Figure 13.12. Photomicrograph of linear array of styli used in electrostatic printers. (Courtesy of Versatec, Inc.)

impact line printers. Around the periphery of the drum are placed bands of character-shaped electrodes corresponding to the character font to be printed. The drum is rotated at high speed past the forms, and a linear array of selector electrodes is placed on the opposite side of the forms. Each electrode is electrically isolated and has an independent drive circuit. When the appropriate character for a given column is aligned with the forms, the corresponding selector electrode is pulsed "on" and a character-shaped charge pattern is transferred to the forms. A complete line is imaged for each revolution of the drum.

Since the characters in a given line are imaged in an arbitrary manner, the forms must be stationary during each imaging cycle and stepped one line between cycles.

The print speed is, therefore, determined both by the drum speed and by the forms advance rate. For example, a printer with a drum speed of 10,000 rpm and a forms advance rate of 4 msec would print at 6K LPM.

The third electrode configuration utilizes a CRT with a pin-matrix face-plate.[15] The matrix of pins is embedded in the faceplate and extends out of the tube into virtual contact with the forms. Each pin is electrically

Figure 13.13. Typical electrostatic printer (5 × 7) matrix format printing. (Courtesy of Versatec, Inc.)

isolated and becomes negatively charged when the electron beam of the CRT is deflected to it (it is assumed that the secondary emission ratio of the pins is less than unity). Since the capacitance of the pins is very small, they are elevated to a high potential by a small charging current.

Imaging is accomplished by programmable deflection of the electron beam onto the pin matrix. When a pin is energized, a dot charge pattern is placed on the forms. If the pin is in physical contact with the forms, charge is transferred directly; in practice, however, there is usually an air film between the pin and forms, and the charge is transferred by an electrical discharge.

The pins are fabricated from short lengths of wire. They are typically 25 to 100 μm in diameter and are packaged with densities of 100 to 400 per centimeter. The high spatial density of pin electrodes permits the use of high-order dot-matrix formats for character formation and greatly increases the quality of the printout.

Since the electron beam can be addressed at very high rates, (e.g., < 1 μsec), extremely fast printing speeds can be obtained, typically in excess of 10K LPM.

13.3.2   Electrographic Printing on Standard Forms

A fundamental limitation of conventional electrostatic printers is their requirement for special dielectric-coated forms. This need arises because the formation and development of the electrostatic images on the forms occur sequentially. Accordingly, the forms must possess a high surface resistivity in order to ensure that the latent charge images will remain immobile until development. Printing can be accomplished on standard forms, however, by either: (1) utilizing a system in which image formation and development occur simultaneously or very nearly so, or (2) employing a reusable dielectric surface on which the electrostatic image is formed and developed.

13.3.2.1   *Contrography*

Simultaneous electrostatic image formation and development can be achieved by maintaining the forms in contact with a liquid toner developer during the imaging process. By combining the imaging and developing functions, the basic techniques of electrostatic printing can be applied to systems using standard forms. This principle is the basis of contrography.[16]

A schematic diagram of a contrographic printer appears in Fig. 13.14. An electrostatic image is placed on uncoated forms by electrical discharges from a linear array of metal styli. Directly opposite the styli, the forms contact a meniscus of liquid toner developer that is created between the forms and a developer roller. The roller serves to both replenish the liquid developer and function as a platen.

When charge is deposited on the forms, the developing process starts immediately; and a visible image is produced in from 1 to 10 msec, depending on the toner particle size and the charging potential used. These development times are much shorter than the relaxation times of standard forms, which are typically 100 msec or greater; therefore, even with standard forms there is no appreciable relaxation of the charge image prior to development.

The liquid toner developer can also be applied to the side of the forms facing the print electrodes.[17] A film of developer is placed on the forms and contacts the ends of the styli at the print station. In this configuration, charge is deposited on the forms by an electrical breakdown of the developer itself, not by gaseous discharges through air as in conventional electrostatic printing. If the forms are charged through an insulating fluid, the resolution of the images can be increased and higher electrode voltages can be used, since the liquid developer serves both to isolate the styli and to prevent electrical breakdown.

The use of relatively high imaging voltages in the range of 1 to 3 kV

CONTROGRAPHIC PRINTER

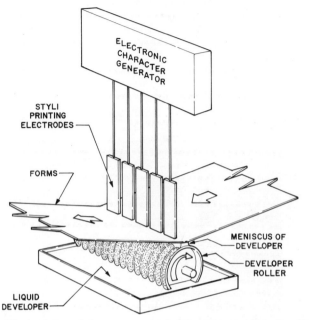

Figure 13.14. Electrostatic printing on uncoated forms by contrography. (Courtesy of Horizons, Inc.)

reduces the imaging time of an individual stylus to as little as 25 nsec. Figure 13.15 is a photomicrograph of contrographic printing of dots at a rate of 3.7 MHz. Rates exceeding 20 MHz have been obtained. These dot-production rates correspond to the printing of full lines of 5×7 dot-matrix characters at speeds in excess of 120K LPM.

13.3.2.2 *Ovography*

Electrographic printing techniques can be applied to systems using standard forms by forming and developing the electrostatic images on a reusable dielectric surface and transferring the developed images to the standard forms. A surface coated with an amorphous semiconductor can be effectively utilized as the reusable dielectric. This technique is the basis of ovography.[18,19]

Amorphous semiconductors have a particularly interesting property—namely, they can be formulated to exhibit a very large change in resistivity when thermally transformed from an amorphous to a crystalline state. The resistivity ranges from 10^{14} Ω-m in the amorphous (nonconducting) state

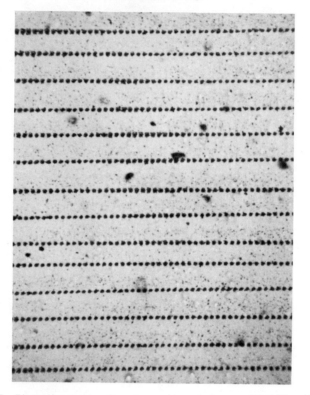

Figure 13.15. Photomicrograph of contrographic printing at 3.7 MHz. (Courtesy of Horizons, Inc.)

to 10^8 Ω-m in the crystalline (conducting) state. Imaging is accomplished on amorphous semiconductors by selectively altering the state of the material on a highly localized basis and therefore spatially modulating the surface resistivity.

The state of the amorphous semiconductor can be controlled by melting it and then varying the cooling rate. A rapid cooling promotes disorder and yields an amorphous structure, but a slow cool gives a crystalline structure.[20] The amorphous semiconductor can be heated by various techniques, including internal ohmic dissipation, electron bombardment, and laser radiation absorption. For printing applications a modulated laser beam has been effectively utilized as a programmable heat source. A high-power, short-duration light pulse ("write" pulse) places the material in the amorphous state. A pulse of approximately equal energy but of lower power and longer duration ("erase" pulse) transfers the material back to the crystalline state. Thus the amorphous semiconductor can be reversibly switched from the amorphous to crystalline states by means of suitably shaped light pulses.

Figure 13.16 is a schematic diagram of an ovonic printer. The reusable dielectric takes the form of a rotating drum, coated with a thin layer (20–60 μm) of an amorphous semiconductor. The resistivity of the drum is modulated by laser pulses which are electrically controlled by an electro-optic light modulator. The modulated laser beam is mechanically scanned across the drum in synchronization with the drum rotation. The drum surface is initially switched to the crystalline (conducting) state and is then selectively switched to the amorphous (nonconducting) state by "write" pulses. After exposure, the drum consists of nonconducting regions corresponding to the image to be reproduced. The drum is then sensitized by uniformly charging it with a high-voltage corona-charging device. The charge quickly relaxes in the conducting areas (relaxation time of 10^{-3} sec); but in

OVOGRAPHIC PRINTER

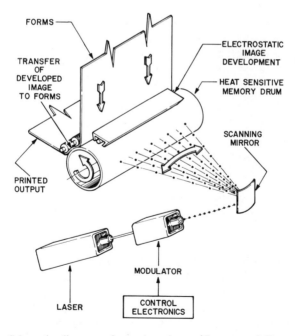

Figure 13.16. Schematic diagram of ovonic printer. (Courtesy of Energy Conversion Devices, Inc.)

the nonconducting areas, the charge remains immobile (relaxation time of 10^3 sec). The resulting charge image is developed by a conventional toning process, and the developed image is electrostatically transferred to uncoated forms and fixed.

Dot writing rates of up to 5 MHz have been obtained with a 1.5-W laser and an amorphous semiconductor with a sensitivity of 0.1 J/cm^2. At this rate, a 136-column printer would output 5×7 dot matrix characters at a speed of 30K LPM.

The resistive image on the drum is unaltered by the charging or toning process; accordingly, after a copy has been produced, the drum can be recycled and additional copies can be made from a single exposure. Before new data can be printed, however, the drum must be "erased." This is accomplished by transforming the entire drum surface back to the crystalline state by exposure to "erase" light pulses. This requires a relatively large expenditure of energy; therefore, ovonic printing is more efficient in multi-copy applications than in single-copy printing.

13.4 ELECTROPHOTOGRAPHIC PRINTING

The basic imaging concepts of xerography can be utilized to produce nonimpact printers that are viable on standard forms. Imaging is accomplished by converting the data to be printed into light energy, which is used to expose a photoconductive drum. The complete printing process parallels that of xerography and consists of five steps:[21]

1. Sensitizing a photoconductive drum by placing a uniform electrostatic charge on it.
2. Exposing the charged drum with light to form a latent charge image.
3. Developing the latent image with toner.
4. Transferring the image from the drum to the forms.
5. Fixing the image to the forms.

To utilize this process in nonimpact printing, an electrically programmable light source must be employed that is capable of exposing the photo-conductor at the required rate. Three types of light sources that are applicable to xerographic printing are: cathode ray tubes, modulated laser beams, and stroboscopic lights. Figure 13.17 schematically represents various xerographic printers utilizing these light sources. (It is interesting to note that these three configurations are the light energy analogs of the electrical techniques used in the electrostatic printers shown in Fig. 13.11.)

Cathode ray tubes controlled by electronic character generators can be utilized to expose a photoconductor in the required image format. However, the large angles at which phosphors emit light (i.e., Lambertian

XEROGRAPHIC PRINTING

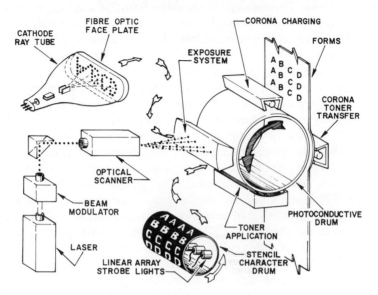

Figure 13.17. Xerographic printing techniques, as labeled: cathode ray tube, modulated laser beams, strobe light.

source) and the light scattering produced by internal glass reflections in the tube faceplate, makes it difficult to obtain both high-resolution and high-intensity images from conventional CRTs. Special CRTs have been developed that circumvent this problem by utilizing faceplates comprised of a vacuum-tight bundle of fused optical fibers. The fiber optic faceplate serves to transmit the image at the phosphor to the outer edge of the faceplate, which is placed in close proximity to the photoconductor. Photometric gains of 50 : 1 can be achieved by using fiber optics in place of conventional lens systems.[22]

The resolution of fiber optic faceplates is very high; it is limited by the size of the individual optical fibers, which are typically 10 μm in diameter or smaller.

Characters can be electronically generated on the face of the CRT in various formats, such as dot matrix, stroke, Lissajou, or scan patterns.[23] The resolution of the characters is limited by the spot diameter of the tube. For the printing of 0.254-cm high characters, spot diameters in the 0.002 to 0.02 cm range are required.

Dot-matrix characters, for example, can be imaged in a line-scan mode by synchronizing the writing rate of the tube with the rotation of the photo-conductive drum. For line-scan times in the 50 to 250 μsec range 5 × 7 dot-matrix characters can be printed at speeds exceeding 10K LPM.

The high-energy densities and directional characteristics of lasers renders their radiation suitable for the controlled short-duration exposure of photo-conductors. A laser-excited xerographic printer would consist of the laser itself, a light modulator, and a light beam deflector. These components, in essence, comprise a light beam analog to a CRT. Imaging is accomplished by using many of the techniques employed in laser displays.[24]

Dot-matrix characters can be produced by amplitude modulating the laser beam and digitally deflecting the beam into the desired character format. The modulation frequencies required are typically 1 to 10 MHz and exceed the rate at which lasers can be internally modulated. Modulation must be achieved, therefore, by external electrooptic or acoustooptic devices.

Imaging is accomplished by single-axis, sequential access deflection of the laser beam in synchronization with the rotation of the photoconducting drum. Laser beam deflection can be obtained by various techniques including rotating mirrors and acoustooptic devices. Of prime importance to printing is the resolution of the deflected beam. The number of resolvable spots that can be produced is determined by the Rayleigh criterion:[25]

$$N = \frac{D\theta}{1.22\lambda} \tag{11}$$

where D is the aperture of the deflector and θ and λ are the deflection angle and wavelength of the light, respectively.

A 136-column printer employing a 5 × 7 dot-matrix format requires a resolution of 1000 spots. At a dot-imaging rate of 2 MHz, a print speed of 12K LPM can be achieved.

The third imaging configuration in xerographic printing utilizes a rotating stencil-character drum with a linear array of stroboscopic lights placed inside it. The drum resembles the print drum used in impact printers and has bands of transparent stencil characters around its periphery. One character band and one strobe light are needed for each column of the printer.

Imaging is accomplished by rotating the character drum past the photo-conductor; when the appropriate character in a column is aligned, the respective strobe light is energized and the image of the character is impressed on the photoconductor. The duration of the strobe light pulse is extremely short (i.e., 1–10 μsec); accordingly, the rotating image is "stopped" on the photoconductor with negligible loss of resolution. Print speeds in the range of 1 to 2K LPM can be obtained by using this technique.

REFERENCES

1. D. S. Swatik, "Non-Impact printing," presented at Albany Conference on Electrostatics, Albany, N.Y., June 8–11, 1971.

2. K. Takagi, K. Tasai, and K. Tsukatani, "On the electrostatic printing sheet for multiple copying," *Fujitsu J.*, **1**, No. 2, 227–260 (Nov. 15, 1965).

3. R. W. Haberle, D. J. J. Lennon, and H. G. Schleifstein, "High speed non-impact printing," U.S. Patent 3,550,153, Dec. 22, 1970.

4. J. W. S. Rayleigh, *The Theory of Sound*, Vol. II, Dover, New York, 1945, Chapter XX, "Capillarity."

5. N. R. Lindblad and J. M. Schneider, "Production of uniform-sized liquid droplets," *J. Sci. Instr.*, **42**, 635–638 (August 1965).

6. J. Zeleny, "Instability of electrified liquid surfaces," *Phys. Rev.*, 2nd ser., **1**, No. 1, 1–6 (July 1917).

7. C. D. Hendricks, R. S. Carson, J. J. Hogan, and J. M. Schneider, "Photomicrography of electrically sprayed heavy particles," *AIAA J.*, **2**, No. 4, 733–737 (April 1964).

8. R. S. Carson and C. D. Hendricks, "Natural pulsations in electrical spraying of liquids," *AIAA J.*, **3**, No. 6, 1072–1075 (June 1965).

9. C. D. Hendricks, "Charged droplet experiments," *J. Colloid Sci.*, **17**, 249–259 (March 1962).

10. J. Sutherland, "High speed non-impact printer using plain paper," Internal Report, Electroprint, Inc., November 1971.

11. G. L. Pressman, "Aperture-controlled electrostatic printing system," U.S. Patent 3,625,604, Dec. 7, 1971.

12. J. D. Cobine, *Gaseous Conductors*, Dover, New York, 1958, p. 163.

13. R. M. Schaffert, *Electrophotography*, Focal Press, London, 1966, p. 329.

14. A. Bliss, S. Marshall, and S. Rutherford, "Computer output printing and plotting with digitally switched arrays on dielectric coated paper," presented at EASCON, Washington, D.C., November 1970, pp. 159–170.

15. D. Kazan and M. Knoll, *Electronic Image Storage*, Academic Press, New York, 1968, pp. 287–290.

16. R. A. Fotland and E. B. Noffsinger, "Contrography, a new electronic imaging technology," *Photogr. Sci. Eng.*, **15**, No. 5, 431–436 (September–October 1971).

17. ———— "Electric recording apparatus employing liquid developer," U.S. Patent 3,623,122, Nov. 23, 1971.

18. P. H. Klose, "Ovonic memory printing," Internal Report, Energy Conversion Devices, Inc.

19. S. R. Ovshinsky, "The Ovshinsky switch," presented at National Conference on Industrial Research, Chicago, Sept. 18–19, 1969.

20. H. K. Henisch, "Amorphous semiconductor switching," *Sci. Am.*, **221**, No. 5, 30–41 (November 1969).

21. R. M. Schaffert, *Electrophotography*, Focal Press, London, 1966, p. 22.

22. N. S. Kapany, *Fiber Optics*, Academic Press, New York, 1967, pp. 219–223.

23. H. R. Luxenberg and R. L. Kuehn, *Display Systems Engineering*, McGraw-Hill, New York, 1968, pp. 266–274.

24. C. E. Baker, "Laser display technology," *IEEE Spectrum*, **6**, No. 12, 39–50 (December 1968).

25. R. Kingslake, Ed., *Applied Optics and Optical Engineering*, Vol. I, Academic Press, New York, 1966, pp. 271–272.

CHAPTER 14

Nonuniform Field Effects: Dielectrophoresis

HERBERT ACKLAND POHL

Department of Physics
Oklahoma State University
Stillwater, Oklahoma
(NATO Senior Science Fellow on leave to
Cavendish Laboratory, University of Cambridge,
Cambridge, England)

14.1 OVERVIEW

Nonuniform electrical fields produce unique, useful, and frequently mystifying effects on matter—even on neutral matter. With nonuniform fields, for example, it is possible to classify and separate minerals, pump liquids or powders, produce images (xerography), provide an "artificial gravity" useful in "zero-g" conditions, clean up suspensions, classify microorganisms, and even separate live and dead cells. And this just starts the list. Applications in biophysics and cell physiology to studies of normal and abnormal cells are at an early but exciting phase. In colloid science the new technique is helping to resolve surface properties. At the molecular level, nonuniform field effects are seeing renewed use in determinations of molecular polarizabilities, in maser operations, and in laser control.

At the broader level, we can readily foresee greater recognition of the role of nonuniform field effects in meteorology and in pollution control.

Historically, the Greeks certainly seem to have observed the effects of nonuniform fields on matter when the behavior of fluff and other light particles near rubbed amber was noted. Gilbert[1] in *De Magnete* remarked on the change in shape of a water drop as electrified amber was brought near. Winckler[2] and Priestley[3] noted the attractive forces on organic liquids long ago. These forces have recently been the subject of much experimental study.[4-26] Rather few quantitative studies seem to have been made, however, until about the middle of this century. Early theory, prettily based on ideal dielectrics and perfect conductors, has led to much overconfidence in our understandings of the behavior of *real* matter in *real* fields. As a consequence many "antitheoretical" or mystifying events are observed. Experiment is still far ahead of theory in many respects.

The problem is complicated by nonuniformities of conduction and polarization in matter assemblies. This leads to situations that are sometimes understandable only with difficulty and are presently impossible to solve before the fact, even with the aid of a large computer effort. This is well shown by two deceptively simple examples.

Example 1. Real Fields

Nothing could be simpler, it would seem, than the description of an electric field between two large, identical, flat and parallel metal electrodes, separated by a highly insulating liquid and having a differing potential. Conventional electrostatic theory leads us to expect a *uniform* field. In reality, this ideality is but a fleeting transient condition, lasting perhaps a few nanoseconds. The field rapidly warps to high inhomogeneity in microseconds. This was shown by the workers at Grenoble,[63] by using the Kerr effect on chlorobenzene. Field warping can continue for minutes after the application of a steady voltage.

Example 2. Insulator-induced Conduction

A second example is that of *insulator-induced conduction*, discovered recently by Cantril and Pohl. It was observed that the current flow was *increased* many fold upon *covering* an electrode with a thick layer of insulator particles (such as quartz or sulfur) in a system in which current was passing through a layer of highly purified dielectric liquid, such as a hydrocarbon fluid. It was demonstrated that this action of an insulating layer—namely, to cause an increase of current—arose because of the resultant field inhomogeneities at the electrode as insulator particles of higher dielectric than the liquid medium came against the electrode. The current flow in insulating liquids is usually of the space-charge-limited type;[29,59] that is, it is nonohmic and dependent on the field to a higher power than unity. Net current flow can therefore be increased by the presence of field inhomogeneities. Insulator-induced conduction is just such an effect.

Since, as we have just seen in the two examples cited, real fields are often far from "ideal" even in simple cases, the study of nonuniform field effects requires careful attention to such matters. The distortion of field conformations from that assumed for ideal dielectrics by phenomena such as space charge, field emission, asymmetric avalanching, corona, and "incipient corona" must always be suspected. For the foreseeable future at least, experiment must lead the theory.

Having remarked that experiment at present overshadows theory in the study of the behavior of real matter in the presence of real but nonuniform fields, we must hasten to say that a great deal is now understood about the theory. Progress will depend on the hand-in-hand development of both experiment and theory.

Much of the past work on the behavior of matter in nonuniform electric fields has focused on studies of matter in the gross. Rather less work has been done with molecular studies. Müller,[23] in a theoretical analysis supported by careful experiment, concluded that nonuniform field effects would not be marked for particles of molecular size. Pohl[4,5] and Lösche and Hultschig[16] independently studied the theory of the magnitude and direction of the effects. The effects of nonuniform fields were shown by the Princeton group[5-10,28] to be both appreciable and useful for macroscopic particles or bodies of liquid. The intermediate range of particle size, exemplified by polymer molecules in solution, was studied by Debye and the Cornell group.[24-27] They concluded that grading by molecular size should be possible, but experimental difficulties were formidable.

As a practical matter, the study of nonuniform field effects on gross matter is made complex by the presence of a number of competing phenomena. Among these, the effect known as *dielectrophoresis*[4] is the main concern of our discussion. Competing with this effect there may be electrophoresis and thermal effects, such as joule heating and attendant convection, conduction, diffusion, as well as the effect to be obtained from the sequence: charging and electrostatic repulsion.

14.2   DIELECTROPHORESIS

The translational motion of neutral matter caused by polarization effects in a nonuniform electric field is defined as *dielectrophoresis*.[4] It is to be distinguished from motion caused by the response to free *charge* on the body in an electric field, termed *electrophoresis*.

The cause of dielectrophoresis is basically simple. Any electric field, uniform or not, exerts a force on a charged body. What is important about inhomogeneous electric fields is that they act on uncharged bodies. From this behavior, a wide variety of potentially interesting and useful events results.

Perhaps the action of a nonuniform field on neutral matter can best be understood by comparing the response of charged and of neutral particles in uniform and nonuniform fields. Referring to Fig. 14.1, let us consider first their behaviors in a *uniform* field. A charged body (e.g., an ion or a particle) is pulled along the field lines toward the electrode of opposite charge. A neutral body, on the other hand, is polarized—it may experience a torque as a result, but it is not impelled to move as a whole toward either electrode. A torque can arise if the body is elongated or anisotropic. The resultant dipole would then tend to align itself to achieve minimum energy in the field.

Referring to Fig. 14.2, where a charged and a neutral body are shown in a *nonuniform* electric field, we see a still different comparative behavior. The charged body behaves much as before. It is still pulled toward the electrode of opposite charge. Now, however, the neutral body tries to move translationally. This happens because under the influence of the field, the neutral body acquires, in effect, a negative charge on the side nearest the positive electrode and a positive charge on that side nearer the negative electrode. Since the body is overall neutral, the two effective charges, positive and negative, are in fact equal. If the externally applied electric field is nonuniform, it diverges across the particle and produces *unequal* forces on the two effective charges. The net effect is an overall force on

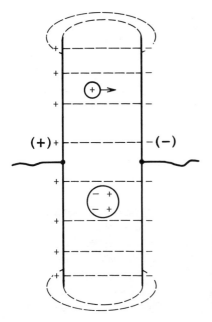

Figure 14.1. Comparison of the behaviors of charged and neutral bodies: In a "uniform" field, a positive body tends to move along the lines of force toward the negative charges on the negative electrode. A neutral body is "polarized" and just "sits."

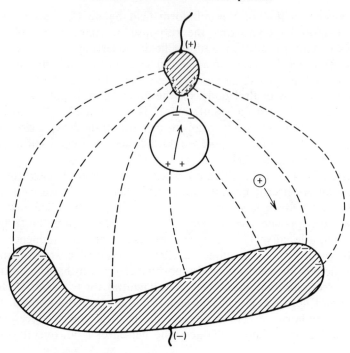

Figure 14.2. In a nonuniform field, the positive body again tends to move toward the oppositely charged electrode. On the other hand, the neutral body is polarized. But its equal amounts of negative and positive charge lie in fields of differing strength. A net force arises which tends to pull the neutral body toward the region of highest field intensity. Note that reversing the sign of the charges on the electrode (i.e., reversing the field) causes the charged body to reverse direction. But the neutral body, having its polarization reversed in the new field, *still* tends to move toward the region of greatest field intensity. In a high-frequency ac field, therefore, the charged object tends to "shudder" or rush alternately to both electrodes, whereas the neutral object, if free to move, would proceed rather steadily toward the "sharper" electrode where the field is more intense.

the neutral body. This generally results in the neutral body being impelled into the region of stronger field. (For exceptions to this paraelectric effect, see later on apoelectric effects.)

One further point in this connection should be made. The force on the neutral, polarizable body by the nonuniform field is in the *same* direction no matter which electrode is positive or negative. The applied field could well be that due to an alternating potential, since the polarization induced in the body is reversed as the field switches. The contrast between the behaviors of charged and neutral bodies in a rapidly alternating field is very noticeable. If the charged object is under the influence only of electrophoresis, it tends to "shudder" about its original position. In contrast, the neutral object, under the influence of dielectrophoresis, tends to move rather steadily toward the region of higher field intensity.

14.2.1 Dielectrophoresis and Electrostriction

As noted earlier, dielectrophoresis[4] is the translational motion of a neutral body caused by its response to an inhomogeneous electric field. The force tending to produce this translational motion is termed the dielectrophoretic force.[4] The word *dielectrophoresis* stems, in part, from the Greek *phoresis*, meaning motion. There is seen in some solid bodies and in certain semisolid ones (liquid crystals, e.g.) still another mechanical response of a neutral to an electric field. This response, *electrostriction*,[30,31] refers to the *distortional response* or strain resulting from an imposed electrical stress. Such electro-strictive strains are used in sound transducers; for example, barium titanate is perhaps the most important of the ferroelectric perovskites used for this purpose.

Other sources[29-31] provide further discussion of electrostriction. It should be noted, however, that historically speaking the two effects—translational and distortional—were both at times referred to as electrostriction with resultant confusion. Modern usage has tended to restrict the scope of the term electrostriction to distortional strain induced electrically. The translational response of bodies is then referred to as dielectrophoresis. The translational response can of course be coupled to a moment arm leading to torsion and in turn to torsional motion of the body with realignment in the field.

14.2.2 Dielectrophoresis in Ideal Dielectrics

The theory for the force exerted by a nonuniform field on a body suspended in a fluid medium can be reduced to relatively simple analytical expressions provided certain simplifying assumptions are made. It is instruc-tive to do just this.

In a static field at equilibrium, the force on a small neutral body is[32]

$$\mathbf{F} = (\mathbf{p} \cdot \nabla)\mathscr{E}_e \tag{1}$$

where $\mathbf{F}$ is the net electric force on the body; $\mathbf{p}$ is the (constant) dipole vector; ∇ is the del vector; and $\mathscr{E}_e$ is the external electric field. For the case where the dielectric body is linearly, homogeneously, and iso-tropically polarizable—that is, if

$$\mathbf{p} = \alpha v \mathscr{E}_e \tag{2}$$

where α = polarizability = dipole moment per unit volume in unit field and v = volume of the body, we can write

$$\mathbf{F} = \alpha v(\mathscr{E}_e \cdot \nabla)\mathscr{E}_e = \tfrac{1}{2}v\alpha\nabla|\mathscr{E}_e|^2 \tag{3}$$

In a conservative field (i.e., "friction-free") the force on the body is the negative of the gradient of its potential energy. Since $\mathbf{p}$ is a constant vector, and $\nabla \times \mathscr{E}_e$ is zero in a static field, we have

$$\mathbf{F} = (\mathbf{p} \cdot \nabla)\mathscr{E}_e = \nabla(\mathbf{p} \cdot \mathscr{E}_e) \tag{4}$$

The energy of the particle U_p is then

$$U_p = -\mathbf{p} \cdot \mathscr{E}_e = -\alpha v|\mathscr{E}_e|^2 \tag{5}$$

In the event that the body is anisotropic, the polarizability α is a tensor quantity. The computation is then more arduous.

To summarize at this point, we note that the force and the energy depend directly on the volume and the polarizability of the body and on the *square* of the electric field intensity. The latter circumstance emphasizes that dielectrophoresis is *independent of the field sign* and could as well take place in an alternating field.

14.2.3   Apoelectric and Paraelectric Behavior

Although it is normal for neutral macroscopic particles to exhibit paraelectric behavior (i.e., to be drawn into the stronger regions of the electric field), this need not always be the case. The paraelectric response is generally the result of what can be regarded as the two-step process: (1) polarization by the field followed by (2) action of the nonuniform field on the *induced dipole* to produce a dielectrophoretic force toward the region of higher field intensity. An oppositely directed force can result, for example, if the body is already strongly polarized (e.g., a permanent dipole or an electret) *and* is spinning. This can produce an apoelectric response, opposite in direction to the conventional paraelectric one.

14.2.4   Apoelectric Cooling

The apoelectric response is expected to be strongest if the axis of the rotation is at right angles to the field direction. It will be negligible if the axis is parallel to the field direction. Clearly, the application of an external electric field to a system capable of making an apoelectric response will increase the energy. This leads to a suggested use of the process as a means of cooling. The prolonged application of an external field normally permits relaxation processes (friction, thermalization, etc.) to operate, thereby decreasing the populations of the apoelectric states and increasing the population of the paraelectric states. The latter have a lower energy in the presence of the field. Bringing such a sample to thermal equilibrium with an external heat bath, and removing first the heat bath and then the external field brings about adiabatic cooling of the sample. The process, which would be analogous to cooling by demagnetization, is an interesting area for future study. Indeed, the apoelectric response of macroscopic bodies, although predicted several years ago[20] does not yet seem to have been observed for macroscopic bodies. On the molecular scale, for example, it is used in a vital step in the maser operation to artificially shift the population of excited states.[33]

14.2.5 Types of Polarization

There are now five modes of polarization known to occur in matter. Four occur on a molecular scale. The fifth, interfacial polarization, involves macroscopic structures.

1. *Electronic polarization* arises from the slight distortion of the electron clouds of the atoms about their nuclei as an external field is applied. Since the fields within atoms are already large (ca. 10^{11} V/m) and man-made fields are feeble $(0-10^8)$ V/m) by comparison, the electronic shift is usually not large (ca. 10^{-8} Å or 10^{-18} m) with respect to the nucleus. In organic solids, electronic polarization leads typically to relative dielectric constants of about 1.8 to 4. In inorganic solids, especially among elements of high atomic number, the electronic polarization can lead to relative dielectric constants which are rather higher (e.g., 12 for silicon and 16 for germanium).

2. *Atomic polarization* arises from the shift of differently charged atoms or groups of atoms with respect to one another. It is well exemplified by the ionic shifts in the ionic solid sodium chloride, when subject to an external field. This type of polarization in many organic molecules causes but a small addition to the total polarization. Typically, a contribution of about 15% of that due to electronic polarization is seen, as judged from a comparison of the low (radio-frequency) and high-frequency (optical) dielectric constants. The atomic polarization can, however, be very large in some solids. That in barium titanate, for example, is the major source of the remarkably large dielectric constant of some 4000.

3. *Orientational polarization* arises from the orientational responses of asymmetric charge distributions between dissimilar parts of molecules. The molecular dipoles (e.g., HF, NO, H_2O) can form permanent dipole moments, which can rotate in an electric field. Orientation polarization can be rather large. It contributes most of the polarization available in liquid water, which has a dielectric constant of about 80 at frequencies

below 10^9 Hz and at room temperature. The dielectric constant at optical frequencies where the molecular dipoles can no longer follow the field is only about 1.8.

4. *Hyperelectronic polarization* is due to the pliant response of thermally excited charges situated on long (polymeric) molecular domains.[34-38] The charges arise in thermally produced pairs as excitons with (most frequently) the individual charges of the pair lying on long separate molecules. Long-range molecular orbital delocalization such as in large aromatic polymers offers particularly suitable domains for the required pliant charge response to the external field. Typically, these domains for the electronic shift can be 100 to 1000 Å in extent. Compared with the electronic shifts of some 10^{-8} Å during the electronic polarization of atoms in a field of some 10^5 V/cm, these hyperelectronic shifts are enormous. Hyperelectronic polarization can be quite large. Certain aromatic organic polymers exhibiting this can have dielectric constants of up to 300,000, whereas polyethylene, exhibiting mainly only electronic but no hyperelectronic polarization, has a dielectric constant of about 1.8.

5. *Interfacial polarization* arises from the migration of charge carriers for some distance through the dielectric. When these carriers are impeded in their motion (as by trapping at impurity centers or at interfaces, or because they cannot be freely discharged or replaced at an electrode), "space charges" appear. Space charge produces distortion of the macroscopic field. To the observer such distortion appears as a polarization and often can be distinguished only with difficulty from real rises in permittivity due to one of the four molecular mechanisms. (See Ref. 30, especially Chapter 31.)

Interfacial polarization occurs more often than is generally appreciated. It is an all too ubiquitous reminder of the way in which *real* matter behaves in real fields, as compared with a textbook dreamworld of ideality. Interfacial polarization in pure homogeneous materials is generally negligible at low field strengths. It can become quite large. A simple example is that of graphite suspended in rubber. Living matter and soils or earth with their conductive regions interleaved with barrier layers can appear to have enormous dielectric constants (up to 10,000 or so). Historically, interfacial polarization was first recognized by Maxwell[39] and by Wagner.[40] A recent review and extension of the theory by von Beek is instructive.[41]

Let us recapitulate. (1) *Electronic polarization* is a general phenomenon that is due to shift of electrons with respect to the nucleus as field is applied. When charge asymmetries arise in matter, even in the absence of an external field, these can respond (2) by exhibiting *atomic polarization* if the charge motions are associated with atoms or groups of atoms in their normal libration distances; (3) by exhibiting *orientation polarization* if the charge motions are associated with rotational motions of groups of atoms; or (4) by exhibiting hyperelectronic polarization if the excitonic charge motions are associated with translational electron (or hole) motions more or less freely within long molecular domains. In the latter instance it is important to realize that neither excitons alone nor the presence of long orbital domains alone is sufficient for the operation of hyperelectronic polarization. Rather, *both* moieties must coexist to produce the effect.

In contrast, *interfacial* or space charge or Maxwell–Wagner *polarization* is a macroscopic phenomenon.

14.2.6   The Polarization due to Diatomic Dipoles

For simplicity we examine here the case of polarization due to diatomic molecules having permanent dipole moment.[42] Even this simple system can display para- and apoelectric states. It can be shown that where the principal rotational quantum number is $J = 0, 1, 2, 3, \ldots$, and the secondary (orientational) quantum number is M, and $J \geq |M|$, the energies $U_{J,M}$ of the first few rotational states of the dipole in an electric field are:

$$U_{0,0} = -W\frac{z^2}{6} \qquad \text{multiplicity} = 1$$

$$U_{1,0} = W\left(2 + \frac{z^2}{10}\right) \qquad \text{multiplicity} = 1$$

$$U_{1,1} = W\left(2 - \frac{z^2}{10}\right) \qquad \text{multiplicity} = 2$$

$$U_{2,0} = W\left(6 + \frac{z^2}{42}\right) \qquad \text{multiplicity} = 1$$

$$U_{2,1} = W\left(6 + \frac{z^2}{84}\right) \qquad \text{multiplicity} = 2$$

$$U_{2,2} = W\left(6 - \frac{z^2}{42}\right) \qquad \text{multiplicity} = 2$$

where $W = h^2/8\pi^2 I$, $I =$ moment of inertia, $h =$ Planck's constant, $z = 8\pi^2 I \mu_e \mathscr{E}_e/h^2$, $\mathscr{E}_e =$ external electric field strength, and $\mu_e =$ dipole moment. This describes the second-order Stark effect for linear molecules.

At this point we note two things of special interest for our discussion. First, the change in energy of this permanent dipole system with field, just as in the case of induced dipoles, is proportional to the *square* of the field intensity. Accordingly, we can assign an effective polarizability α_i to each state, that is,

$$\mathbf{\mu}_i = \alpha_i \mathscr{E}_e$$

This enables us to extend the analysis for particles and for nondipoles to the molecular dipole case.

Second, we observe listed three specific examples ($U_{1,0}$, $U_{2,0}$, and $U_{2,1}$) of apoelectric states. In these states the molecule would be repelled from the regions of higher field intensity. The paraelectric states (e.g., $U_{0,0}$, $U_{1,1}$, and $U_{2,2}$) would be attracted to the regions of greater field intensity. The diatomic dipole systems are therefore of much interest.

For linear molecules and for symmetric top molecules which are rotating about an axis perpendicular to the axis of molecular symmetry, it can be shown[43,44] that the average dipole moment, μ, is given by

$$\mathbf{\mu} = \frac{4\pi I \mu_e}{h^2} \left[\frac{J(J+1) - 3M^2}{J(J+1)(2J-1)(2J+3)} \right] \mathscr{E}_e \tag{6}$$

Here too we may assign an effective polarizability $\alpha_{J,M}$ such that

$$\mathbf{\mu} = \alpha_{J,M} \mathscr{E}_e \tag{7}$$

The situation is again quite analogous to the case for nonpolar molecules where the field-induced moment $\mathbf{\mu}_{\text{ind}}$ is

$$\mathbf{\mu}_{\text{ind}} = \alpha_{\text{ind}} \mathscr{E}_e \tag{8}$$

Equation (3) can accordingly be modified to contain two terms affecting the dielectrophoretic force. One term, α, is due to induced polarization and is always paraelectric. The other term, Ω, is due to "permanent" polarization and may be either apoelectric or paraelectric, namely,

$$\mathbf{F} = \frac{v}{2} [\alpha + \Omega] \mathbf{\nabla} |\mathscr{E}_e|^2 \tag{9}$$

The net force on a given particle depends not only on the field strength (quadratically) but also, in direction and magnitude, on the type and degree of the two polarizations possible (apo- or paraelectric).

14.2.7   The Force on a Small Dielectric Sphere in a Nonuniform Field

One of the well-known cases in electrostatics for which an analytic solution is available is that of a small sphere of an ideal dielectric of relative dielectric constant K_2, in an ideal dielectric fluid of relative dielectric constant K_1, and of infinite extent. The field interior to the

Table 14.1. The Relative Dielectrophoretic Force Factor $K_1[(K_2 - K_1)/(K_2 + 2K_1)]$ for Various Combinations of Dielectrics

K_2 (relative dielectric constant of the sphere)	K_1 (relative dielectric constant of fluid medium)	$K_1\left(\dfrac{K_2 - K_1}{K_2 + 2K_1}\right)$	Remarks
2	1	0.25	Nylon ball in air, say
80	1	0.963	Water droplet in air
4000	1	0.99925	$BaTiO_3$ in air
∞	1	1.00	Metal ball in air
1	2	-0.20	Air bubble in a hydrocarbon liquid
12	2	$+1.25$	Silicon in a hydrocarbon liquid
80	2	$+1.857$	Water droplet in a hydro-carbon liquid
4000	2	$+1.997$	$BaTiO_3$ crystal in a hydro-carbon liquid
∞	2	$+2.00$	Metal ball in a hydrocarbon liquid
2	80	-38.5	Fat or oil drops in very pure water
86	80	$+1.95$	Rutile crystal in very pure water
4000	80	$+75.4$	$BaTiO_3$ crystal in very pure water
∞	80	$+80$	Metal ball in very pure water

sphere boundary, $\mathscr{E}_{iz}$, associated with an external field $\mathscr{E}_e$ parallel to the z-axis in the remote regions of the external fluid can be shown to be[30]

$$\mathscr{E}_{iz} = \left(\frac{3K_1}{K_2 + 2K_1}\right)\mathscr{E}_e \tag{10}$$

Since the specific force (i.e., the force per unit volume) is proportional to $(\alpha/2)\nabla|\mathscr{E}|^2$, the total force on a small sphere is proportional to the sphere's volume $4\pi a^3/3$, where a is the radius of the sphere. Then we have

$$\mathbf{F}_e = \frac{(\text{volume})}{2} \times \text{polarizability} \times \text{local field} \times \text{field gradient}$$

$$\mathbf{F}_e = \frac{1}{2}\frac{4\pi a^3}{3} K_1\varepsilon_0 \left[\frac{3(K_2 - K_1)}{K_2 + 2K_1}\right]\nabla|\mathscr{E}_e|^2$$

$$\mathbf{F}_e = 2\pi a^3 K_1\varepsilon_0\left(\frac{K_2 - K_1}{K_2 + 2K_1}\right)\nabla|\mathscr{E}_e|^2 \tag{11}$$

Here we have introduced the *net force* F_e as that due to the replacement of the dielectric of medium by that of the sphere, hence the term $(K_2 - K_1)$. We have also used the term "small sphere," thereby making the approximation that the field, although nonuniform enough to produce appreciably different force on the positive and negative polarization charges, does not vary strongly enough to sensibly alter the *degree* of polarization and hence the size of the dipole throughout the volume of the sphere. By ε_0 is meant the permittivity of free space, $(36\pi \times 10^9)^{-1} \cong 8.854 \times 10^{-12}$ C/V-m. Equation 11 can also be written as

$$\mathbf{F}_e = \frac{\text{volume}}{2} \varepsilon_1 \left(\frac{\varepsilon_2 - \varepsilon_1}{\varepsilon_2 + 2\varepsilon_1}\right)\nabla|\mathscr{E}_e|^2 \tag{12}$$

where $\varepsilon_1 = \varepsilon_0 K_1, \varepsilon_2 = \varepsilon_0 K_2$ are the absolute electrical permittivities of media 1 and 2, respectively.

The form of Eqs. 11 and 12 indicates that the dielectrophoretic force effect has built-in limitations with regard to the role of the dielectric constants. It does not, as is perhaps at first intuitively inferred, increase without limit as the dielectric constant of the small body becomes large. A glance at Table 14.1 is helpful in appreciating how this varies.

14.2.8 Dielectrophoresis and Electrophoresis Compared

At this time it is opportune to carefully compare the two phenomena—dielectrophoresis and electrophoresis.

DIELECTROPHORESIS

Dielectrophoresis, arising from the polarization of matter and its subsequent tendency to move into regions of different field strength, has the following properties:

1. It produces motion of bodies suspended in a medium. The direction of that motion is independent of the sign of the field. Therefore, either ac or dc fields can be used.

2. It gives an effect that is proportional to the cube of the particle diameter. Thus it is more easily observed on coarse particles and can be observed on molecular levels only under selected conditions.

3. It is an effect requiring quite divergent fields as a rule.

4. It is an effect requiring relatively high field strengths. Commonly 10 V/mm or more is required.

5. It usually requires sizable differences in the relative dielectric constants of the particle and the surrounding medium (e.g., $|K_2 - K_1| \gtrsim 1$).

6. It deposits weights of a coarse sol in direct proportion to the applied voltage during equal times of deposition.

7. It is a relatively gentle effect, all things considered. It is most easily observed with large particles in fluids of low viscosity when fields are high but controlled. It can often be masked by side effects such as turbulence and convection.

ELECTROPHORESIS

Electrophoresis, arising from the direct attraction of an electric field for a charge, has the following properties:

1. It produces motion of suspended particles in which the direction of their motion depends on the sign of their charge and on that of the field. Reversal of the field reverses the direction of travel. Care must be taken in the application of this rule for the distinction between electro- and dielectrophoresis. Occasionally the motion of charged particles toward a sharp electrode occurs even when alternating voltages of high strength are applied. This happens because partial rectification and selective charge injection appear.

2. It is observable with particles of any size—atomic, molecular, colloidal, or even macroscopic.

3. It operates in either uniform or nonuniform fields.

4. It requires relatively low fields.

5. It can be marked even when relatively small charges are present per unit volume of the particles. Unlike dielectrophoresis, it does not depend on the particle radius per se, but on the particle's charge.

14.2.9   The "Bunching" Effect

Clumping or bunching action is frequently observed among suspended particles as they are subject to an electric field,[16] even to an apparently uniform external field. This effect, which is due to dielectrophoresis, arises in the following way. The particles, having polarizabilities different from that of the medium, distort the external field. Each particle, on seeing a nonuniform field at the other, is polarized and mutually attracted.

In the case of the lone small spherical (or ovate) particle in a fluid, the field inside the sphere is uniform ($\mathscr{E}_{iz}$ is parallel to $\mathscr{E}_e$, the external applied field) if both dielectrics are homogeneous, linear, and isotropic. If the sphere is of higher dielectric constant (i.e., if $K_2 > K_1$), then the field lines external to it tend to focus or concentrate at the sphere (Eq. 11). This nonuniformity of the field about the sphere causes two or more spheres to be mutually attracted. This "bunching" effect due to the enhanced local nonuniformities of the field about dielectric particles is frequently displayed in the formation of "pearl chains" or other clusterings in an electric field. Whether $K_2 > K_1$ or $K_2 < K_1$, the pearl chains tend to form along the field lines if the external field is not too strongly divergent. This is true because if $K_2 > K_1$, the particle tends to go into the region of highest field strength (i.e., near the z-axis of the neighboring particle) which is concentrating the field lines about itself. If $K_2 < K_1$, the particle tends to go where the field is weakest. This again means into the region of the z-axis of the neighboring particle, which this time is diverting the field away from itself. (The z-axis here, as before, is parallel to that of the original $\mathscr{E}_e$ field.)

The "bunching" effect is doubtless a factor in such diverse phenomena as the agglomeration of ice, snow, and water particles in the electric field of the earth's atmosphere, as well as in filtration, in xerography, or in the agglomeration of oil or paint particles sprayed in an electric field.

As an example of atmospheric effects we could ask whether falling ice particles, which acquire a charge by some accident of ion disproportionation, might not grow rather differently from neutral ones. Recent evidence suggests that the freezing of supercooled water produces charge separation[60,61] and that the ejection of microdroplets of positive charge leaves the parent drop negative.[62] Charged snow crystals tend to attract surrounding neutral water droplets by dielectrophoresis, and the collection tends to be much stronger at the sharpest tips. The arrival of water droplets and their freezing at crystal-originating symmetry points should tend, it would seem, to produce more feathery snowflakes among the charged ones, whereas the neutral ones would tend, on the average to be more dense and less massive. Experiments to examine this point more closely would be of interest.

14.2.1.0 Dielectrophoresis in Various Field Distributions

Up to this point we have stressed the role of the properties of the materials (i.e., the polarizabilities and conductivities). We have said little about the choices that might be desirable with respect to the electric field distribution in order to optimize dielectrophoretic analysis or separation of particles. Let us now look particularly at the various field geometries that are used.

Suppose that our medium is a simple homogeneous one, and that the electrodes determine the local (idealized) field seen at a given region. This implies, for example, that our test particle is not affected by other nearby particles (dilute suspension) and that our medium does not itself greatly modify the field distribution (as by space charge effects, thermal density gradients, the presence of bubbles, etc.). We can then focus our attention upon the geometric field term $\frac{1}{2}\nabla|\mathscr{E}_e|^2$ or $(\mathscr{E}_e \cdot \nabla) \mathscr{E}_e$ of Eqs. 3, 9, 11, or 12.

Broadly speaking, the electrode systems used in practice to produce nonuniform electric fields consist of either spherical, cylindrical, or isomotive symmetry, or modifications of these.

14.2.11 Spherical Geometry

A popular electrode configuration approximating the spherical geometry is that of a rounded wire tip extending into a hollow formed by an outer electrode. It can be approximately described by a spherical capacitor with a central sphere of radius r and an outer concentric shell of radius r_2. The potential at any point between these spherical shells, with the outer one grounded and the inner at a potential V_1, is

$$V = \frac{V_1 r_1}{r}\left(\frac{r_2 - r}{r_2 - r_1}\right) \tag{13}$$

The field, where $\mathbf{r}_0$ is the unit radius vector, is

$$\mathscr{E}_{\text{spher}} = -\nabla V = \frac{\mathbf{r}_0 V_1 r_1 r_2}{r^2(r_2 - r_1)} \tag{14}$$

and

$$\nabla|\mathscr{E}_{\text{spher}}|^2 = \frac{-\mathbf{r}_0(4V_1^2 r_1^2 r_2^2)}{r^5(r_2 - r_1)^2} \tag{15}$$

14.2.12 Cylindrical Geometry

Electrode configurations that approximate the cylindrical are frequently used. These include the use of a central wire held in a coaxial cylinder electrode, a wire–plate combination and, to a rougher approximation, a wire–wire electrode pair. In the case of cylindrical electrodes with the outer grounded electrode of radius r_2 and the concentric inner one of radius r_1 held at a potential V_1, the potential V at some point r away from the central axis is

$$V_{\text{cyl}} = V_1 \frac{\ln (r/r_2)}{\ln (r_1/r_2)} \tag{16}$$

The field is

$$\mathscr{E}_{\text{cyl}} = \frac{\mathbf{r}_0 V_1}{r \ln (r_1/r_2)} = \mathscr{E}_1 \frac{r_1}{r} \tag{17}$$

and

$$\mathbf{V} |E_{\text{cyl}}|^2 = \frac{-2 \mathbf{r}_0 V_1{}^2}{r^3 \ln (r_1/r_2)^2} \tag{18}$$

14.2.13 Isomotive Geometry

We note in the above that the dielectrophoretic force F_e in the field of spherically or cylindrically symmetric electrodes can vary drastically with the position r of the test object. In spherically symmetric electrodes our idealized field causes a fifth-power variation; in cylindrical electrodes it is a third-power variation, that is,

$$F_{e,\text{spher}} \propto \frac{1}{r^5}$$

$$F_{e,\text{cyl}} \propto \frac{1}{r^3}$$

In such electrode systems it is naturally very difficult to obtain any sort of uniformity of treatment of materials passing by the electrodes. Particles in the stream being treated by the spherical electrode type and which lie at radial distances in a ratio of $r'/r'' = 3$ would experience dielectrophoretic forces differing by 3^5 or 243.

Similar and only slightly less severe remarks apply to the use of cylindrically symmetric electrode systems. Reliable separations of two different materials in such field shapes are difficult. Can we do better?

Can a better field shape be found? Is there a field shape that will more fairly act on particles throughout an extended region? One that will produce the same or nearly the same force on a particle wherever it might ride in the supporting medium?

It turns out that such a field geometry can be provided, at least for limited regions. For clear reasons, we term this an "isomotive" field.[8,20,28,46] Let us see how this might be arranged, by taking an equation for the force such as Eq. 3 as part of a boundary condition on the solution of the Laplace equation. It describes the potential over a region containing no free charge. Then we have

$$\mathbf{F}_e = -\tfrac{1}{2}v\alpha\nabla(\mathscr{E}_e{}^2) \tag{3}$$

is to be constant along a given (radial) direction. The Laplace equation is

$$\nabla^2 V = 0 \tag{19}$$

When we choose the potential V to be independent of the z-coordinate, a solution of Laplace's equation in cylindrical coordinates is

$$V = Ar^n \sin(n\theta) \tag{20}$$

and since

$$\mathscr{E}_e = -\nabla V \tag{21}$$

$$\mathscr{E}_e = -\frac{\partial V}{\partial r}\mathbf{a}_r - \frac{1}{r}\frac{\partial V}{\partial \theta}\mathbf{a}_\theta \tag{22}$$

$$\mathscr{E}_e = -nAr^{n-1}[\mathbf{a}_r \sin(n\theta) + \mathbf{a}_\theta \cos(n\theta)] \tag{23}$$

where $\mathbf{a}_r$ and $\mathbf{a}_\theta$ are unit vectors. Then

$$\mathscr{E}_e{}^2 = \mathscr{E} \cdot \mathscr{E} = n^2 A^2 r^{(2n-2)} \tag{24}$$

and

$$\mathbf{F}_e = -\frac{\alpha}{2}\nabla(\mathscr{E}_e{}^2) = -\alpha(n-1)n^2 A^2 r^{2n-3}\mathbf{a}_r \tag{25}$$

On setting the force equal to a constant F_1 in magnitude along the radial direction, such that

$$\mathbf{F}_e = F_1\mathbf{a}_r \tag{26}$$

we require that $2n - 3 = 0$ or $n = 3/2$ and that $A = (2/3)(2F_1/\alpha)^{1/2}$

The equation describing the surfaces of constant potential which generate the *isomotive field* is then

$$V = \frac{2}{3}\left(\frac{2F_1}{\alpha}\right)^{1/2} r^{3/2} \sin\frac{3\theta}{2} \tag{27}$$

A plot of the lines of constant potential obeying this relation is indicated in Fig. 14.3. If metal electrodes are placed along the heavier lines, the desired field should occur along the $\theta = 0$ axis.

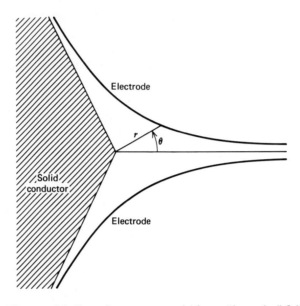

Figure 14.3. Diagram of the lines of constant potential for an "isomotive" field which exerts a constant radial force on a neutral body.

The solution of Eq. 19 suggests that the exponent n is not an integer. But this imposes problems, for θ may be continued past 2π in value. This gives rise to the possible unwanted multivalued solutions of the function $\sin n\theta$. We prevent this and restrict the *defined* region of the function to $(2/3)2\pi = 4\pi/3$, by inserting a conductor in the region $120° \leqq \theta \leqq 240°$. This excludes the forbidden region and makes the solution a valid one.

Now it is clear that the electrode shapes depicted by Fig. 14.3 and Eq. 27 cannot continue indefinitely out along the asymptotic planes $\theta = 0$, 120, 240°, for the dimensions sooner or later must become molecular in size. We are forced to compromise in practice by making the electrodes rather short. A practical design of the isomotive field electrode appears in Fig. 14.4. This is perhaps the simplest of the isomotive electrode systems. The dielectrophoretic force on paraelectric particles in the xz plane in the region OA is such that they will experience everywhere an

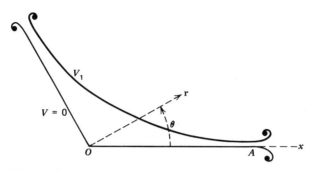

Figure 14.4. Diagram of a practical design of the isomotive field electrode system. The curved electrode is at a potential V_1 (ac or dc). The other is grounded. Paraelectric bodies in the x-plane will experience an essentially constant dielectrophoretic force in the plus x-direction in the region OA.

almost constant force in the positive x-direction. Electrodes in this isomotive design have proved to be capable of giving very satisfactory separations of particle mixtures, where the mixture is of powders of differing dielectric constants.[8,46]

Improved separations can be obtained with the isomotive field by using, where possible, a liquid medium having a dielectric constant intermediate in value between those of the two solids being separated by dielectrophoresis. This improvement occurs primarily because the "bunching effect" described earlier is no longer as active in gathering particles of low dielectric constant to those of higher dielectric constant. The resulting separation is cleaner. An absolute minimum in the bunching of particles of type A, say, onto either themselves or to those of type B will occur when the polarizability of A is identical to that of the liquid medium.

14.2.14 Dielectrophoresis in Real Dielectrics

Perfect or "ideal" dielectrics are nonexistent. They are model materials imagined to have no conduction and are used in dielectric analysis for purposes of simplicity or ease of computation. Real dielectrics do conduct, however, and this often confers substantial difference in behavior. Beginnings in the theory for the time-dependent behavior of forces in mildly conductive media and static fields have been made.[7,45] The broader problem of the behavior of real dielectrics in time spatially varying fields is far more difficult.

In the laboratory or in the field, the actual experimental situation is of course that involving real dielectrics having more or less conductivity, (often in mixtures), having complex shapes and interfaces, and existing in either "static" or sinusoidally time-varying electric fields. Frequently both the medium and the solid are quite conducting, capable of building up charge layers near interfaces. The resultant field distortions are hence complex, and frequently very time dependent. This sounds discouragingly complex. It often is so. But that is partly because too little attention has been given this fascinating and often critically useful technique. Much needs to be done.

14.2.15   Dielectric Relaxation Times

By way of coming to grips with real dielectrics, it is frequently possible to obtain a useful feel for the dielectric behavior of a selected material by learning its "time constant" or "dielectric relaxation time." It is useful especially where the dielectric constant and conductivity do not vary sharply with frequency. Let us consider a linear, isotropic, and homogeneous medium where

$$\mathbf{D} = \varepsilon\mathscr{E} \tag{28}$$

and

$$\mathbf{J}_f = \sigma\mathscr{E} \tag{29}$$

where $\mathbf{D}$ is the "dielectric displacement," $\mathscr{E}$ is the electric field, and $\mathbf{J}_f$ is the "free-charge" (as opposed to the "bound-charge") current. We have assumed ε and σ to be independent of $\mathscr{E}$. From Maxwell's equations we have

$$\mathbf{V} \cdot \mathbf{D} = \rho_f \tag{30}$$

and

$$\mathbf{V} \times \mathbf{H} = \mathbf{J}_f + \frac{\partial \mathbf{D}}{\partial t} \tag{31}$$

From these relations, and because the divergence of the curl is zero, we obtain

$$\mathbf{V} \cdot (\mathbf{V} \times \mathbf{H}) = \mathbf{V} \cdot \mathbf{J}_f + \frac{\partial \mathbf{D}}{\mathbf{V} \partial t} = 0 \tag{32}$$

Combining, we get

$$\mathbf{V} \cdot \frac{\sigma}{\varepsilon} \mathbf{D} + \frac{\partial \rho_f}{\partial t} = 0 \tag{33}$$

Because the material is assumed homogeneous, the last equation becomes

$$\frac{\sigma}{\varepsilon} \rho_f + \frac{\partial \rho_f}{\partial t} = 0 \tag{34}$$

which integrates to

$$\rho_f = \rho_{f0} \, e^{-\sigma t/\varepsilon} = \rho_{f0} \, e^{-t/\tau} \tag{35}$$

where, ρ_{f0} is the free-charge density at time zero.

The time constant τ

$$\tau = \frac{\varepsilon}{\sigma} = \varepsilon_0 \frac{K}{\sigma} \tag{36}$$

for the decay of some initial free-charge density is the analog to the familiar lumped-circuit time constant

$$\tau' = RC \tag{37}$$

The initial free-charge density can be considered to be mathematically equivalent to the voltage-source applied charges residing at the opposite surfaces of a unit cube across which the voltage is initially established. These drain away to recombine within the dielectric. Their recombination rate $1/\tau$ will be proportional to the conductivity of the medium and directly to the Coulombic force ($\propto 1/\varepsilon$) with which they attract.

The relaxation time τ gives measures of (1) the decay rate of the free-charge density in the given medium, (2) the rate at which free-charge is transferred from the interior to the surface, and (3) the rate at which the field within the medium subsides to low levels.

Practically, the relaxation time of materials varies over wide limits. In metals it may be about 10^{-18} sec; in ordinary distilled (but not "deionized") water, about 10^{-6} sec; in fused quartz about 10^{6} sec; and in especially pure polyethylene (British I.C.I. grade), about 10^{10} sec. If water is chosen as the dielectric medium where purity and hence conductivity can vary over wide limits [$\sigma = 0.10$–10^{-7} mho/cm], then we can expect to need to work with frequencies of about 10^{3} Hz and higher (up to 10^{8} Hz) to obtain reasonable responses to dielectrophoretic forces within the aqueous medium.

14.2.16 Theory of Dielectric Forces

We now present a generalized theory of the dielectric force on a body suspended in a fluid. Both the body and the fluid may conduct to some degree, and surface layers of charge accumulation (Maxwell–Wagner-type polarization) may exist, provided there is overall neutrality. In preparation for dealing simultaneously with two dominant frequency-dependent mechanisms (as may exist in biological systems), one affecting conduction, the other affecting polarization, our theory treats with both aspects simultaneously. Formally, at least, this is not "necessary." Even a composite system can be looked at as a "black box" having only in-phase and out-of-phase components of either the permittivity (ε', ε'') or conductivity (σ', σ''). But this is less instructive and offers less insight.

The doubled flexibility of our approach is easily included. It permits easier comparison with experiment such as when we have a "conductive" solid in a "polarizable" liquid (silicon in dioxane, e.g.) or a "polarizable" solid in a "conductive" liquid (cell walls in water, e.g.). This is merely a point of view, but a useful and natural one. It better opens the way for interpretation on a more natural, molecular basis, rather than a cold formalistic one. The development prepares for the possibility that *both* the polarizing and the conduction mechanisms of the system are most easily visualized as separate mechanisms in their frequency response.

For a perfectly insulating spherical particle of radius a and permittivity ε_2 suspended in an insulating fluid medium of permittivity ε_1, the dielectrophoretic force was shown to be

$$F = 2\pi a^3 \frac{\varepsilon_1(\varepsilon_2 - \varepsilon_1)\nabla(\mathscr{E})^2}{\varepsilon_2 + 2\varepsilon_1} \tag{12}$$

Starting with this equation, and assuming the field to be that between concentric spherical electrodes, it is possible to demonstrate[4] that the yield y (expressed as the average length of the chains of particles clinging to the inner electrode) is given by

$$y = \frac{8\pi a^4 C V r_2}{9 r_1 (r_2 - r_1)} \left[\frac{2t\varepsilon_1(\varepsilon_2 - \varepsilon_1)}{\eta(\varepsilon_2 + 2\varepsilon_1)} \right]^{1/2} \tag{38}$$

where C is the concentration of particles in suspension (number/m³) V is the applied rms voltage, r_1 and r_2 are the respective radii of the inner and outer spherical electrodes, t is the elapsed time, and η is the viscosity of the suspending medium.

Application of Eq. 38 to real systems (i.e., to real, not "perfect" dielectrics) shows good agreement in a number of respects. In accord with its predictions, linear response of the yield with voltage and with particulate concentrations, even with aqueous suspensions of living cells, is observed. The linear dependence on viscosity and on the square root of the collection time is noted. No provision is made in Eqs. 12 or 38 for conduction or frequency-dependent effects. An adequate theory should do so. Clearly the permittivities ε_1 and ε_2 of the medium and particle, respectively, should be responsive to these parameters.

Time (frequency) dependence is frequently included in such equations[7,30,32,47,48] by making the noncommital replacement of ε_1 by $\hat{\varepsilon}_1$

and ε_2 by $\hat{\varepsilon}_2$ where the $\hat{\varepsilon}_i$ are now complex quantities and $\hat{\varepsilon}_i^*$ is its complex conjugate, and taking the real part, namely, replacing the term

$$\frac{\varepsilon_1(\varepsilon_2 - \varepsilon_1)}{\varepsilon_2 + 2\varepsilon_1}$$

by*

$$\mathrm{Re}\left\{\frac{\hat{\varepsilon}_1^*(\hat{\varepsilon}_2 - \hat{\varepsilon}_1)}{\hat{\varepsilon}_2 - 2\hat{\varepsilon}_1}\right\}$$

This would give us a reasonable clue about the correct expression to be expected for the case of real dielectrics in real fields. Although we do not derive the equation here, it turns out that a careful analysis of the matter yields the generalized expression

$$\mathbf{F} = -\frac{1}{4}\int_{\mathrm{body}}\mathrm{Re}\left\{\mathbf{V}\left[\xi_1^*\left(1 - \frac{\xi_2}{\xi_1}\right)\mathbf{E}_0^* \cdot \mathbf{E}\right]\right\}d\tau \tag{39}$$

where $\xi_i = \varepsilon_i - i\sigma_i/\omega$, ξ_1 is the complex permittivity of the medium, ξ_2 is that for the body, $\mathbf{E}_0^*$ is the complex conjugate of the (complex) field originally impressed on the volume element $d\tau$ before the body occupied it, $\mathbf{E}$ is the resultant actual field in $d\tau$ with the body present, and ω is the angular frequency. A complete derivation of Eq. 39 is given in Ref. 58.

The foregoing generalized expression looks forward to the later examination of specific systems, be they suspensions of silicon single crystals, of minerals in organic liquids, or of living cells in aqueous media. In each case we can best understand the processes by considering specific mechanisms in molecular terms. Almost always several mechanisms are involved simultaneously. Some can be most easily viewed as field or frequency dependent polarizations, some as frequency or field-dependent conduction. Both types can now readily be included, and in the proper theoretical framework, to describe dielectrophoretic force. The role of Eq. 39 is to provide this framework.

The framework includes the necessarily complex dielectric factor ξ, which we can regard in a dual light—either as a simple sum of two fundamental but complex and frequency-dependent quantities σ and ε (which are the *inherent* conductivity and inherent dielectric constant, respectively) or as a complex sum of two real quantities σ_e and ε_e (the effective—and hence experimentally apparent—conductivity and dielectric constant, respectively). The first pair $\sigma = \sigma(\omega)$ and $\varepsilon = \varepsilon(\omega)$ can be regarded as the interpretive set; the latter pair σ_e and ε_e, as the observable set. Starting with reasonable assumptions about field distributions, polarization mechanisms, conduction

* Note that we take $\mathrm{Re}\{\hat{A}^*\hat{B}\}$, or equivalently, $\mathrm{Re}\{\hat{A}\hat{B}^*\}$, not $\mathrm{Re}\{\hat{A}\hat{B}\}$.

modes, and their locales, experiments on dielectrophoretic force can then be used to test the assumed mechanisms by way of Eq. 39. An example of this is given in Chapter 15, dealing with Biological Aspects of dielectrophoresis.

ACKNOWLEDGMENTS

The author expresses his sincere appreciation to the Research Foundation of Oklahoma State University and to the Department of Physics of the University of California at Riverside for their support during parts of these studies. The author also has appreciated the stimulus and encouragement tendered him during his sojourn at the Cavendish Laboratory of Cambridge University.

REFERENCES

1. W. Gilbert, *De Magnete*, Book II, Peter Short, London, 1600, Chapter II.
2. F. H. Winchler, *Essai sur la Nature; Effets et les Causes de l'Electricite*, Vol. 1, Sebastian Jorry, Paris, 1748.
3. J. Priestley, *The History and Present State of Electricity with Original Experiments*, 2nd ed., London, 1769.
4. H. A. Pohl, *J. Appl. Phys.*, **22**, 869 (1951).
5. —— *J. Appl. Phys.*, **29**, 1182 (1958).
6. —— and J. P. Schwar, *J. Appl. Phys.*, **30**, 69 (1959).
7. —— *J. Electrochem. Soc.*, **107**, 386 (1960).
8. —— and C. E. Plymale, *J. Electrochem. Soc.*, **107**, 390 (1960).
9. —— and J. P. Schwar, *J. Electrochem, Soc.*, **107**, 383 (1960).
10. —— *J. Appl. Phys.*, **32**, 1784 (1961).
11. F. Palmer, *Am. J. Phys.*, **30**, 133 (1962).
12. W. H. Middendorf and G. H. Brown, *Trans. AIEE*, **77**, III, 795 (1961).
13. J. Hart, *Can. J. Phys.*, **32**, 99 (1954).
14. A. Brin and P. Cotton, *Comp. Rend*, **236**, 1485 (1953).
15. J. R. Melcher and M. Hurwitz, *J. Spacecraft*, **4**, 864 (1967).
16. A. Lösche and H. Hultschig, *Koll. Z.*, **141**, 177 (1955).
17. L. D. Sher, "Mechanical effects of a.c. fields on particles dispersed in a liquid, biological implications," Ph.D. thesis, University of Pennsylvania, 1963.
18. E. P. Damm, *J. Electrochem. Soc.*, **115**, 474 (1968).
19. W. F. Pickard, *Prog. Dielec.*, **6**, 105 (1965).
20. H. A. Pohl, *J. Electrochem. Soc.*, **115**, 155c (1968).
21. —— and I. Hawk, *Science*, **152**, 647 (1966).

22. J. S. Crane and H. A. Pohl, *J. Electrochem. Soc.*, **115**, 584 (1968).

23. F. H. Müller, *Wiss. Veröffentl. Siemens-Werken*, **17**, 20 (1938).

24. P. J. W. Debye, *Phys. Rev.*, **91**, 210 (1953).

25. P. Debye, P. P. Debye, B. H. Eckstein, W. A. Barber, and G. J. Arquette, *J. Chem. Phys.*, **22**, 152 (1954).

26. ———, ———, and B. H. Eckstein, *Phys. Rev.*, **94**, 1412 (1954).

27. W. A. Barber, P. Debye, and B. H. Eckstein, *Phys. Rev.*, **94**, 1412 (1954).

28. H. A. Pohl, "Materials separation using nonuniform electric fields," U.S. Patent 3,162,592, Dec. 22, 1964.

29. A Rose, *Phys. Rev.*, **97**, 1538, 1935.

30. A. von Hippel, *Dielectrics and Waves*, Technology Press, Cambridge, Mass., and Wiley, New York, 1954.

31. A. J. Dekker, *Solid State Physics*, Macmillan London, 1962, p. 186.

32. M. Javid and P. M. Brown, *Field Analysis and Electromagnetics*, McGraw-Hill, New York, .1963, p. 145.

33. C. H. Townes, *Science*, **149**, 831 (1965).

34. R. Rosen and H. A. Pohl, *J. Polym. Sci.*, **4A**, 1135 (1966).

35. H. A. Pohl, *J. Polym. Sci.*, **17C**, 13 (1967).

36. R. D. Hartman and H. A. Pohl, *J. Polym. Sci.*, Part A-1, **6**, 1135 (1968).

37. H. A. Pohl, "Capacitors," U.S. Patent 3,349,302, Oct. 24, 1967.

38. J. E. Wyhof and H. A. Pohl, *J. Polym. Sci.*, Part A-2, **8**, 1741 (1970).

39. J. C. Maxwell, *Electricity and Magnetism* Vol. 1, Clarendon Press, Oxford, 1892, p. 452.

40. K. W. Wagner, *Die Isolierstoffe der Elektroteknik*, H. Schering. Ed., Springer, Berlin, 1924, pp. 1ff.

41. L. Van Beek, *Prog. Dielect*, **7**, 69, (1967).

42. P. Debye, *Polar Molecules*, Chemical Catalog Co., 1929, p. 8.

43. J. H. Van Vleck, *The Theory of Electric and Magnetic Susceptibility*, Oxford University Press, London, 1932, p. 152.

44. J. C. Davis, Jr., *Advanced Physical Chemistry*, Ronald Press, New York, 1965, p. 333.

45. C. T. O'Konski, *J. Phys. Chem*, **64** 605 (1960).

46. H. A. Pohl, *Sci. Am.*, **203**, 107 (1960).

47. R. W. P. King, *Electromagnetic Engineering*, Vol. 1, McGraw-Hill, New York, 1945.

48. G. J. Schwarz, *J. Chem. Phys.*, **39**, 2387 (1963).

49. E. R. Peck, *Electricity and Magnetism*, McGraw-Hill, New York, 1953.

50. K. S. Cole, *Cold Spring Harbor Symp.*, *Quantum Biol.*, **8**, 110 (1940).

51. H. P. Schwan, *Adv. Biol. Med. Phys.*, **4**, 147 (1957).

52. ——— and K. S. Cole, *Med. Phys.*, **3**, 52 (1960).

53. J. B. Miles and H. P. Robertson, *Phys. Rev.*, **40**, 583 (1932).

54. G. J. Schwarz, *J. Phys. Chem.*, **66**, 2636 (1962).

55. H. P. Schwan et al., *J. Phys. Chem.*, **66**, 2626 (1962).

56. M. Pollak, *J. Chem. Phys.*, **43**, 908 (1965).

57. C. T. O'Konski, *J. Phys. Chem.*, **64**, 605 (1960).

58. Joe S. Crane, Ph. D. Thesis, Oklahoma State University, Stillwater, 1970.

59. M. Lampert and P. Mark, *Current Injection in Solids*, Academic Press, New York, 1970.

60. B. J. Mason and J. Maybank, *Quart. J. Roy. Soc. Meteorol.*, **86**, 176 (1960).

61. J. Latham and B. J. Mason, *Proc. Roy. Soc.* (*London*), *Ser. A. Math. Phys. Sci.*, **260**, 523 (1961).

62. R. J. Cheng, *Science*, **170**, 1395 (1970).

63. N. J. Felici, *Rev. Gen. Elect.*, **76**, 786 (1967).

CHAPTER 15

Dielectrophoresis of Biological Materials

HERBERT ACKLAND POHL

Department of Physics
Oklahoma State University
Stillwater, Oklahoma
(NATO Senior Science Fellow on leave
to Cavendish Laboratory, University of
Cambridge, Cambridge, England)

JOE S. CRANE

Department of Physics
Cameron State Agricultural College
Lawton, Oklahoma

15.1 OVERVIEW

In recent years it has been found that many techniques developed in the physical sciences are applicable to problems of biological interest. One of these is dielectrophoresis. Through the application of alternating nonuniform electric fields, investigators have learned that biological particles can be made to exhibit movement (much like that of inanimate particles) which is dependent on their electrical characteristics. This technique can be used in two ways—as a method of determining the electrical characteristics of the particles, or as a method of physically separating particles of different characteristics. For example, under proper conditions of high frequency and low suspension conductivity, live yeast cells can be selectively extracted from a mixed suspension[1] of live and dead cells.

In a typical experiment involving suspensions, the ranges of the parameters are: particle diameter, 1 to 50 μ; frequency of applied field,

10^2 to 10^9 Hz; applied voltage, 5 to 30 V for closely spaced (3 mm) electrodes; suspension conductivity, less than 10^{-1} mho/m. Low frequencies and high conductivities lead to excessive heating and other disruptive effects. An important point is the strong dependence of the results on frequency and conductivity. The previous simple theoretical treatment, which works well for insulating materials in insulating media, attributes motion to a difference in permittivities only and has no provision for the inclusion of these other two variables. A more general treatment has been given which includes these parameters.[18] The predicted results using a spherical four-medium model agree with yeast cell behavior.[19] However, the approach has not been checked against other systems, and at present the applicability of the special model to them requires further study and, doubtless, modification.

In this chapter we describe some of the ways in which dielectro-phoresis has been applied to biological systems, presenting some typical results for the yeast cell system, briefly outlining the general theory, and alluding to possible future applications of the dielectrophoresis of biological materials.

15.2   EXPERIMENTAL RESULTS

The reaction of biological particles to rather uniform electric fields was probably first studied in detail by Muth,[2] who subjected fat particle emulsions to high frequencies and noticed pearl-chain formation (the end-to-end attachment of the particles producing a formation similar in appearance to a chain of pearls). Liebesny[3] also observed these formations for erythrocytes in high-frequency fields. Heller and his co-workers[4] studied the responses of various organisms to high-field strengths in the frequency range 10^5 to 10^8 Hz. They observed pearl-chain formation, orientation, preferential movement, rapid rotation, and frequency optima for alignment. Schwan and his co-workers[5,6] presented theoretical treatments for pearl-chain formation and orientation of biological particles. Schwan[7] also made a comprehensive review of the electrical properties of tissue and cell suspensions. It must be stressed that prior to the work of Pohl and Hawk,[1] no special consideration had been given to the gathering, collection, or control of living organisms by the action of *nonuniform* electric fields acting on them as neutral particles by way of their polarizabilities. Such motion produced by a nonuniform field acting on a polarizable neutral body is called dielectrophoresis, as described in the previous chapter. The reports by Schwan of extraordinarily high dielectric constants (10^2 to 10^4) for cell and tissue suspensions, promoted Pohl and Hawk[1] to apply dielectro-phoresis to such a suspension. Under the conditions of high frequency and low suspension conductivity, they obtained the separation of live and dead cells.

A detailed study of yeast cells has been made and a preliminary report[8] indicated that the responses of live and dead cells alike were dependent on the applied voltage, the frequency, the conducitivity of the suspension, and the concentration of cells. The difference between the two types of cells is revealed in their dependence on frequency and conductivity. Figure 15.1 shows the amount of cells collected at a "pin"

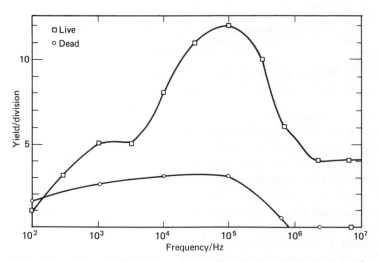

Figure 15.1. Comparison of yields for live and dead yeast cells: $V = 20\,\text{V}, \sigma = 3 \times 10^{-4}\,\text{mho/m}$. Points indicated by squares represent live cells; circles are dead cells.

electrode ("yield") as a function of frequency for both types of cells. The dependence on conductivity is illustrated by Fig. 15.2, which is a frequency plot for various conductivities.

In general, before dielectrophoresis is applied to a particular test system, several decisions must be made concerning experimental approach. The important considerations are the configuration of the electrodes which will produce the nonuniform field, the selection of a critical parameter which reflects changes in the dielectrophoresis and thus measures the effect of the field, and the determination of the variable quantities on which the chosen critical parameter depends.

A nonuniform field will be produced by any electrode design other than parallel plates. The number of possible configurations can be greatly reduced by requiring that the electric field be easily calculable. This desire for mathematical simplicity leads to the standard geometries of

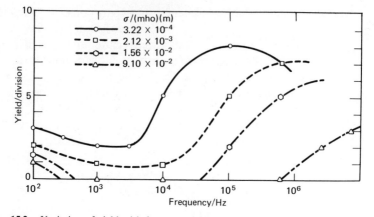

Figure 15.2. Variation of yield with frequency and suspension conductivity: $V = 20$ V.

concentric spheres and concentric cylinders. Other configurations have been devised which can be approximated as cylinders or spheres but which are easier to construct and offer experimental advantages. These include a wire perpendicular to a flat plate (pin–plate), a wire parallel to a flat plate (wire–plate), two coaxial wires end-to-end (pin–pin), and two parallel wires side-by-side (wire–wire). One other calculable shape is that which gives a constant dielectrophoretic force, the "isomotive field." The choice among these five designs for a particular experiment depends on the experimental characteristics desired. Probably the most popular type is the pin–pin arrangement, owing to its ease of construction and observation. A typical system appears in Fig. 15.3.

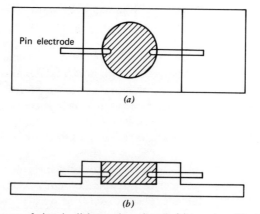

Figure 15.3. Diagram of pin–pin dielectrophoresis cell: (*a*) top view, (*b*) side view.

Once a field shape has been chosen, the next problem is to select a dependent variable that will give an indication of the effect of the field. Several such critical parameters exist, including the dielectrophoretic force, the velocity of a test particle, and the rate at which materials is built up at one of the electrodes. The force has been used as the critical parameter on several occasions[9-11] and is convenient when dealing with a single, large particle. The force can also be used for single microorganisms in a variation of the Millikan oil-drop experiment, where the dielectrophoresis force is balanced by gravity. The velocity of a particle has not served as the dependent variable, but there are no obvious reasons for thinking that it would not be useful in certain systems. The most widely used critical parameter is the rate of collection of particles at an electrode.[12-14,22,23] It is most often expressed as a "yield," which is the amount collected within a specified time, and it is best suited to systems which are suspensions of large numbers of particles. For use with very dilute suspensions of cells, a very precise counting procedure is available. It consists of collecting cells for a short period (1 min, say) at a "driving" voltage (20 V, say), cutting the applied voltage to a "holding" value (2 V, say), and counting the number of cells touching the electrode wire along a prechosen length. Using a mechanical stage, the count density of collected (single) cells can be checked along several regions of the wire-wire electrode and the results averaged. Cell suspensions containing only about 100 to 1000 cells/mm³ can be conveniently studied in this manner (see Figs. 15.1 and 15.2).

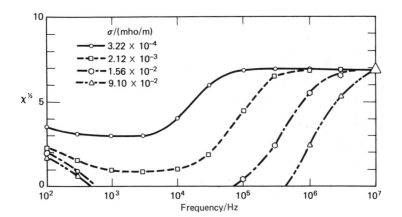

Figure 15.4. Theoretical variation of $\chi^{1/2}$ with frequency at various experimental conductivities for two-shell model.

After the dependent variable has been selected, the various independent variables need to be ascertained so we can provide the equipment and procedures that will allow these quantities to be varied. For the case of a suspension of particles with the yield as the critical parameter, the independent variables are: the size and shape of the particles, the conductivities and permittivities of both the particles and the suspending medium, the concentration of particles, the frequency and strength of the applied field, and the elapsed time. The quantities that can be most easily varied and measured are the suspension conductivity, the particle concentration, the frequency, the field strength, and the elapsed time.

The use of dielectrophoresis with a variety of microorganisms shows so far that each appears to have a collection rate versus frequency curve or "yield spectrum" which is rather characteristic of each organism and of its condition. The "yield spectrum," as it is loosely called, can be varied considerably of course, by altering the conductivity of the medium so that it is not the "invariant fingerprint" that might have been desired. Nevertheless the yield spectra of a wide range of types of microorganisms are of some interest. Several are shown very schematically[26] in Figs. 15.5, 15.6, and 15.7.

Clearly, each microorganism can exhibit up to three characteristic relaxation times. Each relaxation time very probably reflects some basic electrical polarization process associated with the cellular physiology and morphology. We have reason to believe, for example, that the relaxation peaks near 10^4 Hz are associated with the outer Helmholtz ionic double layer. It is just above this frequency region that the dielectrophoretic responses of AgBr solutions,[25] and $PbHfO_3$, and silicon particles[11] are

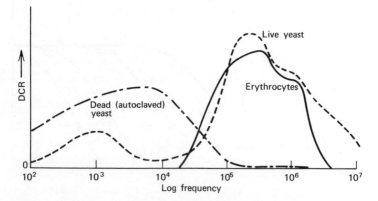

Figure 15.5. "Yield-spectra" or dielectrophoretic collection rates (DCR) of live and dead yeast cells as a function of frequency of the electric field (diagrammatic—$\rho \simeq 2 \times 10^3$ Ω-m).

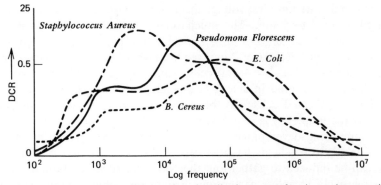

Figure 15.6. "Yield spectra" or dielectrophoretic collection rates of various microorganisms as a function of frequency of the electric field (diagrammatic—($\rho \simeq 2 \times 10^3$ Ω-m).

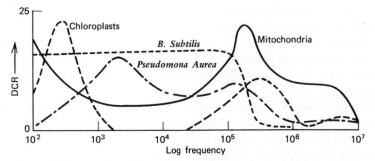

Figure 15.7. "Yield spectra" or dielectrophoretic collection rates of various microorganisms as a function of frequency of the electric field (diagrammatic—$\rho \simeq 2 \times 10^3$ Ω-m).

no longer much affected by specific ions. It is also a prominent region of response in all the living microorganisms we have examined thus far. It is especially marked in those such as mitochondria[26] having high surface-to-volume ratios. The calculations we have made for yeast cells using the spherical four-medium model also indicate that the ionic double layer might well respond at about this frequency.[19] However, the identity of this relaxation, and of the ones above and below it in frequency are not yet firmly established. More work needs to be done in this area.

A remark about the size and ease of dielectrophoretic collection of the various organisms may be in order. The larger cells such as chlorella[29] (ca. 15 μm diameter), yeast (*Cereovisea saccharomyces*, ca. 5 μm), human blood cells[27] (erythrocytes, ca. 8 μm), and canine blood platelets[28] (ca. 3 μm) collect rapidly. At a modest voltage the 0.1-ml suspension may be essentially cleared up in 2 to 5 min. The much smaller organisms such as rat liver mitochondria[26] (ca. 0.5×3 μm), spinach chloroplasts[30] (ca. 0.7 μm), and *Escherichia Coli*[26] (ca. 0.6×2 μm) require rather longer times (10–25 min). *Bacillis subtilis*[26] (ca. 1×4 μm) cells, which are long and thin, live up to their name by coming in sideways at high frequency and end-on at low frequency. Collected cells may be made alternately to stand up and lie down as the frequency is switched back and forth.

An interesting development using the orienting and gathering effect of nonuniform electric fields has just begun. We consider that if loose single cells can be induced to gather and then be "cemented" or bound together by a relatively harmless cementum then the possibility exists to form a "synthetic flesh." Furthermore, if the cells can be preferentially oriented in so doing and can thereby be in position to perform directionalized chemical or physical changes (e.g., pump or do work), then the possibility exists to form a "synthetic tissue."

Now certain aqueous polymer solutions while near 0°C are very fluid, although they gelatinize to a rather firm solid upon warming to about 37°C (e.g., Pluronic–127F resin). Using this interesting property of this rather nontoxic mixture, we have succeeded in making model cellular matrices of both types of solid using spherical cells (yeast) for the one and elongated cells (*B. Subtilis*) for the other.[29] A next interesting step will be to use dielectrophoresis to align unidirectionally functional cells and cement them in a similar manner. It may prove possible to treat sheets of cells on one side to produce unidirectional function. The possibilities are all very interesting. Clearly much remains to be done.

One further point should be made about size and dielectrophoretic collectability. Since it is clear that the dielectrophoretic force *increases* with particle volume (i.e., is proportional to the cube of the particle radius) and that thermal diffusional effects (or osmotic forces, as they are sometimes called) are greater as particle size *decreases*, a critical size limit could exist for collection. That is, since $F_{diel} \propto a^3$ and $F_{osm} \propto 1/a$, there will be some critical size range a_{crit} below which easy success cannot be had in gathering particles of radius a, against diffusional interference. Calculations using reasonable input parameters[23] indicate that this critical minimum particle radius will be about 0.1 μm. It was therefore of much interest to be able to push down through the size ranges and find that even mitochondria (ca. 0.5-μm diameter) could be collected. A preliminary attempt using tobacco mosaic virus (TMV—ca. 0.3×0.015 μm, cylindrical) by one of us[29] while at the Cavendish Laboratory, University of Cambridge, indicates that collection may just be possible. A suspension of TMV in deionized water was subjected to a nonuniform field, using platinum wire electrodes, and was inspected using either phase contrast or crossed Nicols prisms in a polarizing microscope. Only a marginal effect was obtained. This indicates, as we would expect from theoretical considerations, that the limit for the application of dielectrophoresis to suspension of fine particles may not be far away.

A most useful aspect of dielectrophoresis is its application to the rapid comparative study of normal and abnormal cells, or of changes in cellular responses as nutrients, poisons, and so on, are added to the medium. It is necessary, of course, to work with rather dilute ionic solutions (less than about millimolar) so that the conductivity does not overly interfere. It is also often necessary to protect the cells against osmotic damage by working with them in an isoosmotic medium. This is readily done by using say quarter-molar glucose, sucrose, or glycine solutions. When these precautions are taken, very interesting comparisons of cell physiology can be made. Research in this area is quite active.

To sum up the experimental observations to date, we note that a great variety of microorganisms can be collected, although there may be a lower size limit somewhere in the size range of viruses. Moreover, up to three characteristic relaxation peaks can exist in the yield–frequency curve and these curves can be shifted by the conductivity of the medium. The process of dielectrophoresis can be used to distinguish cells by type, age, or physiological condition. As yet the distinctions are not sharp, but they are sufficiently effective to indicate that improvement of techniques will improve matters to a very significant degree. The process can be used to remove and harvest living cells from fresh water. At present the methods have been developed only for small-scale handling. Their extension to a larger scale needs development and investigation.

15.3 THEORETICAL ASPECTS

Four generally accepted mechanisms have been suggested as possible contributors to the electrical behavior exhibited by organic materials in aqueous suspensions. As discussed by Schwan,[7] these are relaxations of the dipole rotation of constituent molecules, gating mechanisms controlling conduction through cell membranes, Maxwell–Wagner-type relaxation of conduction and polarization at various interfaces, and relaxation of a surrounding ion atmosphere. Most explanations rely basically on one of the four. The present theory has provisions for including all four mechanisms, their relative importance being determined by the experimental conditions. That is, the dominant effect at high frequencies is likely to be dipole rotation, whereas at lower frequencies the gating mechanism or ion layer might become more important. The entire problem can be approached using a Maxwell–Wagner-type formulation, with dipole rotation and the gating mechanism accounted for by allowing the appropriate conductivities and permittivities to vary.

As is discussed in Chapter 14, the conductivity and permittivity of a system relate the conduction current and the displacement to the electric field by

$$J = \sigma E \quad \text{and} \quad D = \varepsilon E \tag{1}$$

In the general treatment we make two fundamental assumptions. The first is that the conduction current and the displacement may arise from entirely different mechanisms and, therefore, σ and ε must be considered to be independent of each other. An example of this situation is the comparison of fresh water to seawater. Both have a relative permittivity of 81 but have conductivities of 10^{-3} and 4 mho/m, respectively.[24] Obviously the permittivity arises from the water and the conductivity from the salt concentration. The second assumption is that there may exist a time lag of response between the applied field and the resulting current and displacement. This can be described, if the applied field is sinusoidal, by making σ and ε complex. That is,

$$\sigma = \sigma' - i\sigma'' \quad \text{and} \quad \varepsilon = \varepsilon' - i\varepsilon'' \tag{2}$$

where $i = \sqrt{-1}$.

Because of the nature of Maxwell's equations, the imaginary part of σ behaves physically like a permittivity and the imaginary part of ε is seen as a conductivity. Thus the real quantities

$$\sigma_e = \sigma' + \omega\varepsilon'' \quad \text{and} \quad \varepsilon_e = \varepsilon' - \frac{\sigma''}{\omega} \tag{3}$$

can be considered to be the "effective" conductivity and the "effective" permittivity, respectively, where ω is the angular frequency of the field. These are the quantities which are measured by typical bridge measurements. Since each is a combination of both conduction and polarization effects, the difficulty in extracting expressions for σ and ε separately from such measurements is obvious.

There is a complex quantity, here called the complex dielectric factor ξ, which relates the total current (the conduction current J plus the displacement current $\partial D/\partial t$) to the electric field. It is given by [15]

$$\xi = \varepsilon - \frac{i\sigma}{\omega} = \varepsilon_e - i\frac{\sigma_e}{\omega} \tag{4}$$

It should be emphasized that the terms ε and σ are complex, whereas the terms ε_e and σ_e are real. The latter form, involving the effective values, is often given as a definition of the complex permittivity; but we prefer to restrict that designation to the proportionality between D and E as in Eq. 1.

As we see later, the dielectrophoretic force depends on ζ and so this formulation immediately introduces a dependence on conductivity and frequency as well as permittivity. A further frequency dependence is produced when the real and imaginary parts of σ and ε are made to vary with frequency, which must be the case if the time-lag assumption is to have any meaning. That is, as the frequency of the field is increased, the responses will lag farther and farther behind, decreasing the in-phase, real components and increasing the out-of-phase or imaginary components of the conduction and polarization. The form of the frequency dependence will depend on the type of mechanism involved, and it is here that allowances for dipole rotation relaxation and conduction gating mechanisms can be included. The simplest form is a single-relaxation mechanism as first introduced by Debye.[16] In this form, the complex permittivity is

$$\varepsilon = \varepsilon_\infty + \frac{(\varepsilon_s - \varepsilon_\infty)}{1 + i\omega\tau_\varepsilon} \tag{5}$$

where ε_s and ε_∞ are the real parts of ε at very low and very high frequencies, respectively, and τ_ε is the relaxation time for the mechanism. A similar expression would hold for σ.

15.4 DIELECTROPHORETIC FORCE

For sinusoidal fields, it is shown by King[15] that the time-averaged electrical energy in a volume τ is given in terms of the amplitude E of the varying electric field by

$$U = \frac{1}{4} \int_\tau \varepsilon_e E^2 \, d\tau$$

When a body in the field is free to move, the force can be determined by finding the negative rate of change of energy with position. The development of the force equation (not given here) is involved; it is similar to that derived by Schwarz[17] for the nonconducting case. The result is[18]

$$F = -\frac{1}{4} \int_{body} \text{Re} \left\{ \nabla \left[\zeta_m^* \left(1 - \frac{\zeta}{\zeta_m} \right) E_0^* \cdot E \right] \right\} d\tau \tag{6}$$

where F is the force on the body, ∇ is the vector del operator, m refers to the surrounding medium, * indicates complex conjugate, ζ is the complex dielectric factor of the body, E_0 is the electric field that would exist in the system if the body were not there, and E is the field inside the body. The integration is performed over the body only, since elsewhere $\xi = \xi_m$ and the integrand is zero.

In order to evaluate this expression, we must know (1) the relation between the field inside the body and the original impressed field, (2) the expressions for the dielectric factor, and hence (3) the permittivities and conductivities for all parts of the body. This is no easy task, particularly if the body consists of several regions of differing characteristics and is not of a simple shape such as a sphere or an ellipsoid. For any but the simplest of systems, a comparison between theoretical calculations and the experimental measurements requires many simplifying assumptions, and only qualitative agreement should be expected.

15.5 APPLICATIONS OF THEORY TO YEAST

As a check of the preceding theoretical result, calculations have been made for the system of an aqueous suspension of yeast cells.[19] In the case of the dielectrophoresis of the yeast cells, the critical parameter was not the force but the buildup of cells at an electrode (i.e., the yield). Since the experimental data were in this form, the relation between the force and the yield was necessary. For the approximately radial field near a pin electrode, the yield can be shown[19] to be proportional to the applied voltage, the cell concentration, the square root of the elapsed time, and the square root of a quantity called χ which contains the combination of the ξ's from the force equation. The voltage, the concentration, and the time dependences agree with the experimental results.[8,20]

In order to check the $\chi^{1/2}$ dependence, a model of a yeast cell was chosen and appropriate values were given to the various conductivities and permittivities. The spherical four-medium model, which was kept as simple as possible while retaining the essential electrical characteristics, was composed of three concentric spheres, a central region representing the cell membrane, and an outer ion atmosphere, all lying in an infinite medium. Each region was assigned suitable complex conductivities and permittivities, whose corresponding relaxation times were based on experimental values found in the literature.[7,21] The value of $\chi^{1/2}$ was then computed as a function of frequency and medium conductivity and was found to give good qualitative agreement with experimental results. For example, compare Fig. 15.2 with Fig. 15.4, which gives the computed value of $\chi^{1/2}$ for the same conductivities. Note that as the suspension conductivity increases, the yield at the middle frequencies is lowered and the high-frequency peak is shifted to higher frequencies. This is also true for the calculated $\chi^{1/2}$.

Remember that the yield is proportional to $\chi^{1/2}$; thus the two graphs should be similar in shape if the theory is correct, the model accurate, and the assigned values correct.

As was stated earlier, the general theory has been compared with experimental results in only this particular case. It is hoped that other comparisons can be made in the near future, to give the theory a firmer experimental foundation.

REFERENCES

1. H. A. Pohl and I. Hawk, *Science*, **152**, 647 (1966).
2. E. Muth, *Koll. Z.*, **41**, 97 (1927).
3. P. Liebesny, *Arch. Phys. Therapy*, **19**, 736 (1939).
4. J. H. Heller et al., *Exp. Cell Res.*, **20**, 548 (1960).
5. M. Saito and H. P. Schwan, *Proc. Conf. Biol. Effects of Microwave Radiation*, **1**, 85 (1960).
6. ———, ———, and G. Schwarz, *Biophys. J.*, **6**, 313 (1966).
7. H. P. Schwan, *Advn. Biol. Med. Phys.*, **4**, 147 (1957).
8. J. S. Crane and H. A. Pohl, *J. Electrochem. Soc.*, **115**, 584 (1968).
9. I. L. Hawk, "Effects of non-uniform electric fields on real dielectrics in water," M.S. thesis, Oklahoma State University, 1967.
10. C. Feeley, "Dielectrophoresis of solids in liquids of differing dielectric constant and conductivity" M.S. thesis, Oklahoma State University, 1969.
11. K. W. L. Chen, "Dielectrophoresis of solids in aqueous solutions," M.S. thesis, Oklahoma State University, 1969.
12. H. A. Pohl and J. P. Schwar, *J. Appl. Phys.*, **30**, 69 (1959).
13. ——— *J. Appl. Phys.*, **29**, 1182 (1958).
14. ——— and C. E. Plymale, *J. Electrochem. Soc.*, **107**, 390 (1960).
15. R. W. P. King, *Electromagnetic Engineering*, Vol. 1, McGraw-Hill, New York, 1945.
16. P. Debye, *Polar Molecules*, Chemical Catalog, New York, 1929.
17. G. Schwarz, *J. Chem. Phys.*, **39**, 2387 (1963).
18. H. A. Pohl and J. S. Crane, "Dielectrophoretic Force," *J. Theor. Biol.*, **37**, 1 (1972).
19. ——— and ———, "Theoretical Models of Cellular Dielectrophoresis," *J. Theor. Biol.*, **37**, 15 (1972).
20. ——— and ———, "Dielectrophoresis of Cells," *Biophys. J.*, **11**, 711 (1971).
21. H. Falkenhagen, *Electrolytes*, Oxford University Press, London, 1934.
22. H. A. Pohl and J. P. Schwar, *J. Electrochem. Soc.*, **107**, 383.
23. ——— *J. Electrochem. Soc.*, **107**, 386 (1960).
24. S. Ramo, J. Whinnery, and T. Van Duzer, *Fields and Waves in Communication Electronics*, Wiley, New York, 1965, p. 337.
25. Chi Shih Chen, H. A. Pohl, J. S. Huebner, and L. J. Bruner, "Dielectrophoretic precipitation of silver bromide suspensions," *J. Colloid Interface Sci.*, **37**, 354 (1971).

26. Chi Shih Chen, private communication.

27. K. L. Wiley "A comparison of normal and abnormal cells using dielectrophoresis," M.S. thesis, Oklahoma State University, 1970.

28. J. Rhoads and R. Buckner, private communication.

29. H. A. Pohl, private communication.

30. I. P. Ting, K. Jolley, C. A. Beasley, and H. A. Pohl, "Dielectrophoresis of chloroplasts," *Biochem, Biophys. Acta*, **234**, 324 (1971).

CHAPTER 16

Electrostatics in the Power Industry

JOHN H. MORAN, *Chief Electrical Engineer*
Lapp Insulator Division—Interpace
Le Roy, New York

16.1 OVERVIEW

Nowhere are electrostatic phenomena more important to us than in the generation, transmission, distribution, and use of electric power. And here, as in other areas, some electrostatic effects are indispensable, whereas others are detrimental and must be suppressed. For example, the same electrostatic forces successfully used to remove charged particles from smokestacks of generating stations are responsible for deposition of matter on insulators, thereby lowering their insulating value. Power generation and transmission conductors and allied equipment operate at from considerable to very high voltages. Their surfaces can be highly charged, and thus a power system is a giant collection of electrostatic phenomena.

This chapter reviews the use of electrostatics in the power industry, and the associated problems. Since the material is descriptive, the newcomer to electrostatics is invited to go right ahead and read the full chapter.

We now have but two major means of generating power. One is in hydroelectric plants, using the energy of falling water. The other is in plants where some sort of fuel makes steam at high pressure and temperature, which then drives giant turbogenerators. Since a large fraction of usable hydro sites are already developed, the very large growth that is certain to come will depend on fueled steam plants. The steam plant is energized by burning fossil fuels—coal, oil, or gas—or by energy from nuclear sources.

The mineral content of coal produces what is known as fly ash. Electrostatic precipitators (Chapter 9) are notably successful in trapping this pollutant before it enters the environment.

Nuclear-fired plants do not and cannot be allowed to produce emissions of any kind. But again, electrostatics comes into the picture, for both monitoring and controlling of the nuclear reaction are imperative. The devices used are inherently electrostatic. A major means for monitoring nuclear radiation is to measure its effect on charged capacitors. A small capacitor, properly constructed, can retain its charge for a long time; but if it is placed in an area through which radiation passes, there is an increase in the rate of discharge. These devices serve as direct-connected monitors for the nuclear source itself; they are also used as pocket dosimeters to monitor the amounts of exposure to which individuals are subjected.

It is not practical to build generators for very high voltage. The maximum generated voltage is 23,000 V, the power being at "low" voltage and very high current; but power in this form cannot be transmitted very far. It must be put through a substation, where step-up transformers change it to high voltage and relatively low current. Transformers, although functionally electromagnetic, are also vitally concerned with electrostatics. They have distributed capacitance, and they must be designed so that the windings can withstand the added stress of sudden surge or a lightning stroke.

For generators, transformers, lightning arresters, and circuit breakers, there is the problem of getting conductors into and out of the steel cases in which they are housed. Bushings are used. Consider a 500-kV line-to-line service (290 kV line-to-ground). If a conductor were to be protected by a "bulk" bushing—simply covering it with insulation—the insulation would be 62 in. thick. This is intolerable. Capacitor-type bushings are used instead. By proper electrostatic design, only about 11 in. is required. Also, the length of the air end of a bulk bushing would call for about 212 in., whereas the capacitor bushing needs only 145 in.

Electrostatics enters the transformer and bushing situations by way of fault detection. By use of comparative capacitances, these are checked out electrostatically before installation to detect any possible flaws.

The generated power, after transformers raise it to a voltage high enough for substation purposes, is led on its next step through the substation, but still at relatively low voltage and high current. The currents call for aluminum tubing 4 to 5 in. in diameter. These bus runs are supported above ground by a series of insulator stacks. A modern substation, with its numerous vertical insulating supports, looks like a large stand of trees. These supports are "station posts" which over the past half-century have gradually been perfected by engineering, technical, and manufacturing "know-how."

As recently as 1955, when substation voltages were 138 kV and below, station insulators were about 5 ft tall. Basically, they were two 30-in. sections, one on top of the other. This was as much height as could be effectively used, because of wet weather conditions. When attempts were made to use insulators three or four sections high for, say, 345 kV, rain water collected by the upper sections would course down and flood out the bottom section. Flashover resulted, leading to power outage, which was catastrophic.

The major breakthrough came when electrostatic forces served to repel the rain water by way of "rain shields." The invention was the subject of a U.S. Patent,[11] and it came just in time for the phenomenal expansion of power use at higher voltages. It employs electrostatic forces between a conducting skirt and ground, and water flowing down over an insulator section is propelled away from each section. Thus more sections can be stacked up without the lower ones being flooded out.

16.2   LIGHTNING ARRESTERS

A lightning arrester—more properly called a lightning diverter—must pass the effects of surges or lightning to ground, thus preventing damage to generators and transformers. Over the years, a number of types of arresters were evolved. Modern arresters have the same general form: a gap in series with a nonlinear resistor element. The arrester remains passively connected with the power source it guards until called into action by an abnormal voltage stress on the line. The gap then flashes over, since its withstand ability has been exceeded, and the voltage present on the bus structure is impressed on the nonlinear resistance element. This element (basically silicon carbide) has the following property: the greater the voltage, the lower is its resistance. Figure 16.1 shows the basic parts.

Although the basic device appears extremely simple, its characteristics must be carefully balanced. It has to prevent any current flow at normal voltage, but it also has to automatically act as a switch when abnormal voltage appears. The gap is designed so that once it flashes over, the gap itself elongates, cools, and helps to interrupt the power current that follows the stroke, when the current passes through zero. Normal conditions are then restored.

The series of metallic elements making up the gap electrostatically divides the impressed voltage, and reliable operation depends on effecting the right voltage distribution. The arrester, as an electrostatic protective device, is absolutely vital to the operation of power systems.

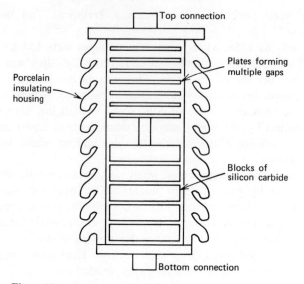

Figure 16.1.   Basic parts of a lightning arrester.

16.3   SWITCHES

For substation maintenance, switches are provided for isolating, connecting, and reconnecting the various parts. A considerable number is required. The voltage to ground they must sustain is determined by the size and shape of the switch, in conjunction with the station post insulators supporting it. The voltage withstand across the open gap is a function of the size, shape, and positioning of the metal parts. In both cases the electrostatic voltage distributions play major roles. Much has been written about switches, and most of it falls in the realm of electrostatics.

16.4   CIRCUIT BREAKERS

Faults and failures are bound to occur in power systems, and circuit breakers must stand ready to interrupt and clear short circuits and faults, isolate them, and restore the integrity of the system. In operation, the breaker is largely an electromagnetic device. But before operation, it must withstand the voltage from conductor to ground and after operating it must withstand the voltage across its open contacts—these characteristics are determined electrostatically.

16.5 TRANSMISSION

For now, and possibly for all time, the transmission line is the one means for transport of large amounts of power. The prevailing three-phase type of transmission accounts for the various configurations and combinations of three wires seen in so many places. Not long ago 115 kV, line-to-line, was a common upper limit. As power increased and lines grew longer, such voltages increased from 115 to 230, to 345, to 500, to 765 kV, and the end is not in sight. The handling of the electrostatic problems involved has called for a high order of engineering expertise.

Two major considerations dictate the size and shape of a transmission line conductor; its ability to carry the current, and its ability to do this at high voltage without permitting corona or static on the conductor. In earlier days it was easy to meet both needs, just by using larger conductors. But there are practical limits to this solution. The weight of conductors, the effects of wind, and the tendency to ice up, all play a part. When current went beyond what the largest single conductor could carry without corona loss or radio noise, a new answer had to be found: bundling.

What would otherwise be one too-large conductor now became two, spaced apart but in parallel. Electrostatically, the effective diameter of the bundle is much larger than that of a single wire. With still higher voltages, the number in a bundle has gone up. For 765-kV transmission, a three- or four-conductor bundle is used, with four preferred (Fig. 16.2).

Dealing with the four-conductor bundle of Fig. 16.2, let g_0 be the voltage gradient (V/in.) at conductor surface, and let V be the line voltage. Let the arrow rotate counterclockwise through the angle θ, thereby defining a point on the surface of the conductor. The voltage gradient next to that surface, as affected by distance above ground, is

$$g_0 = \frac{2}{Nd \ln 4h/\sqrt[N]{N\,dA^{N-1}}}\left[1 - (N-1)\left(\frac{d}{A}\right)\cos\theta\right] \tag{1}$$

For the gradient as affected by nearness of tower parts, these being at radius R, it is

$$g_0 = \frac{2}{Nd \ln 2R/\sqrt[N]{N\,dA^{N-1}}}\left[1 - (N-1)\left(\frac{d}{A}\right)\cos\theta\right] \tag{2}$$

Another useful equation reveals even more directly than Eqs. 1 and 2 the effect of capacitance on these gradients:

$$g = \frac{1.8U_{ph}C_b}{NP_t}\left[1 + \frac{2(N-1)\sin \pi/N}{S^1}\right] \text{ kV}_{\text{eff}}/\text{cm} \tag{3}$$

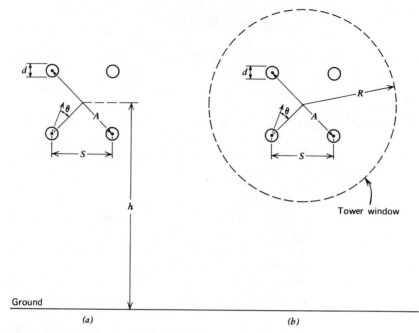

Figure 16.2. Voltage gradient for a bundled conductor.

where C_b = effective capacitance per unit length of conductor (pF/cm)

      N = number of conductors in bundle

      P_t = radius of the conductors

      $S^1 = S/P_t$ = relative distance between conductors in bundle

      S = distance between the conductors in the bundle (cm)

      U_{ph} = voltage between conductor and ground (kV)

The problem of powerline interference with communication systems has grown rapidly, especially in heavily populated centers. Line discharges first caused "static" on radios; and the advent of TV, especially color TV, accentuated the problem. Line corona must be kept suppressed, as must discharges that may show up on inadequate line hardware. A different problem recently found with bundled lines is an audible noise. Apparently, it arises from vibrations set up in the subconductors by electrostatic forces between them.

Transmission line behavior is largely affected by the load the line feeds. In many other circuits, as in lines at low voltage is houses or plants, there is a voltage drop, from the feed-in to the load end. But the high capacitance of a transmission line can, with a light load, cause a *rise* in voltage at the load (called the Ferranti effect). Overvoltage at the receiving end can damage equipment there if this is not taken into account and compensated for by installing inductive reactors. Oppositely, if a line feeds a heavily inductive load, line capacitance alone may not prevent a voltage drop, and capacitors are installed along the line. Material in the references covers these matters.

16.6   INSULATORS

The suspension of lines high up on towers calls for insulators much different from post-type substation insulators. A suspension-type insulator (Fig. 16.3) has two pieces of metal separated by a dielectric. It constitutes a capacitor, and a string of these devices amounts to a batch of capacitors in series. When clean and dry, the voltage distribution over the string is purely electrostatic, but it becomes a mixture of electrostatics and electromagnetics when wet or contaminated or both. The behavior is also affected by proximity to the tower window through which the line passes. Series and parallel capacitance effects now come into play.

High-field stress problems arise where the insulators support the conductors, calling for connectors and clamps. Stress relief to suppress corona is achieved by use of grading rings (Fig. 16.4) to cover sharp points and to give field stress relief.

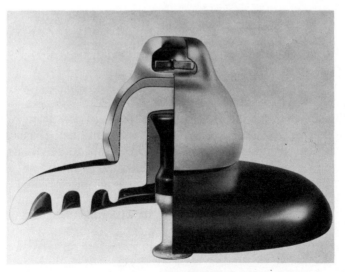

Figure 16.3.   Partial cut-away view of a typical suspension insulator showing the relation of the metal parts to the insulating porcelain.

Figure 16.4. String of suspension insulators supporting a two-bundle conductor arrangement requiring the use of "butterfly" grading rings to eliminate corona discharges from relatively sharp points on the hardware joining the insulators and the conductors. A different type of grading ring can be seen at the top end of the insulator string used to eliminate corona discharges at that point.

16.7   CABLES

Could our numerous long-distance high-voltage powerlines, now overhead, be replaced by underground cables? At this time, the answer is an emphatic *no*. Such a cable is a conductor surrounded by insulation, the whole being protectively encased in a metal sheath. This is a high-capacitance combination; it is too high for any but the relatively short-length runs used underground in some situations, and used then only by necessity. By using a gas, such as sulfur hexafluoride (SF_6) having a dielectric constant near unity, instead of 5 or 6 for solid insulants, the high capacity can be much reduced. Some research interest is being shown in the cryogenic cables for long-distance work of the future, using vacuum for insulation and superconductivity for the conductors. Time will tell where this can be feasible. Of course, common to all ideas is the awareness that electrostatic stress at the conductor surface requires adequate attention, or else breakdown will occur.

16.8   DIRECT CURRENT

All the foregoing has pertained to alternating current. The use of high-voltage direct current for transmission eliminates some problems on the one hand; however, it introduces its own problems. On the good side, the Ferranti effect drops out, together with any need for voltage regulation obtained either by line capacitors or inductors. Also, transmission by dc is simpler and cheaper. Thus it is quite inviting.

However, the equipment needed to change the generated ac into dc is extremely complex and expensive. Moreover, the dc must be inverted to ac for load use. An inverting station also is quite complicated and expensive. Tapping off power at many points along a line seems to be out of the question. There is a further complication, concerned with fault interruption. Alternating current goes through zero twice per cycle and is then interrupted in simple fashion. A dc line has no zero, and fault protection remains a goal that is not yet attained with sufficient reliability.

16.9   PROBLEMS OF VERY HIGH VOLTAGE

Each year cars and trucks and huge vans drive under powerlines crossing highways many millions of times, and no thought is given to it. But look ahead to lines possibly operating at 1000 kV and above. A large aluminum-body semitrailer truck has a flat tire or for some other reason has to be stopped below such a line. The driver gets out. The tires insulate the body from ground. From line to trailer body and from it to ground we have two capacitors in series, with the voltage divided in the ratio of the two capacitance values. Electrostatics now becomes dangerous, for if anyone were to touch the aluminum body, a nasty shock could ensue.

16.10   DISTRIBUTION

For safety, the use of power in the home continues to dictate the traditional 115 to 230 V levels. With some exceptions, factory use calls for the same levels in offices, but may go to 440 V for motors. Virtually no electrostatic effects occur at these levels.

The situation regarding the distribution lines overhead, which bring the power to our load centers, is changing. These lines cannot use the very high voltages demanded in transmission; neither can they operate down at house levels (the conductors would be far too large and costly). Although 4160 V used to be considered a proper distribution voltage, it has had to climb; 34.5 kV is now considered appropriate, and 69 kV is allowed in some areas. This puts us into a range where electrostatics is no longer to be ignored. A linesman working on live lines can be affected without getting close enough to invite the blinding flash that kills. Charges induced on him can affect his control of movements, possibly leading to accidents. Special training is mandatory.

16.11   POWER FACTOR CORRECTION

From the standpoint of line loss, lines should operate at unity power factor to deliver the same power at least current. Motors, on the whole, in home or factory, constitute a somewhat inductive load. To correct line power factor to near enough to unity, capacitors are nowadays installed across the lines. Motors are electromagnetic; capacitors, electrostatic. So here, on a tremendous scale, we have one of the great number of examples of an intimate marriage of electromagnetics and electrostatics.

16.12   PROTECTION

In earlier days the operator in a powerhouse could watch some simple meters, perhaps also hear an alarm buzzer, and take appropriate action in case of trouble. Our giant power systems can no longer be handled that way. The sensing of various kinds of trouble, together with the need to initiate corrective action, has called for elaborate sensing and relaying setups for use when a major power failure occurs. Each protective system is based on a number of devices, and each of these is simple; but the combination is so complex that in many cases the computer must replace man in handling the sensed inputs and delivering the proper (corrective) outputs.

And now the question can be raised, will the protective computer need protection? The answer can be yes. In a case on record, a computer used for accounting in a factory happened to fail. Several hundred feet away, in a high-voltage laboratory, an impulse generator was used to simulate the effects of surges and lightning. The energy discharged electrostatically induced such great charges in the computer location that all efforts to isolate and shield the computer were ineffective, and the computer had to be moved several miles away. This instance was promptly made known to a number of power companies. Surges and lightning discharges are of extremely short duration; but again, a computer operates at very high speeds. If a failure in a power system knocked out its own protective control computer, chaos would result.

16.13   HIGH–VOLTAGE LABORATORIES

The services rendered to millions by our ever-growing and ever-more-complex power systems, and the benefits of those services, are beyond calculation. Yet we would not have reached the power status of today, nor its extremely high reliability, without having had the benefit of electrostatics in the testing area. Lightning, which is Nature's electrostatics, has called on man with his own electrostatics to win a war against Nature.

Around 1920 the idea arose of building devices to duplicate the effects of lightning. The idea was simple: merely use specialized transformers to put insulators and test lines to the test. But that was not enough. It was gradually found that a lightning stroke had a very rapid rise in voltage, followed by a slower fall. The many methods used cannot be detailed here, but there is fascinating reading in the references given.

Through the work of numerous investigators it was determined that many damaging strokes approximate to a certain waveshape and that a similar test waveshape was adequate for testing the ability of equipment to defy the stroke. This is the "1.2 × 50" wave, meaning, that the voltage rises from zero to crest value in 1.2 μsec and falls to half the crest value in 50 μsec. The wave is represented in Fig. 16.5.

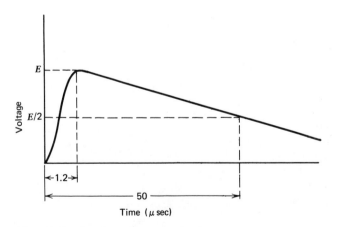

Figure 16.5.   Impulse voltage waveshape.

Use then was made of the Marx circuit, in which numbers of capacitors are charged in parallel; by proper deployment of series gaps, they discharge in series. Thus the high voltages and energies of strokes can be simulated. The earlier small artificial lightning machines were adequate for testing small strings of insulators, arresters, circuit breakers, and transformers of their day. Now there are hundreds of impulse generators ranging from very small units of a few kilovolts up to 10-million V generators using 300 kJ or more of energy, all tailored to perform their own functions.

As transmission lines approached the 345-kV level, it became mandatory that testing for switching surge effects be fully adequate. The impulse generators met the need.

16.14   CONCLUSION

The world's greatest electric field is its own field: that between the positive atmosphere and ground. Its greatest dynamic manifestation of electrostatics is the lightning of a large-scale thunderstorm. Its greatest man-made example of electrostatics—with respect to both size and variety—is seen in interconnecting power systems, spanning much of a continent. And it is good to know that the effects of lightning—Nature's prime threat to those systems—have been conquered by fighting back electrostatically with impulse generators, thereby bringing us an extreme reliability of power service we could not otherwise have had.

REFERENCES

1. *Electrical Transmission and Distribution Reference Book*, 3rd. ed., Westinghouse Electric & Manufacturing Co., East Pittsburgh, Pa., 1944.
2. *EHV Transmission Line Reference Book*, Edison Electric Institute, 750 Third Avenue, New York, N.Y. 10017, 1968.
3. F. W. Peek, Jr., *Dielectric Phenomena in High Voltage Engineering*, McGraw-Hill, New York, 1920 (out of print).
4. A. D. Moore, *Electrostatics*, Science Study Series, Doubleday Anchor, Garden City, New York, 1968.
5. Archer E. Knowlton, Ed.-in-Chief, *Standard Handbook for Electrical Engineers*, 8th ed., McGraw-Hill, New York, 1949.
6. John D. Ryder, *Networks, Lines and Fields*, Prentice-Hall, Englewood Cliffs, N.J., 1949.
7. W. G. Hawley, *Impulse Voltage Testing*, Chapman & Hall, London, 1959.
8. J. Zaborsky and J. K. Rittenhouse, *Electric Power Transmission*, The Rensselaer Bookstore, Troy, N.Y., 1969.
9. D. Vitkovitch, *Field Analysis*, Van Nostrand, New York, 1966.

10. A. R. Von Hippel, *Dielectrics and Waves*, M.I.T. Press, Cambridge, Mass., 1954.

11. C. D. Fiero and J. H. Moran, U.S. Patent 2,884,479, 1959.

12. *Transactions of the Institute of Electrical and Electronic Engineers*, 345 E. 47th St., New York, N.Y. 10017.

13. *Proceedings of the Institution of Electrical Engineers*, Savoy Place, London, WC 2R, OBL, England.

CHAPTER 17

Atmospheric Electrostatics

BERNARD VONNEGUT
Atmospheric Sciences Research Center
State University of New York at Albany
Albany, New York

17.1 EFFECTS OF HUMIDITY

Even those with only a passing acquaintance with electrostatic phenomena recognize that electrostatic behavior depends markedly on the state of the atmosphere. A prime example of this is its sensitivity to the amount of moisture present. When the relative humidity is high, electrostatic phenomena, such as the charging of people as they walk across a rug, or the electrification of garments and blankets, very nearly disappear. Laboratory experiments in electrostatics that behaved admirably in the winter become unreliable or do not work at all when the humidity increases in the spring.

Although some have erroneously attributed this sensitivity to humidity of electrostatics to an increase in the electrical conductivity of the air, we now know that these effects are primarily the result of an increase in the electrical conductivity of the dielectric substances involved in the electrical phenomena. Measurements show that as the relative humidity increases above about 50%, layers of moisture are adsorbed on most solids, causing a pronounced increase in the conductivity of dielectric surfaces. The conductivity of even very good electrical insulators increases by orders of magnitude as the result of the formation of a thin layer of moisture on their surfaces that permits the flow of electricity. Although frictional and contact electrification take place even under conditions of high humidity, the electrical charges accumulated in this way drain off in a fraction of a second; thus electrical effects, which are obvious under conditions of low relative humidity, almost disappear when the humidity is high. The effect of high humidity in increasing electrical conduction of surfaces of dielectrics

390

is very apparent even with clean polished surfaces, but it is even more noticeable in the case of roughened surfaces, particularly those contaminated with layers of dirt, which often contain hygroscopic salts.

Although the effects of humidity on electrostatic phenomena can be attributed primarily to the dramatic changes produced in the resistivity of dielectrics, it can also affect charge-separation processes. The author has observed, for example, that as the humidity is increased the charges acquired by granulated sugar particles colliding with a metal electrode decrease in magnitude, go to zero, and increase again with the opposite polarity.

It is perhaps worth noting that, contrary to widespread belief, the relative humidity of the atmosphere out of doors is not very much less in the wintertime than it is in the summer. Even during the coldest weather when the temperature is far below freezing, the relative humidity of the air may be well above 50%. The great difference we observe between electrostatic behavior in the summer and winter is attributable not to the change in the relative humidity of the atmosphere out of doors, but to the change of the relative humidity that occurs as the result of our heating systems. For every 10°C we increase the temperature of the atmosphere by heating, the relative humidity drops by a factor of 2. Therefore, even if the outside humidity is 100% when the air temperature is $-20°C$, when we heat it up in our homes and buildings to $+20°C$, the relative humidity drops to less than 10%. At such low humidities most dielectrics become excellent insulators, and electrical charges accumulated by contact and friction between surfaces persist for long periods of time.

17.2 EFFECTS OF TEMPERATURE

Another atmospheric variable that sometimes has an important effect on electrostatic phenomena is temperature. In contrast to metals, which almost invariably show a small decrease in resistivity with decreasing temperatures, dielectrics behave in the opposite way, and their resistivity increases rapidly with decreasing temperature. As a result, the resistance of most dielectric materials can be expected to increase appreciably as the temperature is lowered, assuming that the relative humidity remains unchanged.

This large decrease in electrical conductivity with decreasing temperature is well illustrated by observations on the ability of internally charged plastic blocks to retain their charge.[26] If electrons of about 1-MV energy are directed at the surface of a methyl methacrylate sheet, they will penetrate to a distance of about 0.5 cm; if the radiation is continued, the electrical stresses produced by the accumulation of electrons within the plastic will eventually become sufficient to cause dielectric breakdown to take place.

If the blocks are irradiated with electrons for a period of time just short of that required to produce spontaneous breakdown, it is possible to take the block out of the apparatus and to initiate a spark within the plastic by striking the plastic with a sharp object, such as a nail or a prick punch. A beautiful dendritic pattern is produced by such a discharge. Ordinarily, in order to create such a pattern at room temperature, it is necessary to initiate the spark within a period of less than about 10 min following irradiation. After longer periods of time, so much of the charge leaks off that the discharge can no longer take place. When the same experiment is performed at a much lower temperature by refrigerating the block with dry ice before and after irradiation, the charge within the plastic persists for much longer periods of time, and the irradiated block retains its charge even after a period of several months. If the block is allowed to rise to room temperatures, the charge, of course, leaks away rapidly.

17.3 CONDUCTION THROUGH AIR

Although charge transfer by conduction through and over the surface of dielectric substances is usually the primary cause of charge leakage in electrostatics, it is not the only mechanism. Careful experiments carried out with high-quality insulators at very low relative humidity or experiments in which charged objects are supported by the atmosphere itself disclose that the air is an electrical conductor, although a rather poor one.[4] When a charged object, whether positively or negatively electrified, is supported in the free atmosphere, it loses charge exponentially at a rate corresponding to a half-life of 10 or 20 min.

All charge conduction processes occur as the result of the movement of charged particles. In the case of conduction by the atmosphere, it has been determined that the transport of charge is attributable to the motion of positively or negatively charged particles, called ions, which are free to move in the atmosphere under the influence of electrical fields. Under usual atmospheric conditions, the ions primarily responsible for the conductivity of the atmosphere are clusters comprised of a few molecules carrying either a single positive or negative elementary charge; these clusters move with an average drift velocity of about 10^2 m/sec at sea level conditions in an electric field of 10^2 V/m.

Measurements disclose that under normal atmospheric conditions over land they are being produced at a rate of about 10^7 ion pairs per second per cubic meter as the result of ionization processes caused by natural radioactivity and cosmic rays.[51] Approximately 30 eV is required for the production of each ion pair.[19] Near the ground most of the ion-pair production is the result of radioactivity. At increasing altitudes the rate of ion-pair production by cosmic rays becomes large in comparison to the rate of production by radioactivity, and at very high altitudes ionization is almost entirely due to cosmic rays and short-wave radiation from the sun.

A population of small ions is maintained at an equilibrium level by a balance between the rates of production of new ion pairs and the rates of recombination by collision between ions bearing opposite polarities and of attachment to large aerosol particles. Assuming that there are equal numbers of positively and negatively charged small ions, we can write the following equation describing ion population:

$$\dot{n} = q - \alpha n^2 - \eta n N$$

where n = number of small ions per unit volume

q = production rate, typically $2.2 \times 10^7/(\text{sec})(\text{m}^3)$

α = recombination coefficient, typically 1.6×10^{-12} m³/sec

η = coefficient for small ion combination with large particles, on the order of 10^{-12} m³/sec

N = number of large particles per unit volume

(To be precisely correct, the $\eta n N$ term depends on the polarity of charge on n and on N. This requires separate equations for positive and negative small ions and three terms in each equation for negative, neutral, and positive large particles.)

An interesting consequence of this equation in the special circumstance occurs when a very high concentration of ion pairs is suddenly produced by an ionizing event, such as an electric spark or a flame. In this case, the rate at which the ions combine with aerosol particles can be neglected. The steady-state small ion population density, called n_∞, is found by inspection to be $\sqrt{(q/\alpha)}$. For the values of q and α just given, $n_\infty = 3.7 \times 10^9$ ions/m³. The value of n as a function of time may be obtained by direct integration of the differential equation (Ref. 18, p. 24). We see that following the introduction of a very large number of small ion pairs, however high, the ion population will quickly return to the equilibrium value. Figure 17.1 shows how rapidly the small ion concentration decays to the steady-state value after suddenly being raised to 10^{15} ions/m³.

Although the small charged clusters of molecules, which we have been discussing, are primarily responsible for the electrical conductivity of the normal atmosphere, they are by no means the only charged particles that can move under the influence of electric fields and cause the transfer of charge. We have already mentioned that the small, fast ions in the atmosphere disappear by attachment to aerosol particles in the atmosphere. These particles, which range in size from only 10 nm or so up to particles of

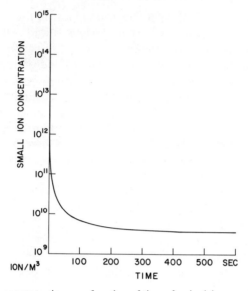

Figure 17.1. Ion concentration as a function of time after ionizing event.

many microns in diameter, occur in concentrations of the order of 10^8 particles/m^3 in clear, unpolluted air and in concentrations as high as 10^{11} particles/m^3 in the vicinity of urban areas.[20] When fast ions become attached to charged particles of this sort, they acquire the single negative or positive elementary charge carried by the fast ion and thereby become ions themselves. Because the aerodynamic drag on these particles is vastly larger than that of the small molecular clusters, these molecules have mobilities of the order of 10^{-7} m/(sec)(V)(m). This means that their contribution to the conduction of the atmosphere is usually small compared with that from the fast ions normally present.

Atmospheric ions are sometimes produced by processes other than radiation or radioactivity. For example, strong electrification can take place when there is blowing dust, sand, or snow. As the result of collisions of these particles with objects on the earth and with each other, highly electrified particles of a few microns and larger are produced, and these carry many elementary charges. It has been discovered that when water drops splash or when small bubbles burst in fresh or salt water they eject strongly charged droplets into the atmosphere.[2] These particles, being highly charged and free to move, contribute to the conductivity of the atmosphere. It has been shown that when volatile, liquid-charged particles such as these evaporate, the charge becomes concentrated on the residual surface.[10]

Under some conditions highly charged large particles in the size range of 10 to 50 μ can be produced, having mobilities of as much as 10^{-4} m/(sec)(V)(m), rivaling fast ions. Observations indicate that high-velocity aerosol jets of steam, oil, sand, and other substances produce strong electrification.[11] Measurements near erupting volcanoes reveal that such natural phenomena can result in the introduction of highly charged particles into the atmosphere.[1]

Man-made activities sometimes produce large concentrations of ions by other mechanisms as well. For example, Cottrell electrostatic precipitators[53] discharge large numbers of charged aerosol particles into the atmosphere. Combustion processes, high-tension lines, electrical apparatus, x-ray equipment, and artificial radioactivity can also result in ion production. For the most part, however, the concentration of ions resulting from these local sources decreases very rapidly with distance and time so that they play a very minor role. Except in the special cases where there may have been large-scale fallout from nuclear explosions or nuclear energy installations, ions from man-made sources are of little importance in affecting atmospheric conductivity.

In conclusion, we can say that man exercises an effect on atmospheric conductivity primarily by the large-scale introduction into the atmosphere of atmospheric aerosols. According to recent observations, there has been a steady decline in atmospheric conductivity in the air over the North Atlantic, amounting to at least 20% over the past half century, which can be attributed primarily to the results of pollution.[6] In very local cases the effects may be even larger, and downwind from large-scale industrial operations the author has made measurements indicating a reduction of fast ion population of as much as a factor of 10 as a result of aerosols released into the atmosphere.

Since most of our familiarity with the electrical properties of matter is derived from experiences with metals, semimetals, and electrolytes, it is worth pointing out the enormous difference between the concentration of charge carriers in liquid and solid conductors and in the atmosphere. In the conductive metals, such as copper and silver, the charge carriers are electrons present in concentrations in excess of $10^{28}/m^3$. In electrolytes, such as salt water, the concentration of positive and negative ions is a few orders of magnitude less, but still many times as many as the approximately 10^9 ion pairs/m^3 normally present in the atmosphere. In the case of metals and electrolytes, we are accustomed to specifying the very useful properties, which we call conductivity or resistivity. These properties are a measure of the ease with which electricity passes through these substances, and we can readily determine them by observing the current that flows per unit area of cross section when we apply a potential gradient.

In making such a measurement on a metal, except for the temperature rise that may be inadvertently produced, there is little cause to worry about any modification of its electrical properties produced by the measurement. The concentration of charge carriers is unchanged, for as many are supplied at one part of the conductor as are removed at another.

In the case of making electrical measurements on electrolytes, there is the danger of altering the substances in the course of the measurement. However, the population of the charge carriers is so extremely large that modest currents of microamperes or milliamperes necessary for the measurement have little significant effect on the total ion population.

In the case of similar measurements of the conductivity of the atmosphere, there are very real problems, for even a tiny current of 1 pA represents a current of 10^6 ions/sec, more than the total ion population of a liter of atmosphere.

If we wish to specify an electrical conductivity for the atmosphere, we can do so only if we use a measuring technique that will not substantially alter the population of the ions that confer on the air its electrical conducting properties. This is customarily done by the use of a Gerdien tube,[13] which consists of a cylindrical electrode mounted inside of a cylindrical tube through which the air is drawn by a fan or a blower at such a velocity that the ion population is not altered significantly by the electrical current caused by an impressed voltage. The electrical "conductivity" measured by such an apparatus under the usual conditions is of the order of $10^{22} \ \Omega^{-1}/m$.

Such conductivity, measured for either the positive or negative charge carriers, is a very useful quantity for determining the flow of current that may take place in the case of electrostatic apparatus when this apparatus is ventilated by the free atmosphere in such a way that it does not substantially alter its ion population. Under such conditions it is useful to think of the electrical relaxation time of the atmosphere, the time period in which, as a result of conduction, a charge will decay by a factor of e. This is of the order of 10 or 20 min in the lower atmosphere, in contrast to relaxation times for the metals and electrolytes of 10^{-6} μsec or less and for distilled water of about 1 μsec.

Great care should be exercised in using the concept of conductivity or relaxation time for the free atmosphere, for this value may be much too large in case there is not free movement of the air to furnish a continuous supply of ions. For example, although a charged object exposed out of doors may lose charge with a time constant of 10 or 20 min, an object in a closed chamber may have a far longer time constant. An example of this is the charged fused quartz fiber used in an electrometer mounted in a small air-filled chamber. This lost only 30% of its charge in three months.[47] The loss of charge, other than that caused by conduction of the fiber, was limited by the rate of ion-pair production in the fraction of a milliliter of air surrounding the fiber.

It follows from our discussion that if we are to maintain charged bodies or electric fields in the open atmosphere, we must continue to supply charge to compensate for the current that is flowing as the result of atmospheric conductive properties. Strictly speaking, there can be no "electrostatics" for more than a period of the order of 10 or 20 min because ions normally present in the atmosphere will act to neutralize any charges or fields.

17.4 DIELECTRIC BREAKDOWN OF THE ATMOSPHERE

The intensity of the electric field that can be maintained in the atmosphere is limited by the dielectric breakdown of air. If the electric field becomes too high, a chain reaction ionization process occurs, causing a small region of the atmosphere to become highly ionized and conductive by the formation of either a corona or spark discharge. Experiments show that for regions having the dimensions of millimeters or centimeters, the breakdown processes occur at sea-level pressures when the field exceeds about 3 MV/m. This breakdown value varies linearly with atmospheric pressure; thus in some cases high-voltage apparatus that functions properly at low altitudes may spark over and fail if it is taken up on a mountain or in an airplane.

One illustration of how significantly dielectric strength depends on the dimensions of the electric field is provided by observations indicating that long sparks can be initiated in a field of 4.5×10^5 V/m (due to Norinder and Salka[31]) and that the initial breakdown process of lightning, which determines the branched character of the spark, is capable of propagating long distances in electric fields at low as 6.5×10^5 V/m (due to Phelps[33]). In the case of very small regions, the electric field can reach values well in excess of the 5 MV/m before breakdown occurs. For example, on the surface of evaporating $50\text{-}\mu$ water droplets, fields can rise as high as tens of megavolts per meter without any ionization processes taking place.[10]

Another illustration is afforded by a sharp metal point of $1\text{-}\mu$ radius, which will not produce point discharge until it is raised to a potential above its surroundings of at least several kilovolts. Simple calculation shows that in this case the dielectric breakdown process does not begin until the electric field at the point is at least two or three orders of magnitude greater than the commonly considered breakdown value of 3 MV/m.

It is helpful in thinking of the electrical properties of the atmosphere to remember that to create fast ion pairs in a gas requires very high specific energies of many electron volts. In the atmosphere such energies are not available at ordinary temperatures from thermal motions, so they must be supplied either by energetic radiation, by the energies available from high temperatures in combustion or volcanism, by the high energies available to charged particles falling through potential differences, or by the energies available from the kinetic energy of large particles. In solids and liquids, the energies required to separate charges are often considerably less and are readily available even at room temperatures.

17.5 FINE–WEATHER ATMOSPHERIC ELECTRICITY

The atmosphere influences electrostatic phenomena not only because of the effects of its electrical conductivity, humidity, and temperature, but also, in a very important way because of the electrical imbalance that is invariably present, particularly in its lower levels. Even in fine weather, when there are no clouds for hundreds of kilometers, there is a weak electric field of the order of 10^2 V/m at the surface of the earth, indicating positive charge in the atmosphere and negative charge on the earth. Integrated over the entire globe, the total fine-weather charge carried on the surface of the earth is of the order of -10^6 C. Soundings made of the potential gradient from balloons and aircraft show that the fine-weather atmospheric electric field decreases with altitude until it essentially disappears, approaching the highly conducting portion of the atmosphere known as the ionosphere. This decrease and disappearance of the electric field of the earth observed with increasing altitude indicates that the charge in the atmosphere is equal and opposite to that carried by the earth.

Since the highly ionized upper part of the atmosphere, known as the ionosphere, is a good conductor of electricity, it is believed by most students of atmospheric electricity to be at very nearly the same electrical potential throughout. To the extent that this is true, we are justified in thinking of the earth and the ionosphere as comprising a gigantic concentric capacitor in which the earth is one electrode and the ionosphere the other, and the lower atmosphere serves as the dielectric separating them. If we sum the potential gradients measured in typical soundings through the lower atmosphere, we find that the total potential difference between the earth and the upper atmosphere is of the order of 300 kV.[5]

17.6 FINE–WEATHER CONDUCTION CURRENT

Because the atmosphere conducts electricity, an electric current flows through the atmosphere as the result of the fine-weather electric field. Measurements show that the fine-weather conduction current flowing to a square meter of the earth's surface is of the order of picoamperes. When summed over the entire surface of the earth, the conduction current is of the order of 10^3 A.[57]

Because the atmosphere can be neither a sink nor a source of charge over any extended period of time, the amount of charge flowing into a volume of the atmosphere must be equal to the current flowing out of this volume of the atmosphere. As a result, it follows that the current density in the atmosphere is almost constant with altitude and that the potential gradient at any level is inversely proportional to the electrical conductivity at that level.[23] Simultaneous measurements of atmospheric conductivity and potential gradient reveal that this is very nearly the case. When there are layers in the atmosphere that have high concentrations of aerosols, the resultant decrease in the electrical conductivity causes the potential gradient in these regions to be abnormally high. In the course of making measurements of the atmospheric potential gradient from an airplane, the author has observed that in smoky layers produced by the burning of the Everglades and in similar smoky layers resulting from industrial operations at Gary, Indiana, the fine-weather gradient can rise to values of about 10^3 V/m, an order of magnitude greater than is usual.

Similarly, if the electrical conductivity of the atmosphere is increased by ionizing radiation, the potential gradient may be abnormally low. Raymond Falconer of the Atmospheric Sciences Research Center at the State University of New York at Albany has reported to the author that the increased ionization resulting from radioactive fallout in the Schenectady, New York, area following one early nuclear test in Nevada was sufficient to reduce the fair-weather field at ground level to less than 10% of its normal value. Subsequently, it has been reported that there has been a systematic reduction in the fine-weather electric field following a period of nuclear testing.[35]

In the first meter or so above the ground, the ionization resulting from the natural and artificial radioactivity in the earth as well as exhalations of radioactive gas sometimes cause an abnormally high conductivity locally. During quiet periods at night it is not unusual to find very low electric fields very near to the surface of the ground as the result of this increased conductivity.[29]

17.7   ELECTRODE EFFECT

Over the ocean and regions of the solid earth surface that do not contain appreciable radioactivity, the so-called electrode effect is often observable. Since neither the ocean nor the earth itself is a source of negative ions, the electrical conduction in this region occurs only to a very small extent through the motion of negative charge upward. Here the conduction current is maintained almost entirely by the movement of positive ions downward. As a result, in this very lowest region of the atmosphere in the absence of atmospheric mixing, the conductivity is only half its usual value, and we commonly observe electric fields about twice as large as those higher in the atmosphere where the current is being carried both by negative ions moving upward and by positive ions moving downward.

17.8   EFFECT OF METEOROLOGICAL CONDITIONS

Since the distribution of aerosols (which can decrease atmospheric conductivity) and the distribution of radioactive substances (which can increase conductivity), depend very much on the circulation of the air, the potential gradient in the atmosphere is highly dependent on meteorological conditions. During a clear day when the sun in shining on the earth, free convection takes place which mixes the lower atmosphere so that it is electrically quite isotropic. Under these conditions, because the electrical conductivity is very nearly uniform in the lower mixing layers, the electrical potential gradient is often very nearly constant. On the other hand, at night when the atmosphere becomes very stable and layers of air pollution or radioactivity can form, the electrical conductivity, and hence the potential gradient, may change markedly from one layer to another.

17.9   NATURAL SPACE CHARGE

All the electrical fields in the lower atmosphere are the result of surface charges residing on the earth's surface and space charges present in the earth's atmosphere. During fine weather in the absence of abnormal electrification, the density of fair-weather space charge seldom exceeds 10^9 elementary charges per cubic meter, and usually it is less than 10^6 elementary charges per cubic meter. It is worth noting that at sea level a cubic meter of air contains of the order of 10^{21} molecules; thus this space charge represents a very slight electrical imbalance. The space charges present during fine weather are, for the most part, the result of the variations in the atmospheric conductivity, which, under the influence of the conduction current, give rise to variations in the potential gradient.

According to Poisson's equation, the space-charge density is related to the change in the vertical potential gradient. We can see from this that, particularly when there is no convection and the atmosphere is stratified, there will be concentrated layers of space charge formed at the interface between layers of contrasting conductivity. For example, when there is a layer of haze in an otherwise fairly clean atmosphere, there will be an accumulation of positive space charge on the upper part of the haze layer and a similar layer of negatively charged ions on the lower part of the layer. When convection makes the air electrically much more uniform, the space-charge density drops to much lower values.

In addition to the regions of the atmosphere carrying net electric charge that result from the fine-weather conduction process, there are also occasions when other phenomena can lead to the development of space charge. For example, when strong winds blow snow or dust, contact electrification takes place that can sometimes result in the formation of volumes of air carrying space charges in excess of 10^9 elementary charges per cubic meter. Usually dust storms result in the introduction into the atmosphere of negative space charge. However, it was recently shown by Kamra[21] that on certain occasions blowing dust can carry positive charge. During dust storms the space charge is sometimes sufficient to cause potential gradients near the earth's surface that are comparable to those that result from thunderstorms.

Another important source of space charge has been elucidated by the studies carried out by Blanchard,[2] in which he demonstrated that when bubbles burst at the surface of bodies of water, they usually eject strongly charged water droplets into the atmosphere. Blanchard has observed that bubbles breaking from bodies of fresh water release negative charges into the atmosphere, whereas similar processes taking place over the oceans release positive space charge. It has been known since the time of Lenard[25] that waterfalls are a strong source of electrification, and the charge from waterfalls (presumably formed as the result of bursting bubbles and splashing) can create electrical anomalies as far as several kilometers downwind from the waterfall. This phenomenon has been discussed in detail by Pierce and Whitson.[36]

17.10   SPACE CHARGE FROM HUMAN ACTIVITY

Human activity can have a pronounced effect on atmospheric electricity by introducing space charge into the atmosphere. We have already discussed the changes in conductivity brought about by air pollution either with aerosols, which decrease conductivity, or with radioactive substances, which increase conductivity. When the conductivity and potential gradient are modified as the result of these effects, there are, of course, corresponding regions of space charge formed. Therefore, during periods of stable stratified atmospheric conditions, human activity may result in the production of considerably higher levels of space charge than might be expected to occur naturally.

Intense concentrations of space charge can be produced directly by human activity—for example, when dust clouds are produced by plowing or by the passage of vehicles over dry soil or sand surfaces. The author once observed strong fair-weather electrical disturbances during a period when large military vehicles were being tested in an area several kilometers upwind from an electrical measuring station. When powders such as insecticides are handled, as in crop-dusting operations, it is to be expected that strong electrical charging will also be experienced.

It has been recognized since the time of Faraday that wet steam escaping from an orifice at high velocity is an intense source of space charge, and we know that steam plumes escaping from steam locomotives, pile drivers, or steam whistles can introduce quantities of space charge sufficient to cause very intense electrical fields. To a lesser degree it is often observed that the exhaust from jet and reciprocating internal combustion engines may also be somewhat electrified and that the space charge resulting from these sources may create detectable perturbations. We readily anticipate that almost any of the many human operations resulting in the dispersion of particles into the atmosphere will, in general, be a source of space charge, for such particles commonly can be strongly electrified during their production. One of the more intense sources of man-made space charge is the Cottrell precipitator. In this apparatus, in which aerosol particles are charged and then precipitated in an electric field, the very small particles, less than 1μ in size, have such low mobilities that they escape into the atmosphere. As the result, a plume of space charge can be readily detected for distances of several kilometers downwind from such installations.

The electric field on high-voltage power wires sometimes reaches such high values that dielectric breakdown of the air occurs. Under these conditions, some of the resultant ions are released into the atmosphere. In the case of alternating-current powerlines, we might expect that because the relative quantities of plus and minus charges are equal, no large quantities of space charge would be introduced. However, the difference in electrical processes between the release of plus and minus charges is, in some cases, sufficiently large that perceptible space charge may be generated and drift downwind. Although no measurements are presently available on the recently constructed high-voltage direct-current powerlines that are being put into use, it is likely that these might be even stronger generators of space charge.

Although the human sources of space charge described previously have been almost entirely the unintentional by-product of other activities, some experiments have been carried out in which space charge has been deliberately introduced into the atmosphere. The equipment investigated for this purpose has been of two kinds. In the first,[46,48] the apparatus consisted merely of a fine horizontal wire, usually of stainless steel, 1 mm or less in diameter and kilometers or more in length, suspended on high voltage insulators 10 m or so above the ground. When such a wire is raised to a dc potential of the order of 50 kV, the resultant strong electric field at its surface causes dielectric breakdown, and there is a copious production of fast ions, which are carried away by the wind. If the air is clean and contains only a low concentration of aerosol particles, the ions produced from the wire rather rapidly move to the earth under the influence of the electric field that they produce. This charge does not remain in the atmosphere for very long, traveling only a short distance before it disappears.

However, in continential atmospheres, which have nuclei concentrations of 10^{10} particles per cubic meter or more, the fast ions released from the wire rapidly become attached to the aerosol particles. This causes their mobility to be greatly reduced. These resultant charged aerosol particles, most of them probably carrying only one elementary charge, move very slowly under the influence of electric fields and can be taken downwind for considerable distances. The natural electric field downwind of such a wire is considerably modified by the emission of several millicoulombs per second of space charge from the wire.

Space-charge measurements taken from an airplane have indicated that the level of space-charge density in the atmosphere can be considerably modified for distances of up to 10 km downwind from a wire space-charge generator.

The density of space charge that can be released from the apparatus just described is limited by the concentration of aerosol particles present in the atmosphere to which the fast ions can become attached. Other apparatus has been devised in which no reliance is placed on the aerosol particles already present in the atmosphere.[50] In this apparatus an aerosol of submicron-size particles is produced in high concentrations, and the fast ions become attached to the particles.

Such an apparatus consists of a gasoline engine, which is made to serve as an aerosol generator by injecting high-boiling-point oil into the hot exhaust system. The submicron aerosol particles are produced when the vaporized oil is caused to condense to an aerosol. This concentrated aerosol is forced by an airplane propeller attached to the engine through a duct and through a fine wire and cylinder corona-charging apparatus similar to that used in a Cottrell precipitator. The concentrated space charge produced by this apparatus produces very intense local electric fields. Grounded objects near the stream of aerosol coming out of this apparatus often emit St. Elmo's fire, visible at night, from points of small radius of curvature. Because vegetation and objects near the space-charge generator go into corona, producing ions that neutralize the aerosol, the charge released from such an apparatus is rapidly dissipated as it moves over land. It has been found that, if the aerosol is released over water surfaces, it persists for much longer periods of time because there are no sharp points to create corona and thus to neutralize the space charge.

Experiments have been carried out with a modified form of the apparatus in which an inductively charged spray of water droplets carrying a charge opposite to that of the aerosol is released. With this technique it is possible to reduce the intense local electric fields near the apparatus. As the stream of electrified particles is carried downwind, the water drops fall out leading to the development of strong electric fields. The intensity of the electric fields so created is sufficient to levitate some of the charged water drops that would otherwise have fallen to the ground. In this manner it is possible to suspend electrified droplets in the atmosphere under the influence of electrical forces.

17.11   ELECTROSTATIC EFFECTS OF FINE–WEATHER ELECTRICITY

Under usual conditions, the electric fields and the electric currents produced by fine-weather electricity are so small that we are not aware of their presence. For example, the conduction current that flows through a person outdoors is of the order of 1 pA, which is many orders of magnitude below the threshold of detectability by the body. Similarly, the electric field produced by fair weather is so small that we are not conscious of any electrostatic effects under normal conditions.

There are circumstances, however, in which consequences of fine-weather electricity make themselves readily evident. A phenomenon of this sort occurs if a long, electrical conductor is placed in the atmosphere so that it spans an appreciable potential difference. For example, if a balloon or a kite is flown on a wire, the wire will span a potential that will be approximately 10^2 V for each meter above the earth's surface. When a kite or a balloon becomes elevated to a height of 50 m, if the bottom of the wire is grounded, the upper part will be at a potential of 5 kV or more with respect to its environment. If sharp points are present, the electric field will be sufficiently intensified to cause point discharge, and a steady current of several microamperes will flow through the wire. As the height of the wire is increased, the potential difference will, of course, increase, and the flow of current will correspondingly increase too. If the wire is not securely connected to the ground, voltages as high as 10^2 or more kV can appear and give rise to electric shocks or other electrostatic phenomena at the lower or upper ends of the wire. Phenomena similar to that occurring with the wire may also take place in the case of tall antenna towers if they are mounted on insulators and not electrically grounded.

Electrostatic phenomena can take place without the existence of point discharge. For example, it has been observed that if a long, ungrounded horizontal wire, such as a telegraph, telephone, or powerline, is suspended on poles on good insulators, under certain conditions it can provide a very strong electrical shock as a result of fine-weather electricity. This can happen in the following way.

Let us assume that the wire is being installed and that it is being supported on the insulators. During the installation process, the wire is almost inevitably connected to ground through the handling by workers or by actually touching the ground. While the wire is connected to ground, it is maintained at ground potential because an induced negative charge flows to the surface of the wire sufficient to cause it to remain at ground potential (even though it may be supported in a region of the atmosphere that would be at several kilovolts with respect to the earth). While the wire is connected to the earth, negative charge can flow from the earth to the wire and thus maintain it at ground potential.

However, if after the wire is installed it becomes disconnected with respect to the earth, the flow of negative charge from the earth to the wire can no longer take place, and in a matter of less than an hour, the induced charge that was on the surface of the wire will have leaked off into the atmosphere. When this happens, the charge on the wire approaches zero, and the electrical potential of the wire approaches that of the space it occupies. If a person connected to ground has the misfortune to touch such a wire, a spark will jump, and a very large surge of current will flow through him to the wire. It is, therefore, desirable as a practical matter that any elevated conductor be well grounded so that it cannot assume high potentials that would make it dangerous.

The fair-weather electric field can give rise to hazards connected with the operation of a helicopter. This can be illustrated by considering the following example in which a helicopter (assumed to carry little or no charge) hovers at a height of 30 m and drops a conducting cable or ladder to the surface. When such a cable comes in electrical contact with the surface or with a person at the surface, a sudden flow of negative charge will take place up the conductor to the helicopter. This flow will transport enough negative charge to the surface of the helicopter to cause its potential to come to that of the earth. When such a surge of current takes place, it can cause dangerous electrical shocks that may startle or injure personnel, and such electrical sparks have even been known to cause the ignition of fuels.

Situations can readily be visualized in which sparks and electrical surges might be produced in the intensified fair-weather electric field at the upper part of a tall building when an ungrounded worker touches a grounded load being elevated by a crane or when a worker who is grounded to the structure touches an ungrounded object being carried by the crane. We would anticipate that the electric shock would not be strong enough to harm the person directly, but it might still be of sufficient magnitude to startle him and to cause an accident. It is desirable therefore, when dealing with elevated objects or when working in the region above elevated objects, that all conductors and personnel be grounded so that they will necessarily all be at the same potential and dangerous shocks cannot occur.

17.12   THE CAUSES OF FINE–WEATHER ELECTRICITY

The conduction current of approximately 1 kA carrying positive charge from the atmosphere to the earth would be capable of neutralizing the 10^6 C of negative charge residing on the surface of the earth in less than an hour. Obviously, then, since the earth and the atmosphere retain their charge, there must be processes taking place that maintain the fine-weather electrical process by bringing negative charge to the earth at the rate of about 1 kA. Although processes, such as bubble breaking at the sea surface and some dust and snow storms, undoubtedly contribute to the negative charge of the earth, most scientists agree with C. T. R. Wilson[55] that the current flowing from thunderstorms is primarily responsible for the maintenance of the negative charge on the earth.

Observations of the current flow above the thunderstorm cloud[14] indicate that the average thunderstorm is transporting negative charge to the surface of the earth at the rate of about 1 A; thus the approximately 10^3 thunderstorms continually in progress over the earth's surface appear to be capable of maintaining the negative charge on the earth.

Additional confirmation to this idea is afforded by the observation that there is a diurnal fluctuation in the charge carried by the earth that appears to be related to the amount of thunderstorm activity occurring over the globe.[52] The charge of the earth, as indicated by the fine-weather field at its surface, appears to reach a maximum during the period of maximum global thunderstorm activity when the sun is shining over the landmasses. The charge decreases during the period when there are fewer storms, when the sun is shining over the regions of the earth covered by water.

17.13   THUNDERSTORM ELECTRICITY

All atmospheric clouds produce electrical perturbations because their electrical properties are different from those of clear air. Usually these perturbations are so small that it is impossible to recognize the external electric fields of clouds against the normal variations of electric field in the surrounding cloudless atmosphere. The notable exception to this is the thundercloud, which can produce electric fields of sufficient size and intensity to result in electric sparks many kilometers in length, which we know as lightning. Almost without exception, thunderclouds are characterized by convective instability with strong updrafts and downdrafts.

Thunderstorms commonly begin under atmospheric conditions that produce many small, fair-weather cumulus clouds showing little or no electrical activity. Suddenly, for reasons that are poorly understood, one of these fair-weather clouds suddenly grows dramatically in size and begins to exhibit strong electric fields and precipitation. And this occurs even though none of the other clouds has exhibited either electrical activity or precipitation. The normal fair-weather field suddenly reverses and increases nearly exponentially with a doubling time of the order of 5 min. In a matter of less than 10^3 sec an innocuous cumulus cloud can suddenly change into a thundercloud producing both heavy precipitation and lightning discharges.

Observations of storms in a variety of meteorological environments indicate that the minimum cloud thickness necessary for producing lightning is of the order of 3 or 4 km. Thunderstorms occur in a variety of sizes, as is shown schematically in Fig. 17.2. The small semitropical thunderstorms observed over the ocean in the Bahamas have cloud bases at about 500 m and cloud tops at about 4 k. Small thunderstorms of approximately the same dimensions are also observed in the American southwest; although they are approximately the same size, these storms have the cloud base at 5 km and the cloud top at 8 or 9 km. Figure 17.2c represents the usual thunderstorm, encountered over continental areas, which characteristically has its base at 2 or 3 km and extends to a height of approximately 10 km. The largest and most active of thunderstorms is illustrated by Fig. 17.2d. These are the exceptionally violent and large thunderstorms, which occur with the greatest frequency in the spring months and which can rise to altitudes as high as 20 km or more and penetrate the tropopause into the stratosphere.

The intensity of the electrical activity, as indicated by the number and repetition rate of the lightning discharges, appears to be closely related to the intensity of the convective activity. Small thunderstorms, whose updrafts and downdrafts seldom exceed 5 m/sec, may produce lightning

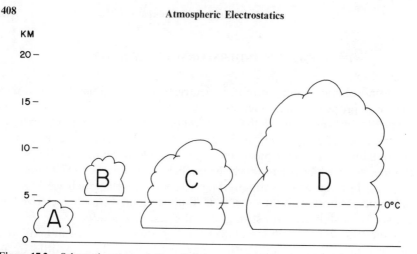

Figure 17.2. Schematic representation of various types of electrical storm clouds: (*a*) warm tropical maritime, (*b*) southwestern United States mountain region, (*c*) ordinary midlatitude continental, (*d*) giant midlatitude continental.

discharges at intervals of several minutes, and some very weak storms may produce only a single lightning flash. The ordinary thunderstorm produces lightning flashes at the rate of 10 to 30/min, and during the life of the storm it is capable of producing hundreds or thousands of flashes. The extremely intense, very large storms, which may have updrafts and downdrafts of as high as 70 m/sec,[41] have been observed to produce flashes at rates as high as 10^3/min, or hundreds of thousands of flashes over the life of the storm.

Estimates based on current flowing in the lightning flashes, which may be of the order of tens of kiloamperes for times of 50 μsec, and field changes and magnetometer measurements show that the average lightning flash transports of the order of 10 or 20 C of charge.[43] There is great variability. Some flashes may involve as little as a fraction of a coulomb, and others are estimated to carry hundreds of coulombs. The momentary flow of charge in the lightning flash is exceedingly large, but the average rate of charge transport by lightning is surprisingly small. For example, flashes carrying 10 or 20 C every 10 or 20 sec, when averaged, amount to a current only of the order of 1 A.

Measurements made outside of a cloud show that the electric field around it is similar to that of a dipole, with positive charge in the upper part of the cloud and negative charge in the lower part of the cloud. In the clear air above the cloud, the electric field is of the order of 10^4 V/m. Under the influence of this field a conduction current of the order of 1 A flows in the clear air above a typical thunderstorm, carrying negative charge from the upper atmosphere to the top of the cloud. As a result of the electric field produced by the negative charge in the lower part of the cloud, point discharge currents, of the order of microamperes, flow from trees and other objects under the storm. The net current flowing in this fashion is also of the order of 1 A and carries positive charge from the earth toward the lower part of the cloud.

Because of the positive ions being released by the point discharge process, a layer of positive space charge is released beneath the cloud that limits the electric field at the surface of the earth to values that are usually of the order of 2 kV/m and seldom exceed 10 kV/m. Approximately 10% of the lightning discharges go from cloud to ground. It is estimated that these discharges carry an average current of about 10^2 mA of negative charge to the earth, whereas charged precipitation falling to the earth carries a similar current of the opposite sign.

17.14   ELECTRIC "CONDUCTIVITY" INSIDE CLOUDS

Most students of atmospheric electricity agree that the cloud is a much poorer conductor of electricity than the cloudless air surrounding the cloud at the same level. The movement of electric charge through the atmosphere under the influence of an electric field, short of breakdown, occurs primarily by the motion of small, fast ions comprising singly charged molecular clusters. It can be shown that, particularly if there is a strong electric field, fast ions inside the cloud rapidly become attached to cloud particles, whether they be small water droplets or ice crystals. As a result, the fast ion population is a function of the electric field and, in fact, of the past history of the electric field within the cloud. Because the ion population is a function of the electric field, it is clear that the conduction of current through a cloud is not a linear function of the electric field.

Since the term "conductivity" implies a simple and reproducible relationship between current and electric field, the concept of the electrical conductivity of a cloud is of dubious value. No one has yet been able to define or devise methods for measuring the electrical conductivity in the cloud, but it does seem obvious that the cloud is a much poorer conductor than the surrounding clear air.

17.15   ELECTRIC FIELD INSIDE THUNDERCLOUDS

Because most workers in atmospheric electricity agree that the cloud is a poorer conductor of electricity than the clear air surrounding it, they also are generally of the opinion that the electric fields inside a thundercloud are probably somewhat larger than they are outside. Gunn[16] and other workers have reported observations in which the electric field measured by an airplane suddenly increased by as much as tenfold when the airplane flew from the clear air into the cloud. Although the workers agree that the electric field inside the cloud is probably greater than that outside the cloud, there is considerable difference of opinion about its actual value. Because of the point discharge and precipitation charging, it is difficult to make accurate measurements of the electric field inside a cloud. Some investigators suggest that the electric field may rise to values as high as 1 MV/m before a lightning discharge is produced;[38] others believe that the field may be only 10% of this value.[31] Since many important electrical effects are proportional to the square of the electric field, it is a matter of some importance to secure better information than is presently available on the electric fields inside clouds.

17.16   CHARGED PARTICLES INSIDE THUNDERCLOUDS

We know because of their lightning and electrical fields that thunderclouds must contain charged regions of tens or more coulombs. Because charge in the form of fast ions rapidly becomes attached to larger particles, there seems to be little question that most of the charges within the thundercloud are residing on water or ice particles ranging in size from cloud drops of 10 or 50 μ in diameter up through snow, rain, and hail particles. Because this charge must reside on the surfaces of these particles, there is reason for believing that the greater proportion of the charge may be residing on the smaller particles. These will have a total surface area that is at least several orders of magnitude greater than that presented by the larger snow, rain, or hail particles. However, we presently have no information on this point. Because of the difficulties of making reliable measurements in clouds, we do not yet know on which particles the principal part of the charge is carried. We are probably safe in assuming that all classes of particles carry some charge, although we do not know the relative amounts carried by each.

Whatever may be the nature of the charged particles within the cloud, many students of thunderstorm electricity are now agreed that because the cloud is a much poorer conductor of electricity than the clear air around it, there is very probably a dense layer of space charge in the outer part of the cloud. This, of course, would account for the increase in electric field that has been observed on flying from clear air into a cloud. Recent theoretical studies have examined the means by which fast ions attracted to a cloud become attached to the cloud particles as they move into the cloud.[3] These studies indicate that the screening layers on the surface of the cloud may be relatively thin (only a few tens of meters thick) and that the space-charge density here may be as high as 10^{12} elementary charges per cubic meter.

17.17   NEUTRALIZATION OF CHARGE BY LIGHTNING

When lightning flashes occur, measurements of the electric field external to the cloud show that there is a disappearance of an electric dipole of the order of 10^2 C-km;[34] thus from a gross point of view there has been a neutralization of charge. Until fairly recently it has been widely assumed by various students of the thunderstorm that the lightning flash itself somehow actually neutralized the individual particles in the cloud carrying the charge, whether they were precipitation-size particles, smaller cloud droplets, or ice crystals.

In recent years, however, it has become less certain whether the lightning flash has sufficient fine structure to extend to all the small particles carrying the charge.[28,49] It has been suggested that, instead of neutralizing the particles, the lightning merely brings the charges close together by injecting a rather coarse dendritic pattern of positive charge into negatively charged regions of the cloud or a dendritic pattern of negative charge into positively charged regions of the cloud. If this were the case, an observer outside the cloud would see the collapse of the large external electric field; but an observer inside the cloud might see that the charged particles that originally gave rise to the discharge were still, for the most part, unaffected, even though a region of high-density space charge had been left in their midst by the lightning discharge.

17.18   THUNDERSTORM ELECTRIFICATION MECHANISMS

Because we presently have no good experimental data on the population of charged particles and the nature of their movements within the thundercloud, there is no general agreement on the relative importance of the various processes leading to cloud electrification.

In order to give a satisfactory description of thunderstorm electrification, it is necessary to explain how the charged particles are produced and how large numbers of particles having positive charge are accumulated in one part of the cloud while large numbers having negative charge are carried so that they accumulate in a different region of the cloud.

Until quite recently it had been widely assumed that the necessary charged particles were formed as the result of a charge-separation process taking place inside the cloud, in which relatively large precipitation particles, such as snow, rain, or hail, acquired one polarity and slower, smaller particles, which fell much more slowly, acquired an equal and opposite polarity. Laboratory experiments and theoretical studies suggested that such charge-separation processes might occur as the result of particle collisions, selective ion attachment, drop breakup, or drop freezing. Following the generation of particles in this way, it was proposed that the movement of charges to form charged regions within the cloud would occur as the result of the gravitational separation when the larger, faster falling particles descended to the lower part of the cloud, leaving the smaller and oppositely charged particles behind.

In recent years there has been a growing appreciation that, in addition to charge-separation processes occurring within the cloud, charged particles might also be formed as the result of ionic conduction processes in the clear air surrounding the cloud and as the result of point discharge from the earth beneath. Lately it has also been recognized that, in addition to the movement of charged precipitation as it falls through the cloud, there would also be extensive motions of charged particles as the result of the large updrafts and downdrafts known to exist within convective clouds.

There seems to be little reason to doubt that most of the mechanisms suggested to explain thunderstorm electrification are capable of playing a part under the proper conditions. The primary problem now appears to be to determine the relative importance of the various mechanisms. The "answer," of course, may be subject to wide variation from one thunderstorm to another.

17.19   ROLE OF ELECTRICITY IN METEOROLOGICAL PROCESSES

It is probable that fine-weather electrification plays a minor or entirely negligible role in the behavior of the atmosphere because it is so weak. On the other hand, more and more scientists are beginning to believe that the electrification of the thunderstorm, which is many orders of magnitude greater, may play a significant role in physical processes taking place inside the thunderstorm.

One of the first suggestions of the possible importance of electricity in cloud behavior was offered by Lord Rayleigh.[37] In this short paper he described experiments indicating that the probability that two colliding drops would coalesce was greatly increased in the presence of an electric field, even one as weak as 10^3 V/m. On the basis of his findings, Lord Rayleigh suggested that

> ... I cannot close without indicating the probable application to meteorology of the facts already mentioned. It is obvious that the formation of rain must depend very materially upon the consequences of encounters between cloud particles. If encounters do not lead to contacts, or if contacts result in rebounds, the particles remain of the same size as before; but, if the issue be coalescence, the bigger drops must rapidly increase in size and be precipitated as rain. Now, from what has appeared above we have every reason to suppose that the results of an encounter will be different according to the electrical condition of the particles, and we may thus anticipate an explanation of the remarkable but hitherto mysterious connexion between rain and electrical manifestations.

Recent field studies,[27] laboratory experiments,[24] and theoretical studies,[9] have indicated that electrical forces may cause a significant increase in collision and coalescence rates within clouds. There are good reasons to believe that, in the case of droplets in the 50-μ size range, electrical forces are acting that may be as much as two or three orders of magnitude greater than those of gravity. According to reported observations,[30] in less than 0.5 min following a lightning discharge there is sometimes an increase in radar reflectivity of three orders of magnitude or more, indicating that the lightning process may be causing a greatly accelerated growth of precipitation particles.

The results of certain laboratory studies[22] have shown not only that electrical forces can have an appreciable influence on the rate of fall of charged precipitation, but that the stability and shape of neutral drops may be greatly modified by strong electric fields. Other laboratory studies have proved that electric fields of the sort to be expected in thunderstorms can induce dipoles in ice crystals suspended in the atmosphere, thereby producing optical effects as the result of changes in the orientation of ice crystal platelets.[45]

Recent years have brought increasing evidence that electric fields can exert a pronounced effect on the rate of growth of ice crystals in super-cooled clouds and on the crystal habit of growing ice crystals; indeed, it appears that under certain conditions electric fields are responsible for ice nucleation phenomena similar to those produced by cloud-seeding techniques. Evidence is rapidly accumulating that electrostatic effects, far from being an incidental by-product of processes taking place in the thunderstorm, may be vitally important in determining the behavior of the cloud.

It has been suggested by C. T. R. Wilson[54] and others that small atmospheric ions, which can be produced as the result of cosmic rays, radioactivity, or dielectric breakdown, might serve as the centers for condensation of water droplets. Although this phenomenon can be readily observed in the laboratory using very clean air, it appears doubtful whether it occurs often in the atmosphere. Investigations disclose that the high relative humidity (ca. 300 or 400%) necessary for condensation on ions is rarely achieved because there is usually a sufficient concentration of larger Aitken nuclei present on which condensation occurs.

The author has reported an observation made in the winter over a hot water pool at Yellowstone Park at approximately 90°C. A horizontal wire placed about 20 cm above the hot water produced a visible water droplet cloud when the wire was put into corona discharge with a high-voltage dc power supply.[39] In this case, condensation apparently was taking place on the ions under the very great supersaturations that occurred when the cold air blew over the heated pool. Such situations are rare in the natural atmosphere, however, and it appears that here small ions seldom serve as nuclei for the condensation of water.

17.20   ACTION OF ELECTRICAL FORCES ON CHARGED REGIONS OF THE ATMOSPHERE

It is easily demonstrated in the laboratory that a small metallic point maintained at a high voltage will go into point discharge and create a surprisingly vigorous draft of air known as an electric wind. Peltier[32] and other scientists have suggested that similar phenomena may take place on a large scale in the atmosphere. We know that the electrical forces in the atmosphere caused by the space charges and electric fields of fine weather would be trivial indeed, corresponding to those that would be expected if the temperature of the air were changed by of the order of 10^{-6} °C. In the strong electric fields in the thunderstorm, it can be estimated that the accelerations that would be experienced by air might be equivalent to temperature differences of as much as 5°C. Conceivably, therefore, in certain circumstances winds resulting from the acceleration of electrified air in an electric field might be important. In general, however, it appears that the space charges and the fields are considerably smaller and that the electrical forces acting on a volume of air are small compared with those resulting from temperature differences which, in a thunderstorm, may be as high as 50°C. For the most part, therefore, it can be concluded that the movements of air in a thunderstorm are the result of atmospheric temperature differences and that electric forces do not play a large role in the motion of air.[7]

Although electrical forces probably have only a minor role in air circulation, electrical energy in severe thunderstorms may be sufficiently large to play a part as the result of electrical heating. Vonnegut[44] has estimated that, in the severe thunderstorms that produce tornadoes, lightning flashes at the rate of 10 or 20 per second may be capable of supplying power of the order of 10^8 kW and has suggested, if a column of air were heated by electrical discharges it might be capable of producing an updraft sufficient to cause a tornado.

17.21   PRODUCTION OF ACOUSTIC WAVE BY ELECTROSTATIC FORCES

C. T. R. Wilson[55] and Colgate and McKee[8] have theorized that a portion of the noise known as thunder may originate not only because of the rapid expansion of air in the lightning channel, but also as the result of the action of electrical forces in the atmosphere. According to this suggestion, when there is a sudden change in the magnitude or direction of the electric field in the cloud due to lightning, there will be a corresponding sudden change in the forces acting on a charged volume of air that could result in the formation of a sound wave. According to recent observations of thunder by Holmes et al.,[17] this effect under some circumstances may be a factor contributing to the acoustic wave.

17.22   MODIFICATION OF THUNDERSTORM ELECTRIFICATION

Scientists have long considered the possibility of modifying the electrical activity taking place in thunderclouds. More than a century ago attempts were made to reduce hail formation by modifying the electrical properties of storms by erecting pointed electrodes. These experiments, based on the idea of reducing the electrification of the storm by point discharge, were inconclusive.

Variations of this idea were used in the early 1900s to try to induce rain by electrification introduced into clouds by the use of balloons flown on wires. Lately it has been proposed that the electrical energies that may cause tornado formation in thunderstorms might be dissipated by firing long wires carried by rockets fired from the ground or from aircraft. It has also been proposed to reduce lightning by dispersing into the thundercloud small pieces of very fine aluminum foil, called "chaff." These bits of metal, by virtue of dielectric breakdown at their pointed extremities, would render the cloud sufficiently conductive to bleed off the charge before the fields became large enough to produce lightning.

Still another suggested means of modifying the electrical property of the storm is to introduce into it small quantities of highly electronegative gases, such as sulfur hexafluoride, which are known to have a pronounced and dramatic effect in capturing electrons and inhibiting electrical discharges. Even with the rather small concentrations required, this "solution" would appear to pose formidable logistic and economic problems.

According to proposals of Workman and Reynolds,[56] the electrification of a cloud is due in large part to the development of potential differences when supercooled water droplets within the cloud begin to freeze. Their laboratory experiments indicated that not only the magnitude of the electrification, but the sign, too, that occurs on freezing can be very greatly influenced by rather small concentrations of ammonia. Following this discovery, they performed several preliminary experiments in which ammonia and ammonium nitrate were introduced into a developing thundercloud in New Mexico with the hope of modifying its electrification. However, no definite conclusions could be drawn from the limited number of experiments performed.

Perhaps the most intensive experimental effort directed toward modifying lightning is that being carried out by the United States Forest Service to modify thunderclouds by the use of cloud seeding.[12] In almost all theories of cloud electrification, the separation of charge depends quite markedly on the nature of the precipitation elements and the convective movements of the cloud. Since it is well established that the use of cloud-seeding materials, such as silver iodide, should cause an appreciable modification of the cloud, a number of experiments have been devised to see if it is possible to modify the electrical activity by this means.

From recently reported data it appears that cloud seeding may be effective in reducing the number of lightning discharges in which a long-duration current flows. These discharges appear to be the ones primarily responsible for setting forest fires. It is quite probable that, as we learn to modify the various physical properties of the cloud, we will in this way be able to exercise a measure of control over the electrification processes.

17.23   ELECTRIC FIELDS IN ENCLOSED BUILDINGS

As the electric field intensifies beneath a developing thunderstorm, it seldom produces electrostatic perturbations within closed buildings, houses, or vehicles because they serve as Faraday cages that shield the interior from the effects of external fields. Even structures made of "nonconductors," such as wood, masonry, or canvas, have electrical relaxation times of less than a second; thus for slow-field changes the charge has ample time to flow to neutralize the external field. As a result, the electric field from a thundercloud does not, as a rule, extend into human habitations.

In contrast, however, when the electric field changes rapidly (within less than 1 msec), as it does when there is a lightning flash, the conductivity of a wood or masonry structure may be insufficient to provide electrical shielding, and strong electric fields may exist for periods of the order of the electrical relaxation time of the structure. In such cases, potential differences may arise sufficient to cause electric sparks between conductive objects or people inside a building, even though they may be some distance from the actual lightning spark.

Even inside an enclosure constituting a good Faraday cage, the atmosphere can produce an electric field by virtue of its space charge. When air containing net charge of either polarity enters a building, it produces an electric field that is proportional to the volume density of space charge and the linear dimensions of the enclosure. Such an electric field inside a building results in a potential difference. Since the potential difference between the center of the enclosure and its walls depends on the product of the electric field times the dimensions of the building, the potential difference depends on the square of the dimensions of the building.

In fine weather, when the space charge is usually of the order of 10^6 elementary charges or less per cubic meter, the effect of the space charge is quite small in structures of ordinary size. For example, in a cubical room 3 m on an edge, the electric field at the walls resulting from the space charge would be only 10^{-2} V/m, and the potential difference between the walls and the center would be less than 1 V. During thunderstorms, when space-charge densities may be as high as 10^{11} elementary charges per cubic meter, these voltage differences may become quite considerable, being of the order of kilovolts in the cubical room just cited. In very large structures, such as large hangars which may be as high as 10^2 m, we can see that, with 10^{11} elementary charges per cubic meter, potential differences of the order of 1 MV might be produced.

17.24   PROTECTION AGAINST LIGHTNING

17.24.1   Faraday Cage

When accumulations of charge inside large cumulus clouds result in electric fields of the order of hundreds of thousands of volts per meter, large-scale dielectric breakdown occurs with the production of lightning. These large electric sparks with estimated energies of 10^9 or 10^{10} J can result in considerable damage to life and property as a result of electric shock, high temperature, and high pressure. Lightning is capable not only of causing fires but also of melting, vaporizing, or exploding structures of metals or dielectrics.

The optimum protection from lightning is provided by a Faraday cage constructed of sufficiently thick metal to conduct safely the currents of many kiloamperes that flow during lightning. An approximation to such a structure is provided by enclosed vehicles, such as buses, trucks, automobiles, railway cars, or all-metal aircraft. With few exceptions, these structures are able to conduct the lightning current without sustaining structural damage. Moreover, although short-duration potential gradients undoubtedly occur within them as the result of electromagnetic effects, people and other living organisms within the structure rarely suffer injury. If the structure of the Faraday cage is electrically continuous and has no points of high resistance, even inflammable mixtures of gas within the structure are usually safe from ignition. Only sensitive electronic equipment, particularly of the solid-state variety, is generally affected by the electrical effects that permeate into the Faraday cage.

It appears on the basis of both experience and theory that maximum safety is provided by a Faraday cage. This is true, however, only if no conductors enter the Faraday cage from the outside. Situations of this sort exist in which conductors, such as powerlines, telephone wires, or radio or television antenna conductors, enter the Faraday cage from the outside. This poses a serious hazard unless the conductor entering from the outside is securely grounded to the Faraday cage. If this is not done, the Faraday cage fails to provide adequate protection against lightning; serious damage, deaths, and injuries have resulted under these circumstances.

17.24.2   Lightning Rod

Particularly in the case of buildings and structures made of dielectric materials, such as masonry and wood, it is often unfeasible to provide complete electrical protection with a conducting, grounded, metallic envelope. In this case protection is provided by lightning rods—a series of electrodes placed in positions on or above the structure to be protected and connected with a heavy conductor to a metal object buried in the earth. The elevated electrodes of such lightning rods or other structures are far more likely to receive the direct discharge than any other points on the structure and, since they are well grounded, they furnish an easy and safe path for the lightning discharge currents to ground. In the absence of the lightning rod and associated grounding conductors, a lightning discharge may take an erratic course through the nonconducting structure, setting fires and causing injury and damage. The primary protective action of the lightning rod system is to provide a good conducting path to earth for the high currents that flow.

Although the lightning rod is not intended to have an electrostatic effect, it undoubtedly produces strong electrostatic influences. A question concerning the action of lightning rods has existed almost since the time of their invention by Benjamin Franklin. The question involves the possible effect of the lightning rod on the space charge and the electric fields about it. Does the lightning rod increase or decrease the probability of lightning?

Measurements indicate that the currents flowing to lightning rods on elevated structures can often reach values as high as several milliamperes in a thunderstorm; and, when the lightning rod is installed on large structures of large radius of curvature, there seems to be little question that often such currents would not have flowed from the earth into the atmosphere, had the lightning rod installation not been present. It is clear that in these circumstances the lightning rod is serving to pass charges that can amount to many coulombs into the atmosphere.

Simple calculations show that such quantities of charge are certainly not negligible when compared with the storm itself and that space charges thus released may have a profound effect on the electric fields nearby. The effect of these space charges released by lightning rods undoubtedly depends very much on how they are located. Under some conditions it appears quite probable that the resultant space charge may serve to screen objects nearby and to reduce the electric field thus minimizing the dangers from lightning. In other circumstances, the space charge released by the lightning rod may serve to intensify the electrification of the storm, thus increasing the possibility of lightning.

It appears that the answer to the old question of the effect of lightning rods on lightning is that, depending on the circumstances, the lightning rod installation may increase or decrease the probability of electric discharge. The circumstances, in turn, depend on such complicated variables as wind direction, wind speed, the location of the storm, and the distribution of other corona-giving structures in the neighborhood.

It is possible to modify the operation of the lightning rod somewhat by altering the corona properties of its exposed tip. For example, if the tip is made blunt and rounded and is given a large radius of curvature, the concentration of the electric field will be reduced, and somewhat higher values of the potential gradient will be necessary to initiate point discharge. If the tip of the rod is made very sharp, the electric field at this point will be greatly intensified, causing the rod to go into point discharge in somewhat weaker electric fields. Similarly, the electrical behavior can be altered by placing radioactive material at the tip of the rod, which will increase the ion population near the tip and cause small but measurable currents to flow even in weak fields far below the threshold necessary for point discharge.

Modifications of this sort have been suggested as a means for improving the performance of lightning rods, but they appear to be of dubious value. Depending on conditions, increasing or decreasing the threshold for point discharge can have a slight effect on the probability of lightning, but such effects are both complicated and unpredictable. It is well established by experience that lightning often strikes sharp and rounded points alike, and this variation in shape should not be expected to give rise to important or readily detectable changes in the lightning rod's behavior (which, of course, is essentially designed to provide a safe path for the flow of charge when the lightning stroke occurs).

Similarly, the use of radioactive material to produce ions near the lightning rod appears to be of very doubtful value, particularly when we consider that, once the lightning rod begins to go into corona, the rate of the ion production is enormously greater than that which would be provided by any safe radioactive source. It should be recognized that, although ionization from corona and radioactive sources results in measurable increases in the conductivity of air in the vicinity of the lightning rod, the electrical relaxation times of this modified air are still very much longer than the time constants involved in the lightning discharge.

Large airplanes in commercial and military operation often carry wicks or points known as static dischargers on exposed portions, such as wing tips and the rudder. Contrary to widespread misconception, these dischargers are not intended to have any effect on lightning, but are designed to bleed off charges accumulated on the airplane by point discharge in a way that minimizes radio interference with the airplane's communication system. Although the presence of these static dischargers unquestionably has some influence on the distribution of space charge about the airplane and on the charge on the airplane, the effect is not large. Moreover, all the evidence suggests that these devices have negligible effect on the probability that the airplane will be struck by lightning.

17.25   UNUSUAL LIGHTNING PHENOMENA

In addition to the common lightning spark, other unusual forms of activity are sometimes seen. In a phenomenon called "crown flash," which has been observed during the day, a bright extension of a sunlit cloud forms and suddenly disappears at the time of occurrence of a normal lightning flash. This phenomenon is probably attributable to the reflection of sunlight by ice crystals in the cloud that have been aligned by the action of strong electric fields.[15]

"Ribbon lightning" refers to a successive displacement of several strokes caused by the wind. "Bead lightning" (also called *Perlschnurblitz*, literally "pearl necklace lightning," and *éclair en chapelet*) refers to a discharge that begins as a normal cloud-to-ground stroke but breaks into a series of luminous sections that continue to glow for an appreciable fraction of a second. Uman[42] has attributed this phenomenon to a magnetic pinch effect.

"Rocket lightning" is a discharge that appears to move so slowly that it can be followed by the eye. This progressive luminosity is probably a series of separate discharges which advance further and further into a cloud, giving the illusion of motion.

Probably the most puzzling phenomenon is "ball lightning" (also called *Kugelblitz* or *éclair en boule*). Usually it appears during a thunderstorm where a cloud-to-ground stroke reaches ground. It is typically a luminous red, blue, or white ball 10 to 20 cm in diameter, and bright enough to be seen in daylight. It floats a meter or so above the ground and moves with a horizontal speed up to several meters per second, usually lasting 10 sec or less. It has also been reported to drop down from the clouds. There is no accepted explanation for this phenomenon, and attempts to duplicate its features in the laboratory have thus far failed. Although some scientists dispute the existence of ball lightning, observations of this phenomenon continue to be reported by creditable observers. A very recent book by Singer[40] provides a competent review of the literature on the phenomenon. Men are challenged by such mysteries, and those who wish to respond to this challenge will consult the book and take up from there.

REFERENCES

1. Robert Anderson, Sveinbjörn Björnsson, Duncan C. Blanchard, Stuart Gathman, James Hughes, Sigurgeir Jónasson, Charles B. Moore, Henry J. Survilas, and Bernard Vonnegut, "Electricity in volcanic clouds," *Science*, **148**, No. 3674, 1179–1189 (1965).

2. Duncan C. Blanchard, "The electrification of the atmosphere by particles from bubbles in the sea," *Progress in Oceanography*, Vol. 1, Pergamon Press, Oxford, 1963, pp. 71–202.

3. K. A. Brown, P. R. Krehbiel, C. B. Moore, and G. N. Sargent, "Electrical screening layers around charged clouds," *J. Geophys. Res.*, **76**, No. 12, 2825–2835 (1971).

4. J. Alan Chalmers, *Atmospheric Electricity*, 2nd ed., Pergamon Press, Oxford, 1967.

5. J. F. Clark, "Airborne measurement of atmospheric potential gradient," *J. Geophys. Res.*, **62**, 617–628 (1957).

6. William E. Cobb and Howard J. Wells, "The electrical conductivity of oceanic air and its correlation to global atmospheric pollution," *J. Atmos. Sci.*, **27**, No. 5, 814–819 (1970).

7. Stirling A. Colgate, "Tornadoes: Mechanism and control," *Science*, **157**, No. 3795, 1431–1434 (1967).

8. ——— and C. McKee, "Electrostatic sound in clouds and lightning," *J. Geophys. Res.*, **74**, No. 23, 5379 (1969).

9. M. H. Davis and J. D. Sartor, "Theoretical collision efficiencies for small cloud droplets in Stokes flow," *Nature*, **215**, No. 5108, 1371–1372 (1967).

10. Arnold Doyle, D. Read Moffett, and Bernard Vonnegut, "Behavior of evaporating electrically charged droplets," *J. Colloid. Sci.*, **19**, No. 2, 136–143 (1964).

11. M. Faraday, "On the electricity evolved by the friction of water and steam against other bodies," *Experimental Researches into Electricity*, Vol. II, 18th series, pp. 106–126, (1844).

12. Donald M. Fuquay, *Weather Modification and Forest Fires*, Ground Level Climatology, American Association for the Advancement of Science, Washington, D.C., pp. 309–325, (1967).

13. H. Gerdien, "Demonstration eines Apparates zur absoluten Messung der elektrischen Leitfähigkeit der Luft," *Phys. Z.*, **6**, 800–801 (1905).

14. O. H. Gish and G. R. Wait, "Thunderstorms and the earth's general electrification," *J. Geophys. Res.*, **55**, No. 4, 473–484 (1950).

15. Maurice E. Graves, John C. Gall, Jr., and Bernard Vonnegut, "Meteorological phenomenon called crown flash," *Nature*, **231**, 258 (1971).

16. R. Gunn, "The electrification of precipitation and thunderstorms," *Proc. IRE*, **45**, No. 10, 1331–1358 (1957).

17. C. R. Holmes, M. Brook, P. Krehbiel, and R. McCrory, "On the power spectrum and mechanism of thunder," *J. Geophys. Res.*, **76**, No. 9, 2106–2115 (1971).

18. H. Israël, *Atmospheric Electricity*, Vol. I, *Fundamentals, Conductivity, Ions*. Translated from German by D. Ben Yaakov and Baruch Benny; Baruch Benny, Ed., Israel Program for Scientific Translations, Ltd. National Science Foundation, Washington, D.C., TT67-51394/1, (1971).

19. W. P. Jesse, "Absolute energy to produce an ion pair in various gases by beta particles from S^{35}," *Phys. Rev.*, **109**, No. 6, 2002–2004 (1958).

20. Christian E. Junge, *Air Chemistry and Radioactivity*. Academic Press, New York, (1963).

21. A. K. Kamra, *Dust Storm Electrification*, Final Report to National Science Foundation, Grant No. GA-18667, (1971).

22. —— and B. Vonnegut, "A laboratory investigation of the effect of particle collisions on the generation of electric fields in thunderstorms," *J. Atmos. Sci.*, **28**, No. 4, 640–644 (1971).

23. J. H. Kraakevik, "The airborne measurement of atmospheric conductivity," *J. Geophys. Res.*, **63**, No. 1, 161–169 (1958).

24. J. Latham and C. P. R. Saunders, "Aggregation of ice crystals in strong electric fields," *Nature*, **204**, No. 4965, 1293–1294 (1964).

25. P. Lenard, "Über die Elektrizität der Wasserfälle," *Ann. Phys. (Leipzig)*, **46**, 584–636 (1892).

26. Charles B. Moore and Bernard Vonnegut, "Effect of low temperature on the leakage of charge from irradiated methyl methacrylate blocks," *J. Polym. Sci.*, **33**, No. 126, 491 (1958).

27. —— and ——, "Estimates of raindrop collection efficiencies in electrified clouds," *Physics of Precipitation*. American Geophysical Union Monogr. No. 5, 291–304, (1960).

28. ——, ——, J. A. Machado, and H. J. Survilas, "Radio observations of rain gushes following overhead lightning strokes," *J. Geophys. Res.*, **67**, No. 1, 207–220 (1962).

29. ——, ——, R. G. Semonin, J. W. Bullock, and W. Bradley, "Fair-weather atmospheric electric potential gradient and space charge over central Illinois, summer 1960," *J. Geophys. Res.*, **67**, No. 3, 1061–1071 (1962).

30. ——, ——, E. A. Vrablik, and D. A. McCaig, "Gushes of rain and hail after lightning," *J. Atmos. Sci.*, **21**, 646–665 (1964).

31. H. Norinder and O. Salka, "Mechanism of positive spark discharges with long gaps in air at atmospheric pressure," *Ark. Fys.*, **3**, 347–386 (1951).

32. A. Peltier, "Observations et recherches expérimentales sur les causes qui concourent à la formation des trombes," *C.R. Acad. Sci. Paris*, **10**, 712 (1840).

33. C. T. Phelps, "Field-enhanced propagation of corona streamers," *J. Geophys. Res.*, **76**, No. 24, 5799–5806 (1971).

34. E. T. Pierce, "Electrostatic field-changes due to lightning discharges," *Quart. J. Roy. Meteorol. Soc.*, **81**, 211–228 (1955).

35. ———— "Radioactive fallout and secular effects in atmospheric electricity," *J. Geophys. Res.*, **77**, No. 3, 482–486 (1972).

36. ———— and A. L. Whitson, "Atmospheric electricity and the waterfalls of Yosemite valley," *J. Atmos. Sci.*, **22**, No. 3, 314–319 (1965).

37. Lord Rayleigh, "The influence of electricity on colliding water drops," *Proc. Roy. Soc. (London)*, **28**, 406–409 (1879).

38. S. E. Reynolds, *Compendium of Thunderstorm Electricity*, U.S. Signal Corp. Research Report, Socorro, N.M., (1954).

39. Vincent J. Schaefer and Bernard Vonnegut, "Electric-field perturbations caused by eruption of Yellowstone geysers," *J. Atmos. Sci.*, **20**, No. 2, 180–182 (1963).

40. Stanley Singer. *The Nature of Ball Lightning*, Plenum Press, New York, (1971).

41. Roy Steiner and R. H. Rhyne, *Some measured characteristics of severe storm turbulence*. U.S. Weather Bureau, National Severe Storms Project Report 10, July 1962, 17 pp.

42. M. A. Uman, "Bead lightning and the pinch effect." *J. Atmos. Terrestr. Phys.*, **24**, 43–45 (1962).

43. ———— *Lightning*, McGraw-Hill, New York, (1969).

44. Bernard Vonnegut, "Electrical theory of tornadoes," *J. Geophys. Res.*, **65**, No. 1, 203–212 (1960).

45. ———— "Orientation of ice crystals in the electric field of a thunderstorm," *Weather*, **20**, No. 10, 310–312 (1965).

46. ————, K. Maynard, W. G. Sykes, and C. B. Moore, "Technique for introducing low-density space charge into the atmosphere," *J. Geophys. Res.*, **66**, No. 3, 823–830 (1961).

47. ———— and Donald A. McCaig, "Simple electrometer employing an electrified, nonconducting fiber," *Rev. Sci. Instr.*, **28**, No. 12, 1097–1098 (1957).

48. ———— and Charles B. Moore, "Preliminary attempts to influence convective electrification in cumulus clouds by the introduction of space charge into the lower atmosphere," in *Recent Advances in Atmospheric Electricity*, Pergamon Press, New York, 317–331, (1959).

49. ———— and ———— "A possible effect of lightning discharge on precipitation formation process," *Physics of Precipitation*, American Geophysical Union Monogr. 5, 287–290, (1960).

50. ————, C. R. Smallman, C. K. Harris, and W. G. Sykes, "Technique for the introduction into the atmosphere of high concentrations of electrically charged aerosol particles," *J. Atmos. Terrestr. Phys.*, **29**, 781–792 (1967).

51. G. R. Wait, "Diurnal variation of concentration of condensation nuclei and of certain atmospheric-electric elements at Washington, D.C.," *Terrestr. Magn. Atmos. Elect.*, **36**, 111–131 (1931).

52. F. J. W. Whipple, "On the association of the diurnal variation of electric potential gradient in fine weather with the distribution of thunderstorms over the globe," *Quart. J. Roy. Meteorol. Soc.*, **55**, 1–17 (1929).

53. Harry J. White, *Industrial Electrostatic Precipitation*. Addison-Wesley, Reading, Mass., (1963).

54. C. T. R. Wilson, "On the comparative efficiency as condensation nuclei of positively and negatively charged ions," *Phil. Trans., A*, **193**, 289–308 (1899).

55. ———— "Investigations on lightning discharges and on the electric field of thunderstorms," *Phil. Trans. Roy. Soc. London, A*, **221**, 73–115 (1920).

56. E. J. Workman and S. E. Reynolds, "Electrical phenomena occurring during the freezing of dilute aqueous solutions and their possible relationship to thunderstorm electricity," *Phys. Rev.*, **78**, 254–259 (1950).

57. T. W. Wormell, "Vertical electric currents below thunderstorms and showers," *Proc. Roy. Soc. (London) A*, **127**, 567–590 (1930).

CHAPTER 18

Electrostatic Nuisances and Hazards

H. FRANK EDEN, *Program Director for Meteorology*
Atmospheric Sciences Section
National Science Foundation
Washington, D.C.

18.1 OVERVIEW

Although the phenomenon of static electrification has been known for hundreds of years, problems arising from it are of more recent vintage. These problems cover a wide variety of physical situations and touch on many human concerns, ranging from large-scale industries to domestic activities; they range in magnitude from major hazards, with potential loss of life and possible damage costing hundreds of millions of dollars, to minor annoyances in the home or office. The problems include explosions during the handling, filtering, refining, or transportation of volatile liquids or gases; dust explosions in sugar mills, granaries, sulfur mills, and coal mines; losses in the manufacture and handling of photographic film and integrated circuits; and minor annoyances such as carpet electrification or the sticking of powders to the walls of containers. Figure 18.1 illustrates some of these hazards and annoyances. Many of the examples come from long-established industries and have been well documented.

It is impossible in one chapter to treat all the topics in detail; thus only certain topics, the greatest hazards and most recent problems shown, are discussed.

Despite the broad range of subjects and physical situations, some general comments can be made regarding electrostatic hazards and nuisances. In all cases there are three criteria for the existence of an electrostatic hazard. These criteria are: (1) charging of the material, or nearby structures, occurs;

425

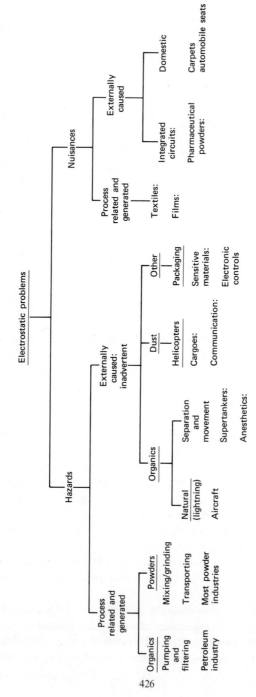

Figure 18.1. Some hazards and annoyances due to electrostatics.

(2) the leakage of such charge is so slow and small that local spark breakdown potential can be reached and a minimal required amount of electrostatic energy stored; and (3) the material ignites, explodes, or is damaged when subjected to the ensuing sparks. Why materials charge electrostatically is the subject of previous chapters. In the case of dust, triboelectric charging and the physics of surfaces are involved; in organic liquids, charge layers form on surfaces and possibly disrupt such layers, and the case of persons, dissimilar surfaces are separated by the action of walking, sliding, or the movement of clothing.

Besides the mechanism of charging, we must consider the second important aspect, namely, the mechanism by which the materials concerned retain the charge. This involves the idea of a relaxation time τ, which depends on the resistivity ρ and the dielectric constant K of the material:

$$\tau = K_0 K \rho$$

where K_0 is the permittivity of free space and equals 8.85×10^{-14} F/cm. Thus

$$\tau \simeq 9 \times 10^{-14} K\rho \text{ sec}$$

where ρ is in Ω-cm.

In general the value of the relaxation time is determined by what leakage paths are available to the charge generated. The leakage path through air depends on the presence of ions and the possible presence of ionizing material. The leakage path through powders involves the resistivity of the powders, which is dependent on humidity and packing. If the resistivities of the leakage paths involved are of the order of, say, 10^{14} Ω-cm, long relaxation times, of the order of 1 sec, will pertain.

The first two criteria may suffice in a nuisance situation such as the sticking of powders to containers or the collection of dust by fabrics.

The third criterion for a potential hazard is that sufficient electrostatic energy be developed at sufficiently high voltage that breakdown can occur and a spark can ensue with enough energy to ignite combustible materials. Typical values of these minimal energy requirements might be taken as 0.1 mJ for organic vapors and 20 mJ for combustible dust. These values are merely representative, since the possibility of ignition depends on the concentration of the material concerned, the oxygen content of the environment, and, in the case of dust, the particle size. Such energy levels are by no means difficult to obtain by simple charging mechanisms. For example, a person can become electrostatically charged to a potential of the order of 10 kV simply by sliding across an automobile seat and stepping out of the vehicle. Since the electrical capacitance of the human is approximately 300 pF, the person could produce a spark of 15 mJ upon discharge. The minimum spark energy required to ignite aviation gasoline vapor in air is estimated from data of Lewis and von Elbe[1] to be about 0.2 mJ; this means that an electrostatically charged person could be a fire hazard in the presence of aviation gasoline vapor.

The following sections discuss some of these in detail. General overviews of the subject can be found in Refs. 2 through 4. The related topic of dangers from lightning strokes is well covered in Ref. 5.

It is appropriate at this point to discuss in general the safeguards that can be applied.

Many potentially hazardous situations involving large quantities of material can be analyzed by laboratory-scale experiments, and estimates can be made including predictions of the roles of additives, controlled humidity atmospheres, or the presence of fast ions to increase atmosphere conduction.

All control systems involve reducing the relaxation time. Bonding and grounding is effective for good conductors. Bonding reduces any potential difference between good conductors, although both may still be at a potential (zero if they are grounded) that differs from other objects. The use of grounded screens through which powder or fluids may flow should not necessarily be undertaken. Electrification may be increased.

Relative humidity in the 60 to 70% range can bring about water films on conducting materials which provide conducting paths.

Ionization of the local atmosphere generally decreases the relaxation time for stored charge. Either radioactivity or point discharge combs can be employed. If the leakage path is not primarily atmospheric, this is a limited mechanism (as, e.g., with flowing liquids or powders).

Various antistatic agents are available for surface or bulk treatments of fabrics or plastics which depend in principle on capturing surface films of moisture. The conductivity of organic liquids can be greatly increased by the addition of a few parts per million of commercial additives.

Warning devices should be used in the process vicinity to indicate when the levels of electrostatic activity are becoming dangerously high. Static detectors may be conveniently divided into two types, namely, those which measure the electrostatic field and variations therein and those whose primary objective is the detection of spark discharges. The former range from very simple devices such as gold leaf electroscopes or neon lamps to sophisticated electrostatic volt meters, electrometers, and field mills. Brief descriptions of these systems and their use are given in Ref. 4. The second class of spark detectors is specifically aimed at detecting rapid changes in the electric field. These devices operate using a capacitive antenna such that when a discharge occurs, there is a change in the equilibrium charge on the antenna. This incremental charge becomes a signal which may be amplified and recorded. Thus it is possible to record the number of sparks and their magnitude over a period of time.

18.2 DUST

The electrostatic properties of dusts can lead both to annoyances and to major hazards. One of the outstanding aspects of electrostatic phenomena in moving powders is the widely differing distance scales on which these phenomena can occur. These range from the microscale, as in the finer aspects of electrostatic copying or in the common annoyance of powders sticking to surfaces, through scales of kilometers or more, possibly to cosmic dimensions. Naturally occurring electrostatic phenomena associated with dust are sometimes seen in dust devils. Remarkable displays of lightning have been observed in the dust clouds of volcanoes such as that which created the island of Surtsey in 1963.

Why powders charge and the differences in charging mechanisms between conductors, semiconductors, and electrophoretic or electrophobic insulators has been the subject of considerable painstaking research (e.g., Ref. 6). Experimental investigation of the electrostatic properties of powders is a notoriously difficult task. This is not because data are difficult to obtain. Moving powders are often very active electrostatically, and high voltage and charge levels can be readily achieved by allowing the powders to impinge on targets or to be collected in containers which approximate Faraday cages. Qualitative data, useful from the point of view of industrial processes, can often be obtained by simple laboratory experiments (Ref. 7). The interpretation of such data in terms of models of the particle behavior or the collection of truly quantitative and repeatable results is, however, subject to considerable error. Thus the magnitude and polarity of the charge acquired by a powder or its container may depend on:

1. The moisture content of the atmosphere and the powder.
2. The particle size distribution.
3. The velocity with which the powder moves and/or impinges on surfaces.
4. The state of surfaces on which the particles impinge.

As a rule, surfaces rapidly become coated with fine particles in the presence of moving powders. This generally reduces charging or causes polarity changes, since impacts are then between different-sized particles of powder rather than between surface and powder. The thin layers of particles adhere very strongly, and a simple calculation shows that electrostatic binding forces can greatly exceed gravitational forces.

Bernard Vonnegut has suggested an arrangement of equipment for triboelectric measurements with blown powder. Nitrogen at about 10 psi enters a gas inlet and swirls the particles up through a hole in a Teflon insulator into a filter. The accumulation of the charged particles on the filter causes a capacitor connected between the metal filter holder and ground to become charged and the potential is recorded with an electrometer. This system affords a convenient way of comparing the relative electrostatic activity of different powders. By varying the gas pressure, the rate of charging can be varied or even the particle size of powder blown up.

Experiments such as these indicate that electrostatically active powders such as starch or cabosil can produce about 1 μC of charge per gram of agitated powder. Assuming reasonable values for various parameters such as particle size and number density for typical powders, the following rough calculation shows that the particles must charge up approximately to breakdown potential.

radius of particles	$r \simeq 10\ \mu$
bulk density	$D \simeq 1$ g/cc
density	$\rho \simeq 3$ g/cc
particle mass	$m \simeq 10^{-8}$ g
total charge	$Q \simeq 10^{-6}$ C/g
charge per particle	$q \simeq 10^{-14}$ C/particle
	$= 3 \times 10^{-5}$ esu/particle

Thus the field f at the edge of the particle is given by

$$f = \frac{q^2}{r^2} \simeq 3 \times 10^{-5} \times 10^6 \text{ stat V/cm}$$

$$\simeq 9 \times 10^3 \text{ V/cm}$$

Similarly, choosing the same parameters for the particles, the following calculation shows that the strength of adhesion of such a particle to a conducting wall can be very high, with the force greatly exceeding the gravitational force on the particle. The gravitational force on particle mass 10^{-8} g $\simeq 10^{-5}$ dyne. The electrostatic force P on a particle charge q in contact with a conducting plane is given by

$$P = \frac{q^2}{4r^2} \simeq 2 \times 10^{-4} \text{ dyne}$$

where q is in electrostatic units and r is in centimeters.

Thus the electrostatic force of attraction is approximately 20 times the gravitational force on the particle.

The relaxation time τ depends on the resistivity ρ and dielectric K constant of the powder; thus we have

$$\tau = K_0 K \rho$$
$$K_0 \simeq 9 \times 10^{-14} \text{ F/cm}$$

Measurements of these variables are also important in assessing a potentially hazardous situation with an electrostatically active powder. In general the value of τ is determined by the resistivity, which can vary over many orders of magnitude. Thus the leakage path through air depends on the presence of ions and the possible presence of ionizing material, and the resistivity of powders is very dependent on humidity and packing.

Again, simple experiments can yield useful data. Resistivity can be determined using a measuring cell with conducting end electrodes. The resistivity (Ω-cm) of the bulk material is computed from measurement of the resistance between the electrodes and the height of the powder. In order to use the method, a known force or pressure must be applied to the top of the sample. Fairly consistent measurements of resistivity as a function of applied pressure can be made using an Instron unit with the 100-kg load cell, which applies constant displacement to the upper powder surface and measures the force applied to the surface. Representative data (taken on various powders subjected to pressures from 0 to 5.6 kg/cm^2, in such a fashion that the pressure was first increased and then decreased) show two gross effects:

1. That the resistivity is very sensitive to increased packing.

2. That a considerable hysteresis may occur. Resistivities of nonmetallic powders may range up to values greater than 10^{14} Ω-cm, which is about the upper limit of conventional measurement techniques. Resistivities of this magnitude ensure long relaxation times of the order of seconds.

In industrial practice where large quantities of powders are concerned, energies stored in the powder or on parts against which the powder has impacted may be considerable and can cause sparks sufficient to ignite the dust. Dust explosions caused by static discharges have occurred in a variety of industries concerned with powders. Table 18.1 lists some major explosions involving various types of dust; these events occurred between 1900 and the present, and at least some have been attributed to statics. The U.S. Bureau of Mines has determined the minimum ignition energy for a variety of dust clouds and layers[9] using apparatus described in Ref. 10. Generally, the energies required are of the order of 10 mJ.

The apparatus consists of capacitors charged to a desired energy level, a spark gap, and a system to produce a small dust cloud which is synchronized with the spark. With such an apparatus it is easy to show that the energy required for ignition is strongly dependent on the duration of the spark. Thus, when one of the points in the spark gap is directly grounded, the minimum energy stored in the capacitors required for ignition is at least an order of magnitude greater than the value quoted in Ref. 9. When the point is grounded through a large choke (several

Table 18.2. Major Dust Explosions Since 1900[a]

		Explosions Attributed to		
Kind of Dust	Total Explosions	Unknown Cause	Electric Spark (Any Origin)	Static Discharge
Plastic dust	35	7	3	3
Paper dust	35	—	2	1
Cork dust	40	8	—	5
Metal dust	81	19	5	3
Starch and corn dust	56	17	5	6
Sulfur dust	34	6	—	6

[a] Data from Ref. 8.

henries), the spark is much more prolonged, and the values of the minimum energy required agree well with Ref. 9, where a large choke was used in the circuit also. The energy required appears to increase with decreasing inductance in the circuit. Possible explanations of the results might involve different heating rates or a shock wave from sharp sparks which blows the combustible material away.

This dependence of effects in dust clouds on the duration of the spark is not limited to ignition. Similar results have been found in experiments to investigate the coagulation of dust by electric spark in attempts to simulate chondrule formation[11] and, as indirect evidence, the ignition of forest fires by lightning stroke, which indicates that lightning strokes of peculiarly long duration and relatively low current flow are the most successful.[12]

We should distinguish here between two different possibilities. If a system involving moving powders has ungrounded metal parts which are contacted by the powders, these will charge to very high potentials and sparks to other grounded conductors are virtually inevitable. However, even when all metal parts are grounded, the airborne powder is still charged and a space-charge situation exists. If the system is large enough, the possibility of breakdown between this space charge of charged particles and the grounds exists in a manner analogous to ground-to-cloud streamers in thunderstorm situations. Little evidence is available regarding whether this occurs and, if it does, whether it has caused explosions. Later, however, we present evidence that such a mechanism may explain recent supertanker explosions. The possibility exists, and it is best guarded against by doing all that can be done to ensure a conducting atmosphere or by monitoring electric fields during the operation.

A further interesting aspect of electrostatic charging due to moving dust is in the case of helicopters. The problem is as yet unsolved in a satisfactory fashion. Helicopters hovering within a few feet of the ground acquire very considerable amounts of electric charge due to triboelectric effects from blown dust, sand, and so on. For such helicopters 1 mJ is considered an acceptable maximum level of stored electrostatic energy. Energies in excess of this value, when released as a spark during cargo operations, may involve hazards to personnel or explosive or volatile cargoes. This potential hazard has resulted in research sponsored by the Armed Services (e.g., Ref. 13). In practice, without some means of reducing the charge on the helicopter, energies of many joules may be realized at potentials of hundreds of kilovolts and with charging currents of hundreds of microamperes.

In attempts to reduce the charge, active corona discharge systems are employed using corona points on the helicopter which are maintained at the proper polarity and potential by means of a power supply and sensor. However, two major problems occur with this system as developed. First, it is difficult to obtain a sufficient discharging current from the points without employing an overly large, expensive, and potentially hazardous power supply; second, the sensors currently used to determine the magnitude and polarity of the charge on the helicopter fail under field conditions.

Helicopters operating at a few hundred feet above the ground and employing a long cable are less susceptible to triboelectric charging (except in rain, ice, or snow). They will, however, be at some potential relative to ground because of the atmospheric electric gradient. Thus in fair weather a helicopter at a 200-ft altitude will be at a potential of the order of 10^4 V relative to the earth. Under thunderstorm conditions the helicopter may reach potentials of the order of 10^5 V relative to ground. This disregards the question of precipitation, which would lead to additional triboelectric charging.

The slurry mixing of sensitive materials poses another problem wherein simple determinations of resistance and capacitance may be of value. Eden et al.[14] have described a manufacturing situation in which a pyrotechnic mixture was slurried with heptane and the slurry dried on a moving belt 1 m wide and approximately 7 m long. The belt was polypropylene approximately 0.1 cm thick. The possible development of static charge on the pyrotechnic mix during the drying process and the characteristics of the dried and drying slurry material from the point of view of initiation by electrical discharge were investigated by means of model experiments. The purpose of the resistivity tests was to measure the electrical relaxation time τ of the mixes as functions of the relative humidity of the atmosphere in which they had been stored and of the amount of heptane present in the slurry. The method involved drying the slurry in a cell in a Buchner funnel with electrodes imbedded and monitoring the resistance of the cell. A relaxation time longer than about 1 msec can permit appreciable electrostatic charge generation in a material and possibly spark breakdown.

The results revealed that the electrical relaxation time at 50% relative humidity is 0.1 sec; at 90 to 100% relative humidity it is 10^{-8} sec. Thus at the low relative humidity, the relaxation time of a specimen is long enough to permit appreciable electrostatic charge separation. When a cake of the mixture was allowed to break contact with a sheet of polypropylene by sliding off it, the cake and the sheet charged with opposite polarities to potentials of the order of 100 V. The potential of the polypropylene sheet may have been an order of magnitude higher. Rough estimates show the energies involved were in the range 10^{-5} to 10^{-7} J.

Ignition level tests showed that all the specimens conditioned for several days at room humidity (i.e., at relative humidity values from 25 to 75%) charred when subjected to sparks of some 20 mJ. There is a difference between the energies required for slight charring and the energies required for complete ignition of the specimen. The experiments presented two clear conclusions: (1) that the relaxation time for these materials when dry was relatively long, and (2) that they could be ignited by electrostatic discharge. The third aspect, the degree of electrical charging which occurs when the material is dried or moved, was indefinite. An experiment that would correctly model possible charging in the actual drying process is difficult to envisage, but it seems likely that some charging would occur and that safety could only be ensured by reducing the resistivity.

18.3   ORGANIC VAPORS

Another serious area of electrostatic hazard occurs in operations involving hydrocarbon vapors. This may include such diverse situations as the use of anesthetics in hospitals, the filling and pumping of gasoline, the filtering of hydrocarbons, the cleaning of storage tanks, and, indeed, any dynamic situation wherein hydrocarbon liquid exists in contact with its vapor. Primarily this is because of the very low energy requirements for ignition of organic vapors and the generally high resistivity which ensures long relaxation times. The lower and upper "explosive limits" for the concentration of flammable vapors in air are roughly 50 to 400% of the stoichiometric proportion for combustion. At the high and low limits, 5 to 10 mJ is needed for ignition, and as we near stoichiometric proportions the spark energy required is about 0.2 mJ.

Ignition of combustible gases may require not only a minimum spark energy but also a critical voltage. Below the critical voltage, ignition requires greatly increased spark energy. The critical voltage for diethyl-ether–air mixtures has been found by Boyle and Llewellyn[15] to be about 1500 V. The corresponding value for gasoline–air mixtures is probably similar or slightly higher. Such energies may be readily attained even by the charging of a human operator. The source of charging may be the motion of the liquid itself either because of free charges on the surface of the liquid or because the liquid is in contact with other materials. In particular, the flow through pipes, the settling of mixed fluids, and filtering provide opportunities for the generation of static charge. Reference 3 supplies examples of computations that can be made to estimate the amount of charge and the potentials generated in such cases. In general liquid hydrocarbons have high values of resistivity in the range of 10^{14} Ω-cm, and thus the relaxation times in such situations are relatively long. A variety of antistatic additives have been developed in the petroleum industry for addition to these insulating liquids to help reduce relaxation times.[16]

Considerable attention has been paid to the control of electrostatic hazards in the handling of organic liquids and vapors.[4] There are two interesting examples wherein the charging and mechanism is not necessarily the motion of the liquid itself but, rather, is due to other operations carried out in the presence of the vapor. These are the classical problems of the use of anesthetics in operating theaters and the relatively new issue of the explosion of tankers at sea during washdown procedures.

Less active anesthetic agents have recently been developed, but initially, when ether was first introduced, candles, gaslights, and coal stoves caused fires and mild explosions of the vapor–air mixtures. Such problems were soon minimized; however, the ignition of the vapors by electrostatic sparks proved a more difficult problem. The situation was discussed in detail in Ref. 17. In 1964 Eichel[3] stated that the estimated number of deaths occurring annually from operating-room explosions may be of the order of 1000 in the United States. In these situations the clothing and shoes of the personnel involved are of consequence. Cotton is now generally considered to be nonhazardous in hospital operating rooms above 65% relative humidity. Outer garments of synthetic materials can build up considerable static charges when moved away from the body. In low-humidity environments and in flammable atmospheres enriched with oxygen, the static levels reached from such sources may be dangerous. Similarly, conductive footwear or conductive floors have been suggested to minimize the accumulation of static on personnel.

The provision of leak paths of suitable resistance to prevent the development of potentials of 1000-V eliminates the electrostatic hazard and provides a reasonable factor of safety. An idea of the leak path resistance that may be required to avoid excessive charge accumulation was given by

Bulgin.[18] In hospital operating rooms, where highly combustible mixtures of ether or cyclopropane and oxygen are possible, he recommended resistance of 10^5 to 10^7 Ω.

The hazard due to personal clothing in such situations can be handled in three basic ways. Increased humidity causes the relaxation time associated with cotton materials to become acceptable, but it is less effective with synthetic materials. Various organic antistatic agents have been developed which can be sprayed on textiles or applied during finishing operations for the cloth. They tend not to be durable and to lose effectiveness at low humidities. In some instances they cause materials to become stiff and uncomfortable. The third alternative is the use of conducting fibers, either metallic or carbon based, in the manufacture of the textiles. For example, Ref. 19 states that primarily synthetic materials containing 0.5% Brunsmet fibers produced only 10 to 20% of the voltage produced by cotton or polyester fabrics in a simple test in which the subject removed a laboratory coat.

Recently there have been reports of tankers having blown up at sea. These ships might be regarded as being of the supertanker class. In the course of 3 weeks in 1969, the 200,000 ton tanker *Merpessa* exploded and sank off the Senegal Coast. Two weeks later the *Macta* exploded in the Mozambique Channel, and the next day the *King Haakon VII* exploded off Liberia. Later two more tankers exploded. From the facts generally available, the following picture emerged. The tankers were traveling empty and were being cleaned when the explosions took place in the tanks. During this process, a high-power water jet was used to wash down the walls with a high-velocity jet of water and sometimes with air venting. These were major explosions involving a loss of life and economic losses of hundreds of millions of dollars. It has been suggested[20] that the electrical charging developed by Lenard splashing during the washing operations may be a cause.

Both theory and experiment indicate that the situation is governed by the equation

$$\frac{dN}{dt} = \frac{Q}{e} - 10^{-3}N$$

where N = number density of large charge carriers
    t = time
    e = electronic charge
    Q = ca. 10^{-11} C/g (for seawater)

These values correspond to $N \simeq 10^5$/cm^3 and space-charge densities of 10^{-8} C/m^3.

The study concluded that the electrification within a supertanker cargo tank may be sufficiently intense for a large-scale spark streamer to develop with a potential for an explosion hazard. The field within a closed container is approximately proportional to the product of the linear dimension of the container and the space-charge density; this charge becomes larger as the size of the cargo tank increases.

Electrostatics and organic fields also present problems in aviation. Lightning and static electricity pose a hazard to high-speed aerospace vehicles. The potential threat of lightning to aircraft and, in particular, the possibility of a catastrophic fuel explosion have long been under investigation.[21,22] The greatest danger of explosion is probably associated with ignition at the fuel vent, although the composite plastic materials for the primary structure are vulnerable.[23]

On the average, a given commercial airplane is struck by lightning once every 5000 to 10,000 hr of flying time. Fuel ignition was the probable cause of the two lightning-related disasters to commercial aircraft—in 1959 near Milan, Italy, involving a Lockheed Constellation, and in 1963 at Elkton, Maryland, involving a Boeing 707. In addition, a number of military aircraft have been destroyed by lightning.[24–26]

There is some danger of accidental fire and explosion in fueling because of improved quality aviation fuel and of the higher flow rates being used to speed refueling. The J.P-4 fuel, a blend of kerosene and gasoline, is more volatile than gasoline and may be more electrostatically active. Desulfurization of fuel has resulted in lower conductivities. Klass[23] reported on a conference at which several explosions were attributed to electrostatic sparks in using such fuel.

18.4   NUISANCES

Electrostatics can play a role in industry which can cause loss of product and difficulty in the production process without the existence of danger to personnel or major facilities. Two well-known examples occur in the production and handling of photographic film and in the textile industry. The photographic industry was at one time at the mercy of static problems. Charge was produced on the back of the film as it passed over rollers or guides and sparks occurred as the surfaces separated. Familiar dendritic discharge patterns on the film were produced. References 27 and 28 discuss the problems in some detail. In the textile industry, the problems similarly involve the transport of materials in belt fashion and the production of charge by the separation of contacting surfaces. Fibrous dust is attracted to the material and handling is made difficult. Nowadays these textile problems are generally reduced by maintaining moist atmospheres.

Problems caused by electrostatic discharges occur in the most modern of industries (e.g., the integrated circuit manufacturing industry). In particular this is a problem in the manufacture and handling of the metal-oxide–silicon (MOS) devices. In these devices an electric field, applied through an oxide-insulated gate electrode, is used to control the conductants of a channel layer in semiconductor material under the gate. The channel is a lightly doped region between two highly doped areas called the source and the drain. The extremely thin oxide separating the silicon and the metallization can easily be destroyed by electrostatic discharges. In monolithic structures as well as in some discrete devices, the external leads are protected by means of resistor networks and zener diodes. These protective devices may be habitated on the common substrate with the MOS circuit and interconnected in a way that protects the circuit.[29] However, the failure rates between shipment of the products from the vendors and final testing in end-use circuits is rumored to be as great as 2%.[30] Failure occurs during shipment, inspection, testing, or assembly into circuits, and it appears that the protective diode devices on the chips are not fully successful.

With the market for MOS devices projected to be of the order of $100 million in the year 1972–1973, such losses are considerable. The probable main cause of gate breakdown is handling by electrostatically charged operators. Another source of static voltages is the styrofoam containers in which the semiconductor products are often shipped.

REFERENCES

1. B. Lewis, and H. von Elbe, *Combustion, Flames and Explosions of Gases*, Academic Press, New York, 1959.
2. R. Beach, "Preventing static electricity fires, III," *Chem. Eng.*, Feb. 1, 1965.
3. F. G. Eichel, "Electrostatics," *Chem. Eng.*, March 13, 1967.
4. National Fire Protection Association, Boston, NFPA, 1966; Section 77, "Static electricity."
5. M. A. Uman, *Understanding Lightning*, Bek Technical Publications, Carnegie, Pa., 1971.
6. W. R. Harper, *Contact and Frictional Electrification*, Oxford University Press, 1967.
7. H. F. Eden, *Aspects of Electrostatic Charging by Powders: The Analysis of Potential Hazards*, Proceedings of the 6th Annual Meeting of the American Institute of Chemical Engineers, San Francisco, December 1971.
8. National Fire Protection Association, *Report of Important Dust Explosions*, NFPA, Boston, 1957.
9. M. Jacobsen et al., *Explosibility of Agricultural Dusts*, U.S. Bureau of Mines Report 5753, 1961.
10. H. Dorsett et al., *Equipment and Test Procedures for Determining the Explosibility of Dusts*, U.S. Bureau of Mines Report 5024, 1960.
11. F. L. Whipple, "Chondrules: Suggestions concerning the origin," *Science*, **153** (1966).

12. D. M. Fuquay et al., "Characteristics of severe lightning discharges that caused forest fires," *J. Geophys. Res.*, **72** (1967).

13. J. Cierra, *Theoretical Analysis of Aircraft Electrostatic Discharge*, Report to U.S. Army Aviation Material Laboratories on Contract DA 44-177-AMC-82(T) DDC No. AD 621 521, 1965.

14. H. F. Eden et al., *Pyrotechnic Mixes: Their Resistivity, Charge Generation and Sensitivity to Spark Discharge.* Arthur D. Little report to Edgewood Arsenal on Contract DA-18-035-AMC-326(A), 1968.

15. A. R. Boyle and F. J. Llewellyn, "Electrostatic ignitability of various solvent vapour–air mixtures," *J. Soc. Chem. Ind.*, **66** (1947).

16. Klinkenburg and Van der Minne, *Electrostatics in the Petroleum Industry*, Elsevier, New York, 1958.

17. P. Guest, V. Sikora, and B. Lewis, *Static Electricity in Hospital Operating Suites.* U.S. Bureau of Mines Bulletin No. 520, U.S.G.P.O., 1953.

18. J. Bulgin, "Factors in the design of an operating theatre free from electrostatic risks," *Brit. J. Appl. Phys.*, Suppl. 2. 1953.

19. R. E. Thompson, "Electrostatic safety in clothing," *Fire J.*, **63** (November 1969).

20. E. T. Pierce, *Waterfalls, Bathrooms and Perhaps Supertanker Explosions*, Stanford Research Institute Report 4454 to O.N.R. Contract Nooo14-71-C-0106, December 1970.

21. B. Vonnegut, *Electrical Behavior of an Airplane in a Thunderstorm*, Arthur D. Little Report of the F.A.A. on Contract FA64WA-5151, 1965.

22. H. S. Appleman, *Lightning Hazard to Aircraft.* Technical Report 179, H.Q. Air Weather Services (M.A.C.) USAF, April 1971.

23. P. J. Klass, "Lightning stroke threat increases," *Av. Week Space Technol.*, Jan. 6, 1969.

24. Lightning and Static Electricity Conference, December 1968, Part II, Conference Paper, May, 1969 Technical Report AFAL, Wright-Patterson AFB, DDC No. AD 693 135.

25. *FAA Report of the Conference on Fire Safety Measures for Aircraft Fuel Systems*, December 1967, DDC No. AD 672 036.

26. A. Few, "Lightning and the new-generation aircraft," *Science*, **168** (May 1970).

27. W. I. Kisner, "Causes and prevention of static markings on motion-picture film," *J. Soc. M.P. T.V. Eng.*, **67** (1958).

28. P. J. Sereda and R. F. Feldman, "Electrostatic charging in fabrics at various humidities," *J. Text. Inst.*, **55** (1964).

29. M. Lenzlinger, "Gate protection in MOS devices," *IEEE Trans. Electron Dev.*, April 7, 1971.

30. "MOS at work: The good and the bad," *Electron. Des.*, **8**, April 1969.

BIBLIOGRAPHY

Barnard, V. H., "Static electricity in textiles," *Am. Dyestuff Rep.*, **44**, P111–113 (1955).

Beach, R., "Industrial fires and explosions from electrostatic origin," *Mech. Eng.*, **75**, 307–313 (1953).

Beck, M. S., and J. H. Hobson, "Electrostatic charge measurement of particulate materials being transported at high velocities—Explosion risk meter for pneumatic conveyors," *Proc. Conf. on Dielectric Materials, Measurements, and Applications*, IEE, London, 1970, pp. 38–41.

Benderly, B., "Static electricity and radiographic films," *Radiogr. Photogr. Med.* (*Kodak Pathé*), **2**, 24–25 (1964); in French.

Carroll, J. M., "Measuring static charge on fabrics," *Electronics*, **25**, 206–210 (1952).

Cleveland, H. W., "Measuring electrification of film," *J. Soc. M.P. TV Eng.*, **55**, 37–44 (1950).

Cole, B. N., M. R. Baum, and F. R. Mobbs, "An investigation of electrostatic charging effects in high-speed Gas–Solids pipe flows," in *Fluid Mechanics and Measurements in Two Phase Flow Systems*, Institution of Mechanical Engineers, London, 1970, pp. 77–83.

Coolige, J. E., and G. Schulz, "Note on static electrification of dust particles on dispersion into a cloud," *J. Appl. Phys.*, **22**, 103 (1951).

Cunningham, R. G., "Electrification of insulating belts passing over grounded rollers," *J. Colloid Interface Sci.*, **32**, 401–406 (1970).

———, "Frictional electrification of belts of insulating materials," *J. Appl. Phys.*, **35**, 2332–2337 (1964).

Dorset, B. C. M., "Static electrification of textile goods and use of antistatic treatment," *Text. Mfr.*, **84**, 468–472 (1958).

Gill, E. W. B., "Static Electrification of Petrols," *Nature*, **173**, 398–399 (1954).

Howard, J. G., "Static electricity in the petroleum industry," *Elect. Eng.*, **77**, 7, 610–614 (1958).

Kisner, W. I., "Causes and prevention of static markings on motion-picture film," *J. Soc. M.P. T.V. Eng.*, **67**, 513–517 (1958).

Kunkel, W. B., "The static electrification of dust particles on dispersion into a cloud," *J. Appl. Phys.*, **21**, 820–832 (1950).

Lenaerts, E. H., "Study of static charge on paper tape for computers," *J. Sci. Instr.* (*GB*), **43**, 377–379 (June 1966).

Owens, J. E., "Methods for preventing static damage to photographic paper," *Photogr. Sci. Eng.*, **13**, 280–283 (1969).

Pierce, E. T., Triggered Lightning and Some Unsuspected Lightning Hazards, A.A.A.S. 138th Annual Meeting. Philadelphia, 1971.

Shafer, M. R., D. W. Baker, and K. R. Benson, "Electric currents and potentials resulting from the flow of charged liquid hydrocarbons through short pipes," *J. Res. Nat. Bur. Std.* (*U.S.*), **69C**, 307–317 (1965).

Smith, J. C., *Static Electricity Generated in Fibrous Materials*, U.S. Nat. Bur. Std. Report 4752, 1956.

"Standard Method of Test for Electrostatic Charge," D2679–69, *Annual Book of ASTM Standards*, **29**, 1111–1115, 1970.

Steiger, F. H., "Evaluating antistatic finishes," *Text. Res. J.*, **28**, 721–733 (1958).

Strickland, A. C., Ed., "Proceedings of the Conference on Static Electrification," *Brit. J. Appl. Phys.* (*Suppl. 2*), **4** (1953).

Vonnegut, B., "Effects of a lightning discharge on an aeroplane," *Weather*, **12**, 8 (August 1966).

CHAPTER 19

Other Electrostatic Effects and Applications

JAMES D. COBINE

Atmospheric Sciences Research Center
State University of New York at Albany
Albany, New York

19.1 INTRODUCTION

Some of the many applications of electrostatics that have not been covered in the previous chapters, for various reasons, are discussed briefly here. Of some of the examples it may be questioned whether they are truly "electrostatic"—especially some of the discharge applications. However, they usually satisfy the requirements that the effects produced are primarily due to charges of one sign, the current present is quite small, and the voltages are usually relatively high. Most of the effects are quite interesting and many have great potential. The individual discussions are necessarily brief, but references are given to allow the interested student to expand his knowledge.

19.2 FIELD EMISSION OF ELECTRONS

Very high electrostatic fields, usually in excess of 10^5 V/cm, can cause the emission of electrons from metal surfaces.[1] Usually this phenomenon occurs under vacuum from fine points where the necessary high field can be produced at reasonable voltages. Cold cathode x-ray tubes make use of the phenomenon. When gases are present, electron emission becomes a very complex process involving space-charge distortion of field distribution and positive ion bombardment of the emitting point. Thermionic emission and emission by positive ion bombardment often are the most important mechanisms. However, the effect of "field" emission by positive ion space-charge fields is often considered an important mechanism at the cathode of the electric arc on low-melting-point metals.

19.3   ELECTROOPTICAL SHUTTER

Certain liquids that are normally isotropic exhibit birefringence in an electro-static field.[2] By polarizing the light entering a cell containing such a liquid and passing it through a second polarizer adjusted to cut off the light, an electrooptical shutter called a Kerr cell is obtained. A suitable voltage producing a field at right angles to the light path "opens" the shutter. Such a shutter can be exceedingly fast in its response to a high-voltage pulse. Photographs of very short duration sparks have been obtained by this method. The Kerr cell can also be used to measure the magnitude of submicrosecond electrical pulses,[3] which are difficult to measure by conventional systems.

19.4   ELECTRON AND ION BEAMS

Electrostatically accelerated and deflected electron beams are well known in their applications in cathode-ray oscillographs. Ion beams have also received considerable attention in connection with isotope separation in the atomic energy field. Neither of these electron and ion beam applications is discussed further.

Electron beams of relatively high power *in vacuo* are especially valuable in metallurgy for melting, purifying, and welding many difficult materials.[4] This process has been found especially useful in melting and purification of titanium.[5] The electron beams used in this technology are electrostatically accelerated through potentials of from 15 to as high as 150 kV. By suitable focusing of the beam (electrostatically or magnetically), power densities at the work surface as high as 10^9 W/cm^2 may be attained. Thus superheating to produce reactions not otherwise obtainable can be employed. Zone refining and the growth of single crystals are important applications of electron beams.

Ion beams have found an important application in the semiconductor art in a process known as ion implantation.[6-8] This is the process whereby atoms of one material are introduced on and into the surface of a solid of another material. This can range from the addition of a relatively small number of "impurity" atoms to literally plating the solid with the new atoms. The process involves the acceleration of ions of the desired atoms in beams variously in the energy range of thousands to millions of electron volts. Mechanical, electrical, magnetic, and superconducting properties of materials are all affected by the presence of foreign atoms in various amounts. In some cases, as in semiconductors, the impurities thus introduced can have a dominant influence. In order to minimize the heating of the surface, the ion beam current density is kept below about 5 $\mu A/cm^2$—much lower than is used in the previously mentioned electron beam technology, where heating is the main process.

A number of processes involving the use of a glow discharge are essentially ion implantation processes wherein the relatively low-energy positive ions can be obtained from the cathode region of an abnormal glow discharge.[9] In the low-pressure glow discharge, an electrostatic field is established at the cathode surface by the presence of a positive ion space charge between the cathode and the luminous portion (plasma) of the glow. These ions occupy and flow through this "cathode drop" region as a steady-state phenomenon. As some move into the cathode others replace them from the plasma, or are produced by the ionizing action of electrons emitted from the cathode by ion bombardment. Thus the cathode and other surfaces at or near this potential are subjected to a continual rain of ions of moderate (several hundred volts) to high (several kilovolts) energy, depending on current and gas pressure. If these ions are of an inert gas or hydrogen, the surface is cleaned. The process of ion bombardment also loosens atoms of the cathode surface,[10] which are expelled and deposited on adjacent surfaces. This is known as sputtering.[11-13] Many types of thin film can be formed on conducting, semiconducting, or insulating surfaces by sputtering.

By a suitable choice of steel and the use of an abnormal glow discharge in nitrogen, strong "case hardening" (nitriding) can be effected more rapidly than by chemical means.[14-16] This is especially useful for intricate or difficult shapes, such as large and small gearteeth and the inner surface of gun liners. The heat produced by the ion bombardment facilitates the diffusion of ions driven through the surface by their incident energy. Boron has been deposited on quartz by a glow discharge in hydrogen–boron trichloride mixture.[17] Metals can also be deposited by the processes in which sufficient vapor, as evaporated from a filament, is mixed with a suitable gas for the glow process. The glow discharge in certain vapors produces various useful chemical reactions. For example, some organic vapors in a glow discharge yield polymers than can be deposited on exposed surfaces, thus forming valuable insulating films.[18]

19.5 DISCHARGE CHEMISTRY

Glow and corona discharges have served for many years to produce chemical reactions.[19] Many interesting and useful applications have been found. In these applications it is important to prevent the corona discharge from causing a complete breakdown of the gas to form a hot, concentrated column arc. When ac is used, this is done by covering the entire exposed surface of one electrode with a dielectric. This material should have sufficient electric strength to withstand the entire applied voltage. Preferably the dielectric should have a high value of dielectric constant. The higher the dielectric constant, the greater the proportion of the applied voltage that is exerted across the reaction zone. If dc were used, the dielectric would have to be partially conducting so that it could act as a ballast resistor and drain off the charge collected at the surface.

Many reactions produced by corona are the result of ozone formed by the discharge in air.[20] Coffman and Browne discussed processes involved in all gas systems, gas–liquid systems, and reactors handling gas-powdered solid (fluidization) systems, and their work should be consulted for more details.[20]

19.6 THE ELECTRIC CURTAIN

It has been found[21,22] that electrically charged particles such as dust and aerosols can be made to move through channels in space by suitably designed alternating electric fields. Alternation can be made to take place without the particles touching the electrodes used to establish the geometric configuration of the electric field. A simple form of "curtain"—perhaps fence would be a better word—consists of an array of parallel cylinders that may follow a curving path, much as a picket fence follows a path. Connecting the cylinders alternately to positive and negative poles of an alternator makes the array function as a curtain. A charged particle brought into this field undergoes periodic motion nearly along a line of force impeded by the viscous drag of the air. Since the lines of force are curved, the particle also experiences a centrifugal force normal to the lines of force.

When there is a "cloud" of charged particles, each individual particle experiences a force due to the space charge of the other particles. A given particle alternates in position according to these forces and its own inertia due to its mass, which contributes to a lag in phase with respect to the electric field. The centrifugal force tends to push the particle away from the nearest electrode, where the field is most intense, pulling it back where the field intensity is weaker. The result is that the particle is always repulsed from the electrode array, which thus acts as a fence or curtain.

Transport of the cloud can be effected by the superposition of a dc field along the fence. Filtering, beam shaping, and funneling of dust clouds can be achieved by suitable choice of field shape, gradients, and so on. Obviously the equation of motion is very complex; however, Masuda[22] found that the paths determined by computer are confirmed by experiment.

19.7   CHARGED PARTICLE SENSORS

Accelerometers, gyros, and other sensing devices have been developed from the levitation of a small charged particle in an electric field. The oil-drop experiment that Millikan[23] used in determining the charge on the electron is its simplest form. It has been found that small charged particles can be held in dynamic equilibrium by alternating electric fields.[24,25] This may be looked on as a closed form of the electric curtain of the previous section. Small metal particles have been studied during combustion by a system of balancing electric fields.[26] The particles were ignited by a laser beam and time-resolved spectrograms were taken during the combustion process. The use of three-phase electric fields, each phase exciting the field in one of the three mutually perpendicular axes[25] in space, permits acceleration sensing along all three axes. The exciting structure is arranged in the form of a hollow cube with each of the three sets of opposite faces connected to the terminals of a Y-connected three-phase voltage source. Such charged particle sensors have been arranged as accelerometers for one, two, or three axes; a single-particle gyro unit can replace three of the more expensive electromechanical type accelerometers, and as free particle devices they can be used to measure gravity.[27] The gravity meter works by tossing a charged particle upward past two photosensors and measuring the time the particle takes to rise and fall a known distance.

19.8   EFFECTS OF ELECTROSTATICS ON FLAMES AND HEAT FLOW

Flames are produced by reacting gases. In flames, positive and negative ions and electrons are produced partly by chemical reaction and partly by thermal ionization occurring in the high-temperature gases. Since free charges are present, a flame can be affected by an electric field,[28,29] and it has a measurable conductivity, usually of the order of 10^{10} mho/m. The various ions move in the field as determined by their respective mobilities.[30] Space-charge layers form in front of electrodes immersed in a flame which distorts the voltage distribution from its theoretical geometric form. Where ions of a single sign are present in any region, the field-directed motions of these ions can produce an "ion wind" in the neutral gas in and surrounding the flame, by momentum transfer. If the electric field impressed on a flame is modulated at audio frequencies, the flame will radiate the sound. It is possible to control and direct the combustion of some fuels, as well as convection, by the use of a suitable electric field.[31,32] An interesting application of this is found where combustion takes place in a gravity-free region. The ability of an electric field to influence convection can have a marked effect on the heat transfer at a surface.[33–35] Corona at a wire heated to 500°F has been found to produce a fourfold increase in the heat-transfer rate.

It has been demonstrated that a high-voltage probe (25 kV) placed about 2 in. from a hot surface can produce a distinct cooling effect[36,37] with currents of from 10 to 60 μA. The effect has been used to control the temperature of the surface of metal near the bead during welding, with resulting metallurgical improvements. It has also served to cool semiconductor elements and in the cooling of optical elements of a high-power carbon dioxide laser system. It is well known[38] that a high-voltage point produces a "corona wind" or electric wind because of the motion of ions leaving the point. This ion motion can drive "cool" gas molecules to the hot surface so that both ions and neutral molecules disrupt the insulating gas layer nearest to the surface. The drawing off of electrons emitted by the combined effects of high temperature and high electric field would also occur with a very hot negative surface. The cooling produced by this process is proportional to the product of the surface work function and the current density. Although the effect is demonstrated with dc, it also occurs where ac is used.

19.9 PHOTOGRAPHIC EFFECTS

Electrostatic charges can be produced on photographic film during handling, while in the camera, and during processing. Safe discharging of the film surface may take place under conditions of high humidity by various leakage processes. This charging is especially bad when the humidity is very low. The charged film attracts dust particles that may adhere tenaciously. What is even worse is the appearance of "static marks" on the developed film.[39] These marks are photographs of the ionized path produced in the air above the film surface as the surface is discharged. These static marks are the same as the so-called Lichtenberg figures,[40,41] which can be used to measure high-voltage surges on transmission lines. There seems to be no easy solution to the problem.

19.10   STATIC ELIMINATORS

Many manufacturing processes cause the generation of electrostatic charges because of the touching and subsequent separation of materials of different composition, such as paper and metal rolls. Fires, product spoilage, and disagreeable shocks to personnel may result from the accumulation of electric charge. The charges seldom cause trouble when the humidity is high. Where operation at a very high humidity is impossible, a number of techniques have been employed in the past. Among these techniques, which were only partially effective and difficult or impossible to control, are (1) an array of sharp points on a grounded metal strap placed close to the charged material, (2) a similar array of points placed at a high potential (5–15 kV) relative to ground and producing a more or less steady corona discharge, (3) an open flame, usually playing on or very near the charged surface.

System 1 requires the material to develop a charge sufficiently great to start a corona discharge—this is erratic; the phenomenon depends on atmospheric condition, and the operator has little control over it. Method 2 works by "spraying" charges of either (as needed) or both signs according to whether dc or ac excitation is used. Better control exists with this method because the potential of the points and their discharge current can be varied. However, there always exists the possibility of the corona at one point developing an arclike discharge, forming a tiny heated channel sufficiently intense to produce ignition of explosive vapors or to fire the material being "protected." The flame of method 3 produces ions of both signs and is a conductor of low conductivity. It may work quite well while the material is moving rapidly, but if the material stops moving, a fire starts immediately!

One of the most effective and safe systems of static elimination consists of incorporating the radio active isotope ^{210}Po in a metal film or other supporting structure, which is placed close to the surface that is becoming charged. This isotope emits only short-range alpha particles, which are capable of ionizing the air sufficiently to discharge adjacent charged surfaces. Since the isotope produces charge separation in the air without any concentrated discharge as corona or other flame, it is safe in even explosive atmospheres.[42] Also important is the absence of potential health hazards such as oxides of oxygen or nitrogen.

In another application, the ionizing source is incorporated in an air gun or blower so that ionized air can be directed onto charged surfaces. Devices for measuring the intensity of static electric fields consist of an isotope ionizing source (tritium, a beta-emitter, is sometimes used) and a meter to measure the current drawn by the field from a partially closed chamber.

Quite a different effect can be produced by an array of points at a high potential as the "static eliminator" mentioned as process 2. The corona discharge thus produced in air has been found very good for treating polyethylene film, bottles, and so on, so that they may be printed.[43] Although some ion bombardment takes place, the most important factor in changing the surface is probably the action of ozone that forms with corona in air. A device utilizing a corona discharge should always be well ventilated, because the oxides of oxygen and of nitrogen form and create a health hazard.

19.11 ELECTROSTATIC FIBER FRACTIONATION

The property of dielectric filaments to align themselves along the lines of force of an electric field has long been used to map electric fields. Uncharged dielectric particles do not move along the force lines in a uniform field, but in a divergent field they move toward the strongest part of the field. Thus even though uncharged, they may move from one electrode to the other in a field such as a wire to a plane or concentric cylinders. This phenomenon can be used in gases and in some liquids to separate particles of different sizes or to purify the fluid. Shaping an electric field into strong and relatively weak regions can be used to separate dissimilar materials such as grain and husks.[44] Somewhat similar schemes can be used in cotton processing to separate short fibers (i.e., those less than ca. 0.75 in.) from the longer fibers. The presence of these short cotton fibers decreases the spinning efficiency, decreases the strength of the material, and affects the appearance of the yarn.[45,46] In this application, the mixed fibers are fed into the low-field region and, as they are carried along by an endless, nonconducting belt, the long fibers move into the strong-field region. The two groups of fibers can then be separately removed by suction.

19.12 ELECTROSTATIC COATING

A number of processes have been developed for coating surfaces by electrostatics. One process is quite analogous to electroplating of metals in that it takes place in a liquid in which positive and negative electrodes are immersed. In this electrodeposition[47] the resin to be used as a coating is dispersed in the liquid (water) and becomes the negative ions that are conducted to the surface to be coated by the electric field. The surface—a wire or foil—is made positive (100–200 V) to attract the negatively charged resin particles. The rather complicated process of transferring the resin to the surface continues until the entire exposed surface is coated. It is easily mechanized for wire by having the wire pass through the bath at a speed suitable for complete coating and then through an oven, where the resin is polymerized to an insoluble thermoset resin that is an insulator. By this system a *single* pass through the bath is all that is required, whereas for a simple dip-and-heat process, from four to six passes are required.

Various resins in powder form may be made to coat surfaces by immersing the object in a "fluidized" bed of the powder.[48] The fluidization is accomplished by passing air up uniformly through the powder by a porous supporting plate. When the air velocity is sufficient to overcome the weight of the powder, it resembles a boiling liquid. Preheated objects to be coated may be immersed or supported in this dry fluid and coated by the thermosetting resin.

Coats as thin as 0.005 to 0.010 in. cannot be obtained except by use of electrostatics.[49] In this process, the powder bed can be quite thin and the air pressure low. Contact electrification takes place in the fluidization process just described, but an applied dc field acts to separate out the charges. For this effect the bed is made about 90 kV negative relative to the *grounded* object to be coated. The object is suspended in the cloud of charged dust and uniformly coated. Thin parts that will not hold a preheat can be coated. In addition, once coated, the charged dust sticks so well that the object can even be stored at room temperature for several days before the final heating.

Electrostatic spray coating[50] is accomplished by using an air-blast spray gun constructed to give a negative charge to the powder by a high-voltage, low-current corona. The part to be coated is grounded. The object being coated can be heated while being sprayed or it can be heated later. Paints can also be applied in this manner with much less loss of paint, since the charged droplets follow the electrostatic lines of force from gun to object, just as the dust particles do.

An electrostatic fluidized bed can also be used to separate ores[51] such as brucite–calcite ore. In this process a field of 800 kV/m was obtained from a 60-kV half-wave rectifier.

19.13   MECHANICAL APPLICATIONS

Devices for mechanically braking and gripping can be actuated at very low power by means of an electrostatic field. Braking is effected by the use of a liquid whose viscosity changes when subjected to an electrostatic field.[52] Winslow discovered that dielectric particles suspended in an oil increased the drag between two plates when an electric field was established by a potential difference between the two plates.[53] In one fluid the viscosity increases as the square of the electric field strength for values of field greater than about 6 kV/cm. The viscosity is nearly constant for frequencies from dc to about 100 Hz and then decreases to a constant value less than a tenth of the low-frequency value for frequencies of the order of 5 kHz.

Exceedingly thin pieces of both ferrous and nonferrous metals can be held electrostatically for machining.[53,54] The workpiece is connected to a low-power voltage source and placed on a base plate of semiconducting material whose purpose is to ensure a uniform distribution of the field over the surface. An oil (which may be field sensitive as previously, but need not be) serves to exclude air and to make good contact between the base plate and the workpiece.

Electrostatically operated signs have been developed that may soon effect a revolution in changeable indoor and outdoor signs. Blackened rectangular aluminum vanes are closely spaced, each way, and are selectively flipped by electrostatic forces. If all are vertical, a solid black panel faces the viewer. When all are horizontal, only a white background shows. By command, only those are flipped to vertical that will show the desired message in terms of letters, cartoons, trademarks, and so on. A variety of controls is available, and the signs can be manual or programmed, operated either locally or from any distance. (One manufacturer of electrostatically operated signs is Display Technology Corp., 10351 Bubb Road, Cupertino, Calif.)

19.14   ELECTROSTATIC—MECHANICAL TRANSDUCERS

An electrostatic field, modulated by signals in the audible range, can stimulate the sense of hearing without an accompanying sound wave in the intervening medium.[55] In practice, the acoustically modulated electrostatic field is coupled to the flesh in the vicinity of the ear (for best results the lobe in the forward portion of the ear) through a high dielectric constant material on the electrode. The electrode (part of a head piece) is made sufficiently massive that little or no energy is coupled to the surrounding air. The electrostatic field produces a force on the flesh near the ear and causes the tissue to vibrate. The force is proportional to the square of the electrostatic field terminating on the flesh. This square relation can introduce a second-harmonic distortion that can be minimized. Applications of promise include training of the deaf and underwater communication with a diver. In working with deaf pupils, a distinct advantage is the elimination of feedback coupling between headsets of student and teacher. In the latter application, excellent reproduction is experienced by the diver without depth sensitivity.

Another transducer works on a different principle to couple an electric signal to a fluid. An organic liquid is forced through a porous membrane by the action of an electrostatic field impressed across the membrane (i.e., by electroosmosis).[56] The electrostatic field drives ions in the fluid through the pores of the membrane; normally the direction is that of positive ions of the fluid moving toward the negative electrode. The direction, of course, depends on the relative abundance of positive and negative ions. The net flow depends on the size and concentration of the positive and negative ions (i.e., on the nature of the fluid). The effect is reversible in that a pressure drop across the membrane induced by mechanical pressure on the fluid will develop an electric field across the membrane. In a cell consisting of a ceramic-frit porous membrane with metal screen electrodes on either side, a pressure difference of 1.2 psi generated an electrostatic potential of 2800 V. Since a special fluid is required surrounding the membrane, the pressure change induced by an electric field is coupled to an external fluid system by flexible diaphragms of compatible material at each end of the cell. The response of such a cell is linear with frequency up to at least 1000 Hz.

19.15   ELECTROSTATICS AND LIVING THINGS

The blue haze over heavily forested areas now seems to be electrostatic in origin. It has been shown that an electric field strong enough to cause high intensity at the tips of pine needles can produce wax fingers, from which come airborne particles of less than 0.6-μm diameter.[57] A wax coating is common to many types of vegetation. It seems possible that the atmosphere's field, especially at its higher values below charged clouds, may account for the haze and for the presence of terpenes and other organic compounds.

One proposed alternative to the usual citrus orchard heating by fires, smoke screens, and so on, with their serious pollution effects, uses electrostatic fields.[58] The high-frequency electric field, produced by a Tesla coil, was located so that it terminated on the leaves of small lemon trees. The presence of the field was demonstrated to protect the plants from temperatures of 13.2°F, which destroyed the unprotected control plants. It seems that the high-frequency currents flowing on the sensitive surface layers of the plants produce enough heat at the right place to prevent harm. Of course some rectification of the high frequency may be present, with a resulting ion current reaching the plant from the air and returning through the ground to the source. It is not clear from this work how the positive and negative ions formed by a discharge current as well as the ozone and various nitrogen oxides[59] that may be present, are separately involved in protecting the trees.

Many experiments have been performed over the course of many years to determine the effects of ions and electric fields on organisms.[60] However, as pointed out by Bachman,[61] it is very difficult to control separately and to distinguish among the effects of an electric field, ions, and the very active chemicals such as the oxides of oxygen and of nitrogen that usually form.

These authors demonstrated a pronounced electric field effect for barley. Initially the electric field increased the growth over that of a control group of plants. However, as the plants continued to grow toward an upper electrode, they reached a critical zone of higher field where the growth rate was markedly reduced. Ion currents of each sign from leaf tips increase markedly for "space fields" in excess of 80 kV/m and are accompanied by increasing amounts of ozone.

Although the fair-weather electrostatic field at the earth's surface is about 100 V/m, this field can reach 50 kV/m for a positive field and 10 kV/m for a negative field. Since the electric field at a point is many times the average field in which it is immersed, corona, ions, and ozone can be present in varying amounts. Thus the "weather," acting through its electric field, can produce a variety of incompletely understood effects on living things. There must be a lot of corona present at the leaf tips of trees in thunderstorm electric fields.

The disturbing effects of ozone may be eliminated by the use of a radio-active material suitable for ionizing the air. Experiments[62] using this technique[63] have been made with rats at ion densities in the range 10 to 60,000 ions/cm^3. This value is from 5 to 30 times the normal air ionization and produced significant physiologic effects. Both positive and negative ions tended to stimulate both heart beat and respirative frequencies. Experiments with mice[64] infected with viral pneumonia have demonstrated that positive ions caused an increase in the death rate. Similar experiments with influenza[65] led to the conclusion that air ions are capable of producing functional changes in a diversity of living forms. It seems that the effects of positive ions are detrimental, whereas the effects of negative ions are beneficial, or at least not harmful.

Experiments[66] on humans inhaling air with an ion concentration of about 10^6 ions/cm^3, or about 1000 times the normal atmospheric value, have been conducted, arranged so that the only ions reaching the subject were through the nose. Reaction time measurements under ionization were ambiguous. In vigilance tests, ions of both signs were equally effective in reducing omissions. Neither polarity affected the heart rate. A reduction in respiration rates occurred for both ions of both the polarization and the control, but the reduction for positive ions was the greater.

Although much work has been accomplished in this field, even more must be done before these complex phenomena are understood and new effects are established.

An improved version of air ion therapy is electro-aerosol therapy (E.A.T.) introduced some 30 years ago. It is estimated that by now, a million patients in 15 countries have been treated, by this method, for bronchial asthma, nasal and sinus conditions, bronchitis, emphysema, and other ailments of the respiratory tract. It was introduced in this country in 1961 by Dr. Alfred P. Wehner, who has published a comprehensive review of the subject that includes 226 references.[67] In E.A.T. the common treatment of inhaling uncharged aerosols is enhanced by first charging the aerosols; there is evidence that the treatment is thus made more efficacious. Research in this area should be strongly supported and extended since this is a matter of interest to possibly 20 million sufferers in the United States.

REFERENCES

1. W. P. Dyke, *Sci. Am.*, **210**, 108 (1964).

2. J. Kerr, *Phil. Mag.*, **50**, 337, 446 (1875).

3. E. C. Cassidy, H. N. Cones, and S. R. Booker, *IEEE Trans. Instr. Meas.*, **IM–19**, No. 4, 395 (1970).

4. R. Bakish, Ed., *Introduction to Electron Beam Technology*, Wiley, New York, 1962.

5. H. Stephan and W. A. Dietrich, *Research/Development*, April 1971, p. 44.

6. J. W. Mayer and D. J. Marsh, "Ion Implantation in Semi-Conductors" in *Applied Solid State Science*, R. Wolf, Ed., Academic Press, New York, 1969.

7. J. W. Mayer, L. Erikson, and J. A. Davies, *Ion Implantation*, Academic Press, New York, 1970.

8. J. F. Gibbons "Ion implantation in semi-conductors, Part I, Range distribution theory and experiments," *Proc. IEEE*, **56**, 295 (1968). (See also: G. R. Brewer, *IEEE Spectrum*, January 1971, p. 23.)

9. J. D. Cobine, *Gaseous Conductors*, Dover, New York, 1958, p. 205.

10. J. B. Newkirk and W. G. Martin, *Trans. ASM*, **50**, 574 (1958).

11. J. W. Nickerson and R. Moseson, *Research/Development*, March 1965, p. 52.

12. L. Holland, *Brit. J. Appl. Phys.*, **9**, 410 (1958).

13. G. Wehner and G. Medicus, *J. Appl. Phys.*, **25**, 698 (1954).

14. P. M. Unterweiser, *Iron Age*, May 2, 1957, p. 91.

15. T. A. Vanderslice and J. D. Cobine, *Reactions of Surface Layers by a Glow Discharge Process*, General Electric Research Laboratory Report 58RL2023, June 1958. See also: H. Knuppel, K. Brotzman, F. Eberhard, *Stahl U. Eisen*, **78**, No. 26, 1871 (1958).

16. C. K. Jones and S. W. Martin, *Metal Prog.*, February, 1964, p. 95.

17. A. E. Hultquist and M. E. Sibert, *Advances in Chemistry Series*, No. 80, American Chemical Society, 1969.

18. S. M. Lee, *Insulation/Circuits*, June 1971, p. 33.

19. G. Glockler and S. C. Lind, *The Electrochemistry of Gases and Other Dielectrics*, Wiley, New York, 1939.

20. J. A. Coffman and W. R. Browne, *Sci. Am.*, **212**, 91 (June 1965).

21. W. Paul and H. Steinwedel, *Z. Naturforsch.*, **8a**, 448 (1953).

22. S. Masuda, *Electric Curtain for Confinement and Transport of Charged Aerosol Particles*, Albany Conference on Electrostatics, 1971, Electrostatic Society of America, June 8–11, 1971.

23. R. A. Millikan, *The Electron*, 2nd ed., University of Chicago Press, Chicago, 1924, p. 57.

24. R. F. Wuerker, H. Shelton, and R. V. Langmuir, *J. Appl. Phys.*, **30**, 342 (1959).

25. ———, H. M. Goldenberg, and R. V. Langmuir, *J. Appl. Phys.*, **30**, 441 (1959).

26. L. S. Nelson, N. L. Richardson, and J. L. Prentice, *Rev. Sci. Instr.*, **39**, 744 (1966).

27. Martin Marietta Corp., *Charged Particle Devices Status*, Report OR–9638, October 1968.

28. J. J. Thomson and G. P. Thomson, *Conduction of Electricity Through Gases*, 3rd ed. Vol. I, University Press, Cambridge, England, 1928, p. 399. See also: H. A. Wilson, *Rev. Mod. Phys.*, **3**, No. 1, 156 (1931).

29. R. J. Heinsohn and P. M. Becker, *Flames, Ions and Electric Fields*, American Society of Mechanical Engineers Publ. 70-WA-Fu-4. This paper has a good bibliography, including many very early works.

30. J. D. Cobine, *Gaseous Conductors*, Dover, New York, 1958, Chapter 2.

31. See Ref. 29.

32. J. Wong and J. R. Melcher, *Phys. Fluids*, **12**, 2264 (1969).

33. M. Robinson, *Int. J. Heat Mass Transfer*, **13**, 263 (1970).

34. E. Bonjour, J. Verdier, and L. Weil, *Chem. Eng. Progr.* **58**, 63 (1962).

35. H. Y. Choi, Jr., "Heat transfer," *Trans. ASME*, February 1968, p. 98.

36. *Design News*, **17**, Oct. 27, 1969, pp. 62–63; **19**, Jan. 18, 1971, Cover; **19**, June 21, 1971, p. 13; **19**, July 5, 1971 pp. 52–53.

37. J. McDermott, *Design News*, **19**, 22 (Sept. 30, 1971).

38. L. B. Loeb, *Electrical Coronas*, University of California Press, Berkeley, 1965, p. 402.

39. N. Goldberg, *Pop. Photogr.*, **68**, 94 (April 1971).

40. J. D. Cobine, *Gaseous Conductors*, Dover, New York, 1958, p. 201.

41. E. Nasser, *Fundamentals of Gaseous Ionization and Plasma Electronics*, Wiley, New York, 1971, Chapter 11.

42. Technical data: 3M Company Nuclear Products, St. Paul, Minn.; Nuclear Radiation Developments, Inc., New York; Simco Co., Inc., Lansdale, Pa.

43. D. R. Mills, *Flexography*, September 1960, p. 26.

44. H. A. Pohl, *Sci. Am.*, **203** No. 6, 107 (1960).

45. M. Mayer, Jr., H. W. Weller, Jr., and J. L. Lafranca, Jr., *Textile Bulletin*, Agricultural Research Service, U. S. Department of Agriculture, March 1965.

46. J. J. Lafrance, Jr., M. Mayer, Jr., and H. W. Weller, Jr., *Text. World*, **116**, 58 (June 1966).

47. J. P. Haughney, *Insulation*, May 1966, p. 55.

48. A. H. Landrock, *Chem. Eng. Progr.*, **62**, 67 (1967).

49. ——— Plastec Note 18, Plastics Technical Evaluation Center, Picatinny Arsenal, Dover, N.J., February 1968.

50. *Ibid.*

51. K. I. Burgess, I. I. Inculet, and M. A. Bergougnou, Paper 41a, presented at the 62nd Annual Meeting of the American Institute of Chemical Engineers, Nov. 16–20, 1969.

52. U.S. Patent 3,047,507, Warner Electric Brake and Clutch Co., Beloit, Wisc.

53. D. DeSimone, Ed., *Education for Innovation*, Pergamon Press, New York, 1968, p. 85.

54. V. E. Vavra, *Machine and Tool Blue Book*. Also: Electroforce, Inc., Fairfield, Conn.

55. M. Salmanshon, Naval Air Development Center Report NADC-AE-6922, Nov. 7, 1969. See also: R. H. Einhorn, *Electron. Des.*, Dec. 26, 1967.

56. R. Lee, *Prod. Eng.*, July 4, 1970, p. 71. See also *Technical Data*, Electro-Dynamics, Inc., Cleveland, Ohio.

57. B. L. Fish, "Electrical generation of natural aerosols from vegetation," *Science*, **175**, 1239–1240 (March 17, 1972).

58. G. E. Horanic and G. Yelenosky, *Proc. 1st Int. Citrus Symp.*, **2**, 539 (1969).

59. C. H. Bachman, D. G. Hademenos, and L. S. Underwood, Jr., *Atmos. Terrestr. Phys.*, **33**, 497 (1971).

60. M. Kroll, J. Eichmeir, and R. W. Scüdn, in *Advances in Electronics and Electron Physics*, L. Marton, Ed., Academic Press, New York, 1964, p. 177.

61. See Ref. 60.

62. C. H. Bachman, R. D. McDonald, and P. J. Lorenz, *Int. J. Biometeorol.*, **9**, 127 (1965).

63. S. Katoka and A. P. Krueger, *Int. J. Biometeorol.*, **16**, 1 (1972), provides other references.

64. A. P. Krueger, S. Kotaka, E. J. Reed, and S. Turner, *Int. J. Biometeorol.*, **14**, 247 (1970).

65. ———, ———, and ———, *Int. J. Biometeorol.*, **15**, 5 (1971).

66. R. D. McDonald, C. H. Bachman and P. J. Lorenz, *Aerospace Med.* **38**, No. 2, 145 (1967).

67. A. P. Wehner, *Am. J. Phys. Med.*, **48**, 3.

CHAPTER 20

The Status Abroad

NOEL J. FELICI
University of Grenoble
Grenoble-Gare, France

20.1 HISTORICAL OUTLOOK

The spectacular electrostatic phenomena of the earth's atmosphere, usually associated with thunderstorms, fascinated man long before he acquired the rational turn of mind he is so proud of. Primitive man's attitude toward the awe-inspiring manifestations of Nature, however, has never been one of mere resignation; for man tried from the very beginning to placate the powers he suspected were behind these forces and to bring them under some sort of control. Magic, as Lynn Thorndike pointed out, was the forerunner of both rational knowledge and technology; no wonder that the beliefs, superstitions, ceremonies, and rites related to thunder and lightning clearly testify to man's innate conviction that he should be able to take advantage of them. The claim of having Jove's thunderbolts under control was repeatedly made by magicians and priests; Numa, a semi-legendary king of Rome, could direct lightning toward a piece of ground to give it *numen* (i.e., to have it hallowed by a god's will revealing itself). Much later, with the fall of Rome approaching as the Goths prepared for a final assault under Alaric's leadership, Etruscan priests proposed to Pope Innocent I (the only authority surviving in the harassed city) that they might arrange for the destruction of the Gothic hordes by lightning strokes. The negotiations between the heathen priests and Innocent went very far, but eventually failed when the Etruscans insisted on being given full credit for their feat of magic.

It has also been argued by several writers that the Ark of the Covenant was designed like a large capacitor, since it was overlaid with gold inside and outside, whereas the shittim wood has good insulating qualities under dry climatic conditions. Thus trespassers would be struck by discharges but the priests would be protected by their vestments laced with gold threads and adorned with chains, brooches, breastplate, and so on, of the same metal.

In Renaissance times, when modern rationality began to flourish, we encounter a typical electrostatic device, well designed for meteorological purposes. On the top of the Duino Castle, in the Friuli, near the Adriatic Sea (a part of Italy stricken with thunderstorms) a halberd was planted on a wooden pole. The sentinel on duty was to touch the halberd occasionally with the tip of his weapon and, if "fire" was seen jumping between the two points, the alarm-bell was rung at once, to warn peasants and fishermen of an impending thunderstorm.

The resounding discoveries of Franklin in America and Lemonnier in Europe, who demonstrated the electrical nature of lightning almost at the same time (although Lemonnier had a slight lead) need not be discussed here. It should be mentioned, however, that although the invention of the lightning rod was a piece of American practicality, many European scientists engaged in the more questionable field of physiological or medical electrostatics.

Experiencing jerks from Leyden jars became the fad of late eighteenth century; tens or even hundreds of people, standing in a ring and holding one another by the hands, were subjected to the discharge of a single capacitor, and the simultaneousness of their jumps delighted the spectators. It was claimed that the electric shock could cure the most various ailments, and strange electrode arrangements were designed to direct the electric fluid to the aching part of the body.

Many people believed in some mysterious link between life and electricity, a tempting idea that set the stage for the epoch-making discoveries of Galvani and Volta three decades later but also enticed respectable scientists into putting forward unwarranted theories. Sigaud de la Fond, having met with a failure when repeating the ring experiment with his pupils in Paris, suggested in a public lecture that persons lacking in sexual ability were insensitive to the electric fluid. A grand discovery indeed, particularly for the eighteenth century, which was even more sex-conscious than ours. The big news made the rounds of town and court—it was Louis XV's court—and the prospect of a quick and decent assessment of everyone's sexual powers was a rake's delight. Poor Sigaud soon felt that the joke was getting him too far afield, but in spite of his prevarication and excuses, one of the most distinguished gentlemen, the Duke of Orleans, appeared at his laboratory with a large retinue, including three eunuch singers from the royal chapel. The ring was formed, with the Duke and Sigaud at the ends, the eunuchs and various attendants in the middle. All jerked in unison, and Sigaud was ridiculed, although he tried to explain away the outcome of the experiment by assuming that eunuchs do not react like naturally impotent persons. Later on he discovered the truth about the earlier experiment: the boy in the ring who was supposed to stop the electric fluid had been standing on a conductive patch of ground and, moreover, tended to perspiration.

Some genuine discoveries of this period did not meet the same enthusiasm but are much more interesting to us, for they gave the first hints of the feasibility of two important processes, atomization and precipitation. When trying to elucidate the nature of the electric fluid, the French physicist Abbé Nollet discovered that electrified liquids escaping from a nozzle experienced a thrust that enhances their rate of flow and, eventually, disperse into a cloud of droplets attracted by grounded objects. Other experiments with lint particules or bits of thread are reported by the same author; this is clearly reminiscent of the well-known flocking process.

Skipping nearly a century, we reach the 1860s, the beginning of another golden age of European electrostatics, when interest focused on the first practical electrostatic generator, the influence machine. Electrostatic influence charging had long been known and devices taking advantage of the "addition" or the "multiplication" of charges had been imagined by Bennett, Volta, Cavallo, and others. But these were not generators in the technological sense (i.e., machines delivering electric power at the maximum rate compatible with size and structural materials and as based on physical principles). As concerns high voltage, the only generators available were the old friction machine and the more recent induction coil; however, these had to be supplied by dirty and expensive chemical batteries.

Under the circumstances, the invention of efficient electrostatic generators by three German physicists, Toepler, Holtz, and Musaeus, was hailed as a great achievement of applied science. How the new machines worked, nobody exactly knew, not even their inventors; experimenters contented themselves by observing that they could get, by merely turning a crank, more power than from a cumbersome induction coil with its costly chemical cells; and this lasted well into the end of the 1880s.

Near the close of the nineteenth century, however, the tide was slowly turning. Primitive power systems began to expand, making electric energy available in most cities, and primary chemical cells could be replaced by more efficient storage batteries. On the other hand, the working mechanism of the influence machine was still a riddle to the physicists; it could deliver up to 10 W of useful power with a 20-in. disc, but nobody knew how to improve that output, or whether it could be improved at all.

A very interesting attempt of the German physicist Hempel in 1885 might have given a clue, but it was misunderstood because no scientist living could explain how the composition and pressure of the gas surrounding an influence machine could act on its power capacity. The decisive role of the dielectric strength was not suspected. Under the sway of the prevailing theories, the explanation was sought in an inductive effect of the medium, not in a limitation related to the breakdown of this medium, although the data at hand, together with the known electric field laws, might have allowed a correct analysis. Thus, for lack of scientific understanding, the influence machines could not keep abreast of the slowly improving electromagnetic sources of dc high voltage, and they must have been doomed in the long run.

An unpublished correspondence between Toepler and Roentgen, the discoverer of x-rays, gives a vivid idea of the competition between the influence machine and the induction coil in the 1890s. Toepler, like a good salesman, boasted of the superior electric output of his multidisc generator, and Roentgen retorted that the x-ray photographic effect is better with an induction coil. In fairness, it should be added that the pulsed voltage of the coil was much better adapted to the requirements of Crookes's vacuum tube (as used by Roentgen) than the constant dc generated by the disc machine, and Toepler was thus playing a losing game.

After the turn of the century, an attempt by Sir Oliver Lodge to energize an electrostatic precipitator with a Wimshurst machine met with complete failure; henceforward, electrostatic generators were discarded altogether and no one cared about them until the revival initiated by Van de Graaff in the United States after 1930.

Before leaving this interesting period, we have to mention the numerous electrostatic gadgets imagined by Sir William Thomson (Lord Kelvin) in connection with atmospheric electricity (the water-dropper*) and trans-Atlantic telegraphy (the siphon-recorder and the replenisher). The siphon-recorder took advantage of Abbé Nollet's observation on the thrust exerted on a liquid by electrification; it is the first example of electrostatics applied to nonrepetitive printing. The replenisher, a tiny self-exciting influence generator, was used to energize the siphon-recorder; later it was applied to an electrometer and also to a domestic gas lighter.

20.2   EUROPEAN ELECTROSTATICS IN THE 1940s

The modern era of electrostatics was initiated by Van de Graaff's belt machine, which suddenly revived interest in a science hitherto regarded as antiquated and fruitless. The impact of Van de Graaff's invention was almost immediately felt in Europe. Nobel prize winner F. Joliot had much interest in it and at the Paris Universal Exhibition of 1937 he displayed

---

* Two water-droppers cross-connected gave an amusing self-exciting electrostatic generator. Thomson also constructed rotating-potential equalizers.

a twin generator of impressive dimensions whose construction he had directed personally. It could deliver sparks 16 ft long and was one of the chief attractions at the Exhibition.

At the same time, Pauthenier's research work on the electrification of tiny material particles by corona began to draw attention. Pauthenier's air-blast generator could not successfully compete with Van de Graaff's, but his physical theories were instrumental in showing the efficiency of electrifying divided matter for many technological processes. Early patents on the electrification of atomized liquid chemicals (filed by Julius and Feldzer in Paris in 1944) and on the electrostatic dusting of crops (Truffaut and Hampe; Paris, 1943) clearly testify to the influence of Pauthenier's findings.

The stage was thus set for the developments that took place in Western Europe immediately after the close of World War II. Joliot, now at the height of his career and reputation, hearing from Prof. L. Néel of the work done in Grenoble by N. Felici and R. Morel on pressurized rotating sector-type generators, gave enthusiastic support, and in 1945 a Laboratoire d'Electrostatique was set up by the Centre National de la Recherche Scientifique (CNRS), an organization of the French Ministry of Education. The following year a private firm was established and was granted a general license of the patent rights held by the CNRS. In 1950 A. W. Bright joined the team of the Laboratoire d'Electrostatique; after returning home, he was to play an important role in the development of applied electrostatics in Great Britain.

Early developments and trends in the main advanced countries outside the United States (Britain, France, Germany, the U.S.S.R., and Japan) reflected the status of these countries immediately after the war, as well as the prevailing mood in scientific and technological matters. Germany and Japan had to work very hard for their economic reconstruction and could muster but a limited number of trained scientists (a consequence of war and emigration), but the astonishing successes of big science in the last years of the war in Britain and France supplied a mood of faith in advanced research and technology. There was also a strong tendency to equate science with good technology, technology with industry, and industry with good business. It was often overlooked that a good scientist might be a poor technologist and that a good technologist might be a poor industrialist and, moreover, an inept lawyer. Thus in the early 1950s Britain and France engaged in several big projects concerned with nuclear energy or futuristic aircraft that met with ultimate failure, while more conventional segments of their economies—agriculture, railroads, automobiles, chemistry, and electrical engineering— made tremendous progress.

The case of the U.S.S.R. was fairly peculiar. Although the Russian mind has shown a twist in favor of unconventional and sometimes far reaching innovations, the needs of the State, whose majesty is hallowed in the Roman sense, have always had top priority. In the worldwide struggle for prestige and power, applied electrostatics could have no more than a third-rank, purely ancillary role. Consequently, the Soviet policy has mainly been one of "wait and see," taking good notice of trends and developments abroad, but not plunging into the field until the new procedures have stood the proof of conventional industry.

20.3   APPLIED ELECTROSTATICS IN BRITAIN AND FRANCE

Developments in Britain and France can be lumped together because of the similarity of their approach to research and development emphasizing genuine innovation and ambitious projects, and also because of the excellent cooperation that was very early established at the scientific and industrial levels. In the following, we briefly examine the main endeavors and their fates.

20.2.1.1   *Electrostatic Dusting of Crops*

The electrostatic dusting of crops began with the work of Hampe in Paris (patented 1943). Hampe's first experiments were performed indoors with an induction coil, and when he heard of the Grenoble generators he felt that an unhoped-for breakthrough had occurred. As a matter of fact, a self-exciting, hand-driven machine enclosed in an iron casing 8 in. long and 3 in. in diameter was constructed in a few months. It weighed (with casing) less than 4 lb, had an overall efficiency of 50%, and could be driven by the hand lever actuating the bellows, thanks to an ingenious gear equipped with pressure-tight packing to prevent loss of the insulating gas (air at 300 psi). The maximum electrical output was 50 kV, 40 μA, with an unusual voltage–current characteristic (the ratio of both quantities was kept constant). This feature, however, was quite compatible with the electric behavior of a corona ionizer and ensured a nearly constant electric gradient between the high-voltage electrode and the grounded objects it faced, even if their distance was rapidly changing. Moreover, the machine deenergized itself if the outer resistance fell below, say, 1000 MΩ, but recovered immediately if this resistance rose again. Absolute safety could be guaranteed, but the useful effect was interrupted for no more than a split second in the case of unwanted contacts.

Field tests that took place in 1946 and 1947 were very convincing. The overall pesticide consumption for a given amount deposited onto the plants was cut by a factor of 5 to 10 when electrification was applied, and the powder particles were much more evenly distributed (because of mutual repulsion). A considerable fraction was attracted to the lower side of the leaves, where pests find shelter (in their early stage of development) and easy access into the vegetal tissues (through the stomates). Even the remanence after a period of rain was improved by electrification, presumably because the particles impinge much faster and develop better mechanical adherence.

Unfortunately, the economic conditions prevailing in the late 1940s, which made this innovation enormously attractive, did not last long enough to permit commercial success. The severe shortages in raw materials and chemicals were rapidly eased, and nobody yet spoke of environmental pollution (Dr. Muller, the inventor of DDT, was awarded the Nobel prize for medicine in 1948). After 1950 European food production expanded at a rapid pace, and markets were soon glutted with surpluses, creating a sore political issue that has been with us ever since. Under such circumstances, it would have been foolish to expect much support from the chemical industry; in fact, it was bent on encouraging farmers to use pesticides as lavishly as possible.

In a nutshell, electrostatic crop treatment collided with so many vested interests that its commercial prospects were very dim. An attempt to revive it with large-scale modern machinery in the early 1960s soon had to be abandoned, although the efficiency of the process was again demonstrated. In fairness, it must be added that the economical assessment of agricultural techniques is often difficult, particularly when they involve (as in the present case) considerable capital expenditures for an unusual sort of equipment which cannot pay off for some years. According to available information, Communist or developing countries have no more taken advantage of electrostatic crop spraying than we did.

20.3.1 Electrostatic Ignition

Electrostatic ignition was also initiated in France, but the main developments benefited by the active support of British and (later) American firms.

Firing the plugs of an internal combustion engine with the discharge of a capacitor is a simple idea with a number of fundamental advantages. In the conventional coil (or magneto) ignition, the rise time of the voltage is comparatively long (10^{-4} sec) and, consequently, ignition fails if the insulation resistance of the plugs is below, say, a few megohms—a common occurrence when the ceramic is fouled by sooty deposits or when its outer surface is damp from nocturnal condensation. Besides, the only useful part of the spark delivered by a coil is its initial, capacitive stage, when the capacitance of the high-voltage lead and of the secondary winding releases its electrostatic energy. It is followed by a much longer inductive stage, dissipating the magnetic energy stored in the iron core in an arc discharge that can be seen as a faint flame, the capacitive spark being bright and bluish. The only effect of the inductive arc is a considerable erosion of the electrodes, and it plays no part in the ignition of the gaseous mixture.

Electrostatic ignition eliminates the useless inductive stage and concentrates all the energy delivered in the true spark. By using a fast-pressurized switch to connect the ignition capacitor to the plug, all the preceding inconveniences are removed. The voltage rise time is extremely short (10^{-8} sec), for the energy is now delivered as a traveling wave with a very steep front; ignition cannot fail whenever the shunting resistance of the plug exceeds the wave impedance of the lead (i.e., a few hundreds of ohms, a very small figure, indeed). Actually, no plug fouling can prevent electrostatic ignition, and an engine could run when immersed in seawater.

When electrostatic generation of the high voltage for charging the capacitor was resorted to, several additional advantages were secured: the output of the generator could match exactly the needs of the igniter, if both were driven by the same shaft, and a constant igniting voltage was thus ensured at every speed; moreover, very high compression ratios ceased to pose a problem, since voltage was plentiful.

In the 1950s a large number (about 100) of experimental igniters were built and tested at the Delahaye plant in Paris, at the Lucas* research facility in Birmingham, England, and later at Littleton, in Colorado, and at Holley Carburetors in Detroit.

All igniters were self-contained, autonomous devices, fitted to the standard explosion engines like conventional spark distributors, the top of the machine being similar to the usual distributor cap. No electrical connection to the battery was needed. A very precise ignition timing was easily ensured by internal pressurization; a rotating finger passed in front of circular array of stationary pins, leaving a clearance of about 0.01 in. Internal insulation was good enough for the device to remain fully energized after standing for several days; at worst, a few revolutions were necessary to start it again, a requirement which had also to be met when the gasoline pump needed priming after being left idle for a long time.

In spite of the very limited charge and energy delivered (ca. 1 μC or 0.01 J per spark), the igniting efficiency was greater than that of a coil, even at low temperatures ($0°F$) and for a lean mixture. Complete screening of the leads to prevent radio interference was quite tolerable and detracted little from the advantages of the device, whereas it was very detrimental to coil ignition.

* Dr. E. A. Watson, the head of the Research and Development Department at Joseph Lucas, Ltd, Birmingham, was an enthusiastic supporter of electrostatic ignition. He had tried it himself with an influence machine before World War I, and many important results were arrived at thanks to his advice and also thanks to the equipment he made available.

In spite of vigorous support from enterprising and competent people, electrostatic ignition was never accepted in the automobile industry. Actually, its proponents were fighting against heavy odds. The device was inordinately expensive by usual standards: the ball bearings it required (to take the thrust from the internal pressure) cost as much as a conventional distributor! But the biggest obstacle it had to face was the automobile industry's usual attitude toward its suppliers. They had to meet certain definite standards for a fixed price—the industry was not prepared to pay more for a fixture it bought from outside just because engine performance could be improved. This spelled the doom of electrostatic ignition.

20.3.2   Electrostatic Generators

Improving electrostatic generators to such an extent that they could success-fully compete with rectifier circuits and generate big blocks of high-voltage dc power has been a physicist's dream for many years. Prof. John G. Trump of M.I.T. considered in his doctoral thesis (1933) the feasibility of huge vacuum-insulated electrostatic generators of the variable-capacitor type, capable of transmitting the hydraulic power of the St Lawrence River to New York City as high-voltage direct current.

A similar attempt on a more modest scale was made in Grenoble (1942–1952), compressed air or nitrogen being used instead of vacua, which had proved to be practically unmanagable. The power outputs from machines of moderate size (1 ft in diameter) were substantial (200–300 W at 1500 rpm) by electrostatic standards, but the mechanical precision and stability required from the active parts, with their numerous interleaving metallic sectors held by insulators, could not be correctly maintained. Moreover, brush commutators would create conductive dust particles that were eventually attracted by the sectors and caused erratic breakdown. Only very small units worked well enough to reach the stage of industrial experimentation; they were mainly used for portable crop sprayers and electrostatic igniters, where their self-starting capability was very useful. The French army also ordered about 100 miniature generators (rotor, 1.8 in. diameter; voltage, 30 kV) as power supplies for sniperscopes (portable infrared telescopes).

A more enduring success was scored by generators with cylindrical insulating rotors in pure hydrogen at 250 psi, studied by N. Felici and E. Gartner from 1951 onward. The stator is a slightly conductive glass sleeve machined to fit the rotor within 0.02 in., commutation of charges being performed by ionizing steel blades. These machines are still manu-factured, although their field of application has considerably shrunk in recent years because of the great strides of solid-state rectifiers and converters;

several thousand units are in operation. Their practical range is 50 to 600 kV, 20 to 3000 W. Rotors more than 2 ft in diameter have performed poorly because the stator and ionizer clearances cannot be kept narrow enough for strong electric fields to be present, and this puts an upper limit on both voltage and power. Although overall efficiency is exceptional (80–90%), excitation requires an auxiliary source (10–30 kV) which can be electronically controlled to ensure high voltage stability (1 part in 10^5). Self-starting of a deenergized unit is impossible, because several kilovolts is needed to ionize the gas at the blades. This makes an auxiliary exciter necessary, which is a serious inconvenience when price and weight should be kept as low as possible.

In the 1960s considerable efforts were made to take advantage of organic liquids both as charge carriers and as insulating media. If the compressed gas used in belt or cylinder generators could be replaced without too much inconvenience by nitrobenzene, the output would be multiplied by the ratio of dielectric constants (36), since breakdown strengths are similar.

Introducing a high-permittivity liquid into an engineering structure, however, raises extremely difficult and fundamental problems because such liquids are very sensitive to electrolytic pollutants. Considerable progress was made in Grenoble (1960–1970) by resorting to ion exchange and electrodialysis to eliminate the minute traces of electrolytes left by conventional purification, and resistivities in the 10^{12} to 10^{14} Ω-cm range were consistently recorded with nitrobenzene ($\varepsilon = 36$) and propylene carbonate ($\varepsilon = 69$). This new technology could be successfully applied to light modulators (Kerr cells) filled with nitrobenzene, but experiments with rotating-disc generators, mainly pursued in Southampton, England, by Prof. A. W. Bright (1966–1969), remained inconclusive. The latter experiments were not entirely successful because deionization could not keep pace with pollution and the actual resistivity of the fluid (ca. 10^{10} Ω-cm) was too low for efficient operation.

Insulating liquids can also be used as charge carriers in place of discs or sectors with a promise of great structural simplicity. The best results were obtained by Dr. P. Secker, of the University College of North Wales in Bangor, Great Britain. Secker's generator uses pure hexane as internal medium; it seems to have overcome one of the problems raised by liquid-stream machines—namely, pole-to-pole breakdown within the insulating ducts carrying the liquid, which can disrupt the ducts and even start a fire.

Gas-flow generators which have been also studied in Britain and in France, can hardly be regarded as specifically European development, since many papers on this subject appeared in the United States in the early 1960s. Two lines of research, however, deserve mention: at Reading University, in England, Prof. W. D. Allen and Dr. P. Musgrove investigated the feasibility of powerful generators, the high efficiency aimed at (80%) being made possible by keeping the gas velocity under, say, 200 ft/sec. In Grenoble, very small supersonic generators, fed by compressed air and having no pretence to energetic efficiency, have been successfully used as built-in voltage sources for portable paint-spraying equipment, the same air flow passing through two successive stages of expansion. In the first (from, say, 80–20 psi gauge) it generates 50 kV, 20 μA, whereas the second (20 psi to atmospheric) is used for paint atomization.

As a conclusion, it may be said that although the long and painstaking efforts aimed at a basic improvement of electrostatic generators were original and interesting (the specific power of a cylinder machine reaches 1 W/cm^2; cf. 0.1 W/cm^2 for Van de Graaff's and less than 0.01 W/cm^2 for disc machines at atmospheric pressure), they have been only moderately successful from the standpoint of industry. This is mainly due to hard competition from the fast improving solid-state devices, whose meteoric expansion took place in between 1950 and 1970 and was accompanied by a precipitous fall in prices. At present, prospects for gas-insulated rotating machines are bleak; new concepts and technologies are badly needed if electrostatic generation is to survive at all.

20.3.3   Electrostatic Powder Coating

Although closely related to the electrostatic dusting of crops (at least from a physicist's standpoint), electrostatic powder coating has had a completely different fate. Applying plastic powders instead of paint to metallic items has distinct advantages—no solvent needed, thicker and more enduring coatings—but it also raises difficult problems. There is no adherence between a cold powder and a metallic surface; thus the items must be heated beforehand above the melting point of the coating material and dipped into a fluidized bed of the same substance. The impinging particles would then adhere, giving a molten layer which still had to be cured in a separate oven.

If electrostatic charging is resorted to, items can be coated with a dry, cold layer of electrified powder, whose stability allows transport to the curing oven without significant losses. Thus the impact of electrostatic techniques is considerably greater than in the case of painting. Most painting work has been so far carried out without the use of electrification, whereas powder coating (except in the case of tiny objects) owes its present expansion almost entirely to electrostatics.

Electrifying a cloud of solid particles by corona is very simple, but the arrangements used for crop dusting are not satisfactory. A high specific charge must be ensured and, above all, powder deposition in the ducts and on the high-voltage electrode must be prevented because the impact of powder clumps on the coated surface would be very detrimental. This problem was elegantly solved by the late Dr. R. P. Frazer, one of the leading British specialists in the field of aerodynamics and liquid atomization, who applied the so-called vortex principle to the charging nozzle. The eddying motion of the air keeps the electrode clean and also diminishes the forward momentum of the powder particles, thus making deposition easier.*

Another approach to electrostatic powder coating, using an electrified fluidized bed whose particles rose and deposited themselves on grounded objects held above the bed, was also developed in Britain and France; its industrial success was rather limited, however, because it could not handle sufficiently large items.

In the early 1960s, several varieties of powdering equipment were manufactured in Grenoble and lent or rented to prospective customers, mainly in the United States. They were returned, however, after a year or so, because no coating powder was available that could be cured fast enough to make the process really attractive. For a fairly long time, it appeared that the prospects of powder coating were very poor indeed. Fortunately, the response of the big chemical industry was decidedly favorable, and later the impending ban on solvent exhausts into the atmosphere gave the process a considerable boost. Thanks to the fast-curing powders that were speedily developed, the economics of powder coating and painting compare favorably. At present, powder consumption is soaring in all advanced countries, and powder coating may be regarded as one of the greatest breakthroughs of electrostatic technology. A side development, so far moderately successful, has been the electrostatic deposition of vitreous enamel for bath manufacture. Prof. Bright's Southampton group has recently carried out extensive research work on the physics of powder and enamel deposition.

20.3.4   Electrostatic Ore Beneficiation

Considerable efforts were devoted to electrostatic ore separation from a fluidized bed in Grenoble in the early 1960s, and in some cases (e.g., phosphate–silica separation) the results were good enough to secure orders for full-scale equipment. Nevertheless, evolution is very sluggish in the mining industry and this development may be considered as dormant.

20.3.5   Conclusion

Like the ill-fated "Brabazon" and "Comet" aircraft or that French submarine designed to be powered by a natural uranium reactor, whose hull never touched seawater, many of the preceding endeavors went awry. This

* The vortex arrangement was also applied by Dr. Frazer to electrostatic paint sprayers.

may be blamed on the overly confident mood mentioned already, which set much store by innovation as such and neglected more practical (especially commercial and legal) aspects. But, as aptly recalled by Dr. Watson, in his spirited 1949 address to the British Institution of Mechanical Engineers "They who seek for gold dig up much earth, and find a little" (Heraclitus).

At present, both Britain (in Southampton) and France (in Grenoble) maintain university chairs and laboratories specially devoted to the teaching of applied electrostatics and to research and development in that field, a legacy of a quarter-century of strenuous efforts, which can hardly be found elsewhere.

20.4   GERMANY AND JAPAN

In 1945, smitten with defeat and destruction, Germany and Japan had little left for innovation as such, until they were well on the way to their amazing recovery. By the end of the 1950s, they began seriously to pay attention to the most important electrostatic innovation of the day from the standpoint of a conventional, mass-producing industry: electrostatic painting. Nevertheless, their approaches were very different.

Germany, already a wealthy country by European standards, bought or rented equipment, mostly of American or French manufacture. Attempts by big firms to develop original techniques were abandoned or remained inconclusive. It is worth mentioning parenthetically that electrostatic engineering, although producing an enormous impact on some industrial processes, has never been a good business opportunity for large firms, because the capital expenditures needed are comparatively modest. The only significant German development of that period is the A.E.G. painting process, which has the peculiarity of being purely electrostatic. Instead of combining electrostatic forces with mechanical (centrifugal, aerodynamical) ones, it performs atomization by unassisted electrostatic pressure, as in Abbé Nollet's experiments. The paint flows between two elongated plates whose edges are separated by a few hundredths of an inch; it escapes as a liquid sheet that is soon atomized by electrostatic instability. Later, the fast-expanding German industry adopted powder coating on a large scale, and small manufacturers tried to market fairly conventional painting equipment.

The ingenious Japanese made use of their proverbial skill from the very start. By 1960 several original types of painting equipment had been developed, some of them by small teams working under adverse conditions and having no more than the bare minimum of floor space and equipment at their disposal. Part of these attempts was aimed at circumventing embarrassing patents, often at the price of a loss in efficiency; others were of more general value. Amusing gadgets of little economic significance could also be mentioned, again testifying to the lively inventive spirit of the Japanese—the magnetic brush developer in electrophotography, and an electrostatic tea sorter, which picked up the broken stems very neatly without lifting the dry leaves. How the machine worked was not clear, but the result deserves mention.

In recent years, Japanese industry has fully realized the tremendous potential of powder coating for a pollution-ridden country, controlling no sources of raw materials and, nevertheless, expanding its production at an incredible pace. On the other hand, research and development efforts in electrostatics involving not only ingenuity but also highly scientific concepts appeared as a new factor in Japan. Prof. Masuda's experiments at Tokyo State University on levitation and control of electrified particles by ac electric fields revealed a new promise of application, and future developments will certainly be watched with much interest.

20.5 SOVIET UNION

Information on work in the Soviet Union is, of course, very scarce, for any data pertaining to economics are considered secrets of State unless officially released. It is necessary to rely on books, journals, and personal contacts. In this context, the pioneer work of Chizhevsky (1897–1964), which extends over nearly four decades (1920–1960) must be mentioned, for it is disclosed in much detail in his monumental book *Aeroionization in the National Economy*. In fact, the author's main pursuit was of a biological and medical nature. Like many eighteenth-century scientists Chizhevsky tried to uncover the hidden links between electrostatic and biological processes. With stupendous perseverance he investigated the effects of natural and artificial atmospheric ions, charged droplets, and so on.

Some of Chizhevsky's most astonishing results (he claimed to have shown that higher organisms cannot survive in a ion-free atmosphere) are probably spurious; but in the course of his experiments, he came across a number of electrostatic problems which he solved in his own way. He described, for instance, an electrostatic sprayer, that was not intended for industrial processes like painting, but merely for dispersing charged droplets or particles of medicinal substances into the air breathed by patients (1933). It is plain from his book that Chizhevsky felt, toward the end of his career, that he missed several rewarding developments of which he might have been a leader if only he had had a more practical turn of mind. Nevertheless, tribute must be paid to this typically Russian character, whose eccentric theories and enthusiasm will always be a reader's delight.

Perhaps as a consequence of Chizhevsky's work, electrostatic painting was adopted in the Soviet Union and some other Communist countries (Eastern Germany, Hungary) as early as in Western Europe. The Russians bought a few pieces of equipment from France and elsewhere and then felt they could do the job themselves, whereas the Germans imported a much larger number of units at the start. Later they too manufactured everything, and so did the Hungarians, Western patent rights being apparently ignored or invalid.

As concerns electrostatic generation (apart from the routine manufacture of belt generators intended for nuclear research, which were quite similar to those built in the United States), a lively interest for new machines was shown in Leningrad and Tomsk (Siberia). Just before the war, academician Ioffe had started an ambitious research program aiming at the electrostatic generation of high-voltage direct current for power transmission, an idea similar to Trump's but more realistic, inasmuch as compressed gas insulation was considered instead of vacua. Of course, this interesting endeavor came to nothing under the pressure of war events. Work was resumed in Leningrad in the 1950s, but the former team under Ioffe's guidance had been scattered and the results, as judged from published papers, were insignificant.

More enduring interest was shown at the Tomsk Polytechnic Institute by Prof. Vorobjov and his co-workers. One of them, A. F. Kalganov, came to work in Grenoble for a year, and Prof. Vorobjov included large developments on electrostatic generation in his comprehensive treatise on high-voltage technology, still a very useful book for anyone who can read Russian. Very recently, the Energetic Institute of Moscow sent one of its members to France for a year's training in electrostatic atomization and related problems.

20.6 ELECTROSTATICS TODAY

Although the research and development efforts devoted to applied electrostatics were very unequal in the developed countries outside the United States, the actual impact of electrostatics in everyday life and industry has not been so different. For instance, there is no need to emphasize the role played by electrostatic copiers, mostly of American manufacture, in almost every walk of life, in Japan as well as in Western Europe. Cheap processes like flocking are also practiced everywhere, although their economic impact is rather insignificant. On the other hand, the tremendous development of the petroleum industry and the mass manufacture of artificial polymers has entailed widespread electrostatic nuisances and hazards, whose abatement has been an important concern of producers and users. In those realms of application, the U.S.S.R. has distinctly lagged behind, as befits a country putting no emphasis on mass production of dispensable goods and hitherto immune to consumerism as it has become familiar to Western Europeans and Japanese, too.

Other differences are more meaningful, Japan, a booming industrial country plagued with pollution, has paid much attention to the electrostatic elimination of unwanted wastes, whereas the French, who still do not react angrily at the view of an ugly smoke plume from an industrial stack, are contenting themselves with the routine application of electrostatic precipitation to such exceptional polluters as coal-fired power stations. Electrostatic ore separation, which scored considerable success at the hands of American firms both in the United States and abroad, has been completely neglected in Japan because of adverse climatic conditions; developments have been modest in Western Europe, most domestic mines being on the verge of abandonment.

Gathering quantitative data on electrostatic applications would entail a tremendous amount of painstaking work which, however, would have but limited interest because most of the applications are intimately intertwined with other techniques or developments and merely follow the same trends.

This does not apply to the electrostatic techniques that have been successful enough to displace more conventional ones or even to create new products or open new markets. Apart from electrostatic printing, two innovations are really significant from this standpoint: electrostatic painting and electrostatic powder coating. It seems worthwhile to give a short survey of their impact in the countries we are dealing with.

20.6.1 Electrostatic Painting

The first equipment for electrostatic painting appeared in Europe in 1950, but customers remained skeptical until 1954, when the first breakthrough occurred in the flourishing bicycle and motorcycle industry, for obvious technical reasons. Home furniture and appliances were soon painted electrostatically too, although other industries (e.g., the car manufacturers) were not interested.

Breaking down the paint consumption according to the main segments of industry gives the following picture, valid for all Western European countries:

Industry	% of Paint Consumed	% of Paint Deposited Electrostatically
Building	60	0
Automobile manufacturing	10	0.3
Other	30	2.5

Thus less than 3% of all the paint used is deposited electrostatically today. It is hoped that this figure will double in a matter of five years, for at present only one-tenth of the amount that could be electrostatically deposited benefits by this technique.

Although the largest industries make little if any use of electrostatic painting, some particular segments have gone very far in taking advantage of it. For instance, 70% of the paint used by the bicycle industry, 50% of that used for home appliances, and 25% of that for furniture is applied electrostatically. For reference, we give the total paint consumption (metric tons/year) in 1970 for a few countries:

> Japan   2,000,000
> West Germany   800,000
> France   600,000

The number of spraying heads in operation is difficult to determine because of the number of competing firms that sell or rent them; a reasonable approximation is 4000 to 5000 for each of the countries mentioned. The corresponding capital investment is about $10 million ($2000 per head). As we mentioned previously, selling electrostatic equipment is not an interesting business for large firms.

20.6.2   Powder Coating

Today the consumption of coating powder is but a tiny fraction of that of paint: 2500 metric tons in 1970 for the whole of Western Europe and 1200 for Japan, as opposed to millions of tons for paint. All projections, however, show that this consumption is sure to soar in the near future. Estimates give 25,000 metric tons in 1975 for Western Europe alone and at least 50,000 in 1980. It should be remembered that almost all this amount will be deposited electrostatically; thus in a few years electrostatic powder coating may excel electrostatic painting in economic significance.

As already mentioned, this breakthrough is mainly due to active support by the big chemical industry. Ten years ago, curing would last half an hour; today, 5 to 10 min at 380°F is sufficient—a usual figure for paint.

Breaking down the powder consumption (metric tons) according to countries gives the following picture (for 1970):

> Japan   1200
> West Germany   800
> France   600
> Scandinavia   400
> Benelux countries 310
> Great Britain   150
> Italy   150
> Switzerland   100

As we might have expected, Japan and Germany, whose industry was expanding fastest, took the leading places, closely followed by France, where the process was first developed. Britain has an unexpectedly low rank because of the economic stagnation of the late 1960s.

The total number of spraying heads for Western Europe is estimated 1255, the capital investment per head being about $6000 including all accessories.

20.7   CONCLUSION

At present, the status of electrostatic techniques in the countries we have considered is much more related to their development level or to the rate of their industrial expansion than to their inventions and innovations in the field. The same lesson could be drawn from almost every segment of industry today, with a few notable exceptions—computers, for instance. Under such circumstances, the mood of disenchantment with applied science, which is beginning to spread in the most advanced countries, is not surprising at all. It must be admitted, however, that we needed a more sober appreciation of technology, and the belief that technological innovation could solve any problem or make any country richer at the expense of others, was childish indeed.

Nevertheless, this necessary reappraisal, now long overdue, will point out some of the most favorable features of applied electrostatics which were completely overlooked when more glamorous subjects diverted public attention.

Electrostatics is one of the least controversial segments of applied science. Not only is its immediate impact clearly beneficial, but the precise assessment of its short- and long-term consequences is much easier than in the case of most technologies (think of DDT!), and its preservational and conservational capabilities are unique (what about the avian fauna if electrostatic spraying of pesticides had been compulsory?). Another remarkable side of applied electrostatics is that it requires imaginative thinking and engineering skill rather than huge research laboratories and heavy capital expenditures. Wherever there have been successful developments, they must be credited to small dedicated teams, not to the heavy machinery of large research organizations. In a sense, electrostatic engineering is too efficient to warrant large financial means or costly projects, and this has detracted much from its potential appeal to big firms or organizations.

But the best of applied electrostatics is still to come, particularly as it concerns such crowded areas as Western Europe and Japan, with serious pollution problems and no significant natural resources. When the present superabundance of raw materials and agricultural products has come to an end, when these lightheartedly expanding nations are choked with refuse on one hand and see their prosperity in jeopardy on the other, no doubt they will turn to electrostatic engineering as one of their last resorts. According to reliable American experts, this turn of the tide is in the offing, a quarter of a century from now. Then applied electrostatics will reveal itself as a gentle, protective, and benevolent science, best suited to reconcile man with his environment and to help him along through the trials he is sure to face in the future.

Index